GEOTECHNICAL INVESTIGATION METHODS

A Field Guide for Geotechnical Engineers

GEOTECHNICAL INVESTIGATION METHODS

A Field Guide for Geotechnical Engineers

Roy E. Hunt, P.E., P.G.

CRC Press
Taylor & Francis Group
Boca Raton London New York

CRC Press is an imprint of the
Taylor & Francis Group, an **informa** business

A TAYLOR & FRANCIS BOOK

The material was previously published in *Geotechnical Engineering Investigations Handbook, Second Edition* ©CRC Press LLC 2005.

CRC Press
Taylor & Francis Group
6000 Broken Sound Parkway NW, Suite 300
Boca Raton, FL 33487-2742

First issued in paperback 2019

© 2007 by Taylor & Francis Group, LLC
CRC Press is an imprint of Taylor & Francis Group, an Informa business

No claim to original U.S. Government works

ISBN-13: 978-1-4200-4274-0 (hbk)
ISBN-13: 978-0-367-39001-3 (pbk)

Library of Congress Cataloging-in-Publication Data

Hunt, Roy E.
    Geotechnical investigation methods : a field guide for geotechnical engineers / by Roy E. Hunt.
       p. cm.
    Includes bibliographical references.
    ISBN 1-4200-4274-2 (alk. paper)
    1. Engineering geology--Handbooks, manuals, etc. 2. Earthwork. I. Title.

TA705.H865 2006
624.1'51--dc22   2006048956

Visit the Taylor & Francis Web site at
http://www.taylorandfrancis.com

and the CRC Press Web site at
http://www.crcpress.com

Contents

Introduction

Purpose and Scope

This book describes and provides the basis for the selection of the numerous methods and procedures for:

1. Exploring the geologic environment and mapping surficial conditions, including rock, soil, water, and geologic hazards; preparing subsurface sections; and obtaining samples of the materials for identification, classification, and laboratory testing.
2. Measurement of material properties (basic, index, hydraulic, and mechanical) in the field and laboratory.
3. Field instrumentation to measure and monitor movements, deformations, and stresses occurring naturally or as a consequence of construction.

Although, in practice, analytical procedures and design criteria are often presented as part of an investigation, they are not included within the scope of this book.

Significance

The investigation phase of any geotechnical study undertaken for development, construction, or any other engineering works is by far the most important phase. Not only must conditions at the project site be thoroughly identified, but for many projects, the regional geologic characteristics must also be determined. For all phases of investigation, there are a large number of methods and devices to choose from, ranging from simple to complex, and usually several are applicable for a given subject of study.

Geotechnical engineering analyses and evaluations are valid only when based on properties truly representing all of the natural materials that may influence the works. Properties of some materials are best measured in the laboratory, while others must be field tested. In some cases, properties cannot be adequately defined by direct testing and the result will be designs that are conservative and too costly, unconservative and risky, or unconservative but based on contingency plans. To monitor ground conditions during construction, field instrumentation is an important element of many studies, where subsurface conditions cannot be adequately defined by exploration and testing. Instrumentation is used also to obtain design data and to monitor changing natural conditions such as slope failures and fault movements.

1

Exploration

1.1 Introduction

1.1.1 Objectives

The general objective of an exploration program is to identify all of the significant features of the geologic environment that may impact on the proposed construction. Specific objectives are to:

1. Define the lateral distribution and thickness of soil and rock strata within the zone of influence of the proposed construction.
2. Define groundwater conditions considering seasonal changes and the effects of construction or development extraction.
3. Identify geologic hazards, such as unstable slopes, faults, ground subsidence and collapse, floodplains, regional seismicity, and lahars.
4. Procure samples of geologic materials for the identification, classification, and measurement of engineering properties.
5. Perform *in situ* testing to measure the engineering properties of the geologic materials (Chapter 2).

1.1.2 Methodology

Three general categories subdivide exploration methodology:

1. *Surface mapping of geologic conditions* (Section 1.2), which requires review of reports and publications, interpretation of topographic and geographic maps, remote-sensing imagery, and site reconnaissance
2. *Subsurface sectioning* (Section 1.3), for which data are obtained by geophysical prospecting, test and core borings, and excavations and soundings
3. *Sampling* the geologic materials (Section 1.4) utilizing test and core borings and excavations

A general summary of exploration methods and objectives is given in Table 1.1.

1.1.3 Scope

The scope of the investigation will depend upon the size of the proposed construction area, i.e., a building footprint, or several to hundreds of acres, or square miles, and the

TABLE 1.1

Exploration Objectives and Applicable Methods

Method	Regional	Surficial-land	Surficial-seafloor	Major structures	Faults[a]	Deep-land	Shallow-land	Subaqueous	Soft-soil depth	Sliding masses[a]	Rock depth	Rock-mass conditions	Disturbed to GWL	Representative	Undisturbed	Deep, offshore	Rock cores	Normal depths	Deep
General																			
Reports and Publications	X	X	X	X															
Topographic Maps	X	X	X	X															
Imagery: Satellite, SLAR	X	X	X	X															
Imagery: Low-Altitude Photos	X	X	X	X				X											
Bathymetry			X	X															
Side-Scan Sonar			X	X															
Underwater TV			X	X															
Reconnaissance	X	X	X					X											
Geophysics																			
Seismic Refraction				X	X	X	X	X			X	X							
Seismic Reflection				X	X			X											
Electrical Methods								X			X								
Gravimeter				X	X							X							
Magnetometer				X	X														
Radar Profiling					X	X					X	X							
Video-Pulse Radar												X							
Boring																			
Wash Boring						X			X	X			X	X	X				
Rotary Drilling					X	X	X	X	X	X	X		X	X	X		X		
Rotary Probe										X	X								
Continuous-Flight Auger							X		X				X	X	X				
Hollow-Stem Auger							X		X	X			X	X	X				
Wire-Line Drilling				X		X		X				X					X		X
Borehole Sensing																			
Borehole Cameras				X			X					X							
Acoustical Sounding								X			X								
Electric Well Log							X	X	X		X	X							
Radioactive Probes				X			X	X	X		X	X							
Ultrasonic Acoustics				X			X				X	X							
3-D Velocity Log				X			X					X							
Test Pits/Trenches				X			X			X	X		X		X				
Miscellaneous																			
Adits				X			X					X							
Bar Soundings										X									
Retractable Plug							X			X					X				
Continuous Cone Penetrometer							X	X	X							X			
Hand Augers							X						X						
Bucket Auger							X			X			X						

a   See also Instrumentation.

experience of the investigator in the area. Do they have prior knowledge or is the area new to them? This text basically assumes that prior knowledge is nil or limited.

1.2   Surface Mapping

1.2.1   General

Objectives

Data Base

For all sites it is important to determine the general geologic conditions and identify significant development and construction constraints. For large study areas it is useful to prepare a map illustrating the surficial and shallow geologic conditions.

Preliminary Site Evaluations

An overview of geologic conditions permits preliminary evaluations regarding the suitability of the site for development. The first step is the identification of major geologic hazards and "constraints" in the study area. Depending upon the construction or development proposed, constraints could include shallow rock or water, or thick deposits of weak soils. Taking into account the hazards and constraints, the optimum location for the proposed construction is selected, and the planning of the site investigation then begins.

Methodology

A geologic reconnaissance study may advance through a number of steps as described briefly in Figure 1.1, including:

- Research of reference materials and collection of available data.
- Terrain analysis based on topographic maps and the interpretation of remotely sensed imagery.
- Preparation of a preliminary engineering geology map (large land areas).
- Site reconnaissance to confirm initial data, and, for large areas, amplification of the engineering geology map, after which it is prepared in final form.
- Preparation of a subsurface exploration program based on the anticipated conditions.

1.2.2   Research Data

Basic Objectives

A large amount of information is often available in the literature for a given location. A search should be made to gather as much data as possible before initiating any exploration work, particularly when large sites are to be studied, or when the site is located in a region not familiar to the design team. Information should be obtained on:

- *Bedrock geology*, including major structural features such as faults.
- *Surficial geology* in terms of soil types on a regional or, if possible, local basis.
- *Climatic conditions*, which influence soil development, groundwater occurrence and fluctuations, erosion, flooding, slope failures, etc.
- *Regional seismicity* and *earthquake* history.

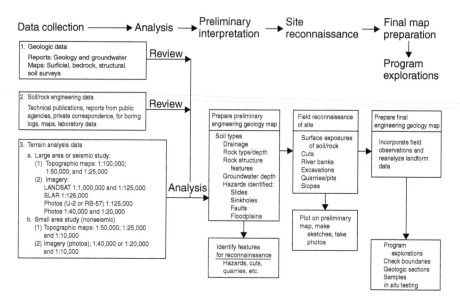

FIGURE 1.1
The elements of the geologic land reconnaissance study.

- *Geologic hazards*, both regional and local, such as ground subsidence and collapse, slope failures, floods, and lahars.
- *Geologic constraints*, both regional and local, such as expansive soils, weak soils, shallow rock, groundwater, etc.

Information Sources

Geologic texts provide information on physiography, geomorphology, and geologic formation types and structures, although usually on a regional basis.

Federal and state agencies issue professional papers, bulletins, reports, and geologic maps, as do some cities. Sources of geologic information include the U.S. Geological Survey (USGS) and the U.S. State geological departments. Agencies for agriculture, mining, and groundwater also issue reports, bulletins, and maps. Information on the USGS and State Agencies can be found on the Internet.

Engineering soil surveys have been prepared for New Jersey (Rogers, 1950) and Rhode Island, which are presented as reports and maps on a county basis. The maps illustrate shallow soil and rock conditions and the soils are classified by origin in combination with AASHO Designation M145–49. The prevailing or average drainage conditions are also shown.

Technical publications such as the journals of the American Society of Civil Engineers, Institute of Civil Engineers (London), the Association of Engineering Geologists (USA), the *Canadian Geotechnical Journal*, and the various international conferences on soil and rock mechanics, engineering geology, and earthquakes, often contain geologic information on a specific location.

Climatic data are obtained from the U.S. Weather Bureau or other meteorological agencies.

Geologic Maps

Geologic maps generally vary in scale from 1:2,500,000 (U.S. map) to various scales used by state agencies, to USGS quadrangle maps at 1:24,000, and vary in the type of geologic information provided. A guide to map scale conversions is given in Table 1.2. On a worldwide basis the availability of geologic maps varies from excellent in modern, developed countries, to poor to nonexistent in other countries or areas.

TABLE 1.2

Guide to Map Scales

Scale	ft/in.	in./1000 ft	in./mile	miles/in.	m/in.	acres/in.^{2}
1:500	41.67	24.00	126.72	0.008	12.70	0.040
1:600	50.00	20.00	105.60	0.009	15.24	0.057
1:1000	83.33	12.00	63.36	0.016	25.40	0.159
1:1200	100.00	10.00	52.80	0.019	30.48	0.230
1:1500	125.00	8.00	42.24	0.024	38.10	0.359
1:2000	166.67	6.00	31.68	0.032	50.80	0.638
1:2400	200.00	5.00	26.40	0.038	60.96	0.918
1:2500	208.33	4.80	25.34	0.039	63.50	0.996
1:3000	250.00	4.00	21.12	0.047	76.20	1.435
1:4000	333.33	3.00	15.84	0.063	101.60	2.551
1:5000	416.67	2.40	12.67	0.079	127.00	3.986
1:6000	500.00	2.00	10.56	0.095	152.40	5.739
1:7920	660.00	1.515	8.00	0.125	201.17	10.000
1:8000	666.67	1.500	7.92	0.126	203.20	10.203
1:9600	800.00	1.250	6.60	0.152	243.84	14.692
1:10000	833.33	1.200	6.336	0.158	254.00	15.942
1:12000	1,000.00	1.000	5.280	0.189	304.80	22.957
1:15000	1,250.00	0.800	4.224	0.237	381.00	35.870
1:15840	1,320.00	0.758	4.000	0.250	402.34	40.000
1:19200	1,600.00	0.625	3.300	0.303	487.68	58.770
1:20000	1,666.67	0.600	3.168	0.316	508.00	63.769
1:21120	1,760.00	0.568	3.000	0.333	536.45	71.111
1:24000	2,000.00	0.500	2.40	0.379	609.60	91.827
1:25000	2,083.33	0.480	2.534	0.305	635.00	99.639
1:31680	2,640.00	0.379	2.000	0.500	804.67	160.000
1:48000	4,000.00	0.250	1.320	0.758	1,219.20	367.309
1:62500	5,208.33	0.192	1.014	0.986	1,587.50	622.744
1:63360	5,280.00	0.189	1.000	1.000	1,609.35	640.000
1:100000	8,333.33	0.120	0.634	1.578	2,540.00	1,594.225
1:125000	10,416.67	0.096	0.507	1.973	3,175.01	2,490.980
1:126720	10,560.00	0.095	0.500	2.000	3,218.69	2,560.000
1:250000	20,833.33	0.048	0.253	3.946	6,350.01	9,963.907
1:253440	21,120.00	0.047	0.250	4.000	6,437.39	10,244.202
1:500000	41,666.67	0.024	0.127	7.891	12,700.02	39,855.627
1:750000	62,500.00	0.016	0.084	11.837	19,050.04	89,675.161
1:1000000	83,333.33	0.012	0.063	15.783	25,400.05	159,422.507
Formula	$\dfrac{\text{Scale}}{12}$	$\dfrac{12.000}{\text{Scale}}$	$\dfrac{63.360}{\text{Scale}}$	$\dfrac{\text{Scale}}{63.360}$	$\dfrac{\text{ft/in.}\times}{0.3046}$	$\dfrac{(\text{Scale})^2}{43{,}560\times144}$

Bedrock geology maps (Figure 1.2) often provide only the geologic age; the rock types are usually described in an accompanying text. There is a general correlation between geologic age and rock type. The geologic time scale and the dominant rock types in North America for the various time periods are given in Appendix A. The formations for a given period are often similar in other continents. For the purpose of mapping, rocks are divided into formations, series, systems, and groups. Formation is the basic unit; it has recognizable contacts to enable tracing in the field and is large enough to be shown on the map. Series are coordinate with epochs, systems with periods, and the largest division, groups, with eras.

Structural geology may be shown on special maps or included on bedrock geology maps using symbols that identify faults, folding, bedding, jointing, foliation, and cleavage. The maps often include geologic columns and sections.

Surficial geology maps depict shallow or surficial soil and rock types.

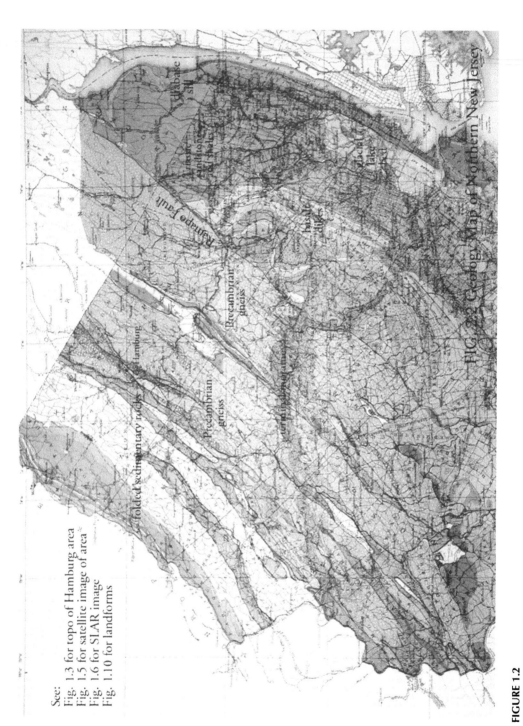

See: Fig. 1.3 for topo of Hamburg area
     Fig. 1.5 for satellite image of area
     Fig. 1.6 for SLAR image
     Fig. 1.10 for landforms

FIG. 1.2 Geology Map of Northern New Jersey

FIGURE 1.2
Geology map of northern New Jersey. (From New Jersey Geological Survey, 1994.)

Folios of the Geologic Atlas of the United States was produced by the USGS until 1945. Detailed maps of bedrock geology, structural geology, and surficial geology for many cities in the United States and other areas of major geologic importance were included.

Soil survey maps, produced by the Soil Conservation Service (SCS) of the U.S. Department of Agriculture, are usually plotted as overlays on aerial photographs at relatively large scales. Prepared on a county basis they show the soil cover to a depth of about 6 ft (2 m), based on pedological soil classifications. They are often combined with symbology describing slopes, shallow groundwater, and soil drainage conditions. Recent maps contain engineering-oriented data prepared by the Bureau of Public Roads in conjunction with the SCS. However, the shallow depth depicted limits their usefulness in many engineering studies.

Flood insurance maps identify 100- and 500-year-old floodplains adjacent to water bodies. These are available from the Federal Emergency Management Agency (FEMA), the USGS, and State Agencies.

Tectonic maps give regional lineations often indicative of faulting.

Earthquake data may be presented as intensity maps, isoseismal maps, various forms of seismic risk maps, or as microzonation maps.

Other useful maps published by the Geological Society of America include the glacial map of the United States and the loessial soils or wind-blown deposits of the United States.

Topographic Maps and Charts

Topographic maps, such as quadrangle sheets, show landforms, drainage patterns, stream shapes, and surface water conditions, all indicators of geologic conditions. Because of their availability and usefulness they should be procured as a first step in any study. They are available from a number of sources and in a variety of scales as follows:

- USGS provides maps covering a quadrangle area bounded by lines of latitude and longitude available in 7.5° series (1:24,000) (Figure 1.3), 15° series (1:62,500), 30° series (1:125,000), and 1° series (1:250,000) for most of the United States, although many of the larger scales are out of print.
- Other countries use scales ranging from 1:10,000 to 1:1,000,000 but coverage is often incomplete. 1:50,000 is a common scale available for many areas, even in countries not fully developed.

Coastline charts, available from the National Ocean Service (NOAA-NOS), provide information on water depths and near-shore topography.

Remotely Sensed Imagery

Remote-sensing platforms now include satellite-borne digital imagery and radar systems, airborne imagery including digital and radar imagery, and aerial photography. In recent years, many new sources of remotely sensed imagery have been, and continue to be, developed. The relationship between the various forms and the electromagnetic spectrum is given in Figure 1.4. Remotely sensed imagery is discussed in the following section.

1.2.3 Remotely Sensed Imagery

Satellite Imagery (Digital Sensors)

Satellite-borne systems obtain images of the Earth's entire surface every 16 days but images are affected by cloud cover:

- LANDSAT (USA-NASA): Landsat satellites have been launched periodically since the first satellite (ERTS) was launched in 1972. Landsat 7, launched in 1999,

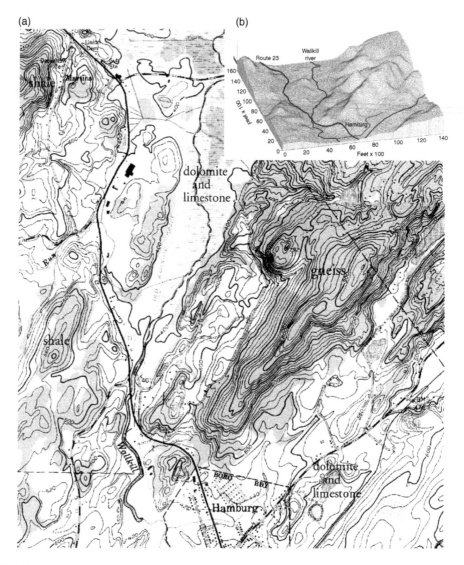

FIGURE 1.3
(a) Portion of USGS Hamburg NJ Quad Sheet, Scale 1″ = 2000′ (1:24000). Area has been glaciated. Shown are general bedrock types. Note relation to landform shown on inset as evidenced by topography. (b) 3-D diagram of topography of Figure 1.3a. (Courtesy of USGS.)

includes a multispectral scanner (MSS) system and an enhanced thematic mapper (ETM) system. The MSS system, with a spatial resolution of 79 m, has four channels that record reflected solar energy corresponding to green and red bands and two near-infrared spectroscopy (NIR) spectral regions. A fifth channel, with a spatial resolution of 240 m, records emitted energy in the thermal infrared region. The ETM system, with a spatial resolution of 30 m, collects reflected energy in three visible bands and three infrared bands. The system has one thermal infrared channel with 60-m spatial resolution. Also included is a panchromatic (black and white) channel with a spatial resolution of 15 m. Swath width is 185×185 km. 3-D stereo-projections can be prepared from sequentially obtained images. (Examples of satellite images are given in Figure 1.5).

- SPOT (France): First launched in 1986, SPOT now includes a high-resolution visible sensor and a multispectral sensor with a spatial resolution of 10 m.

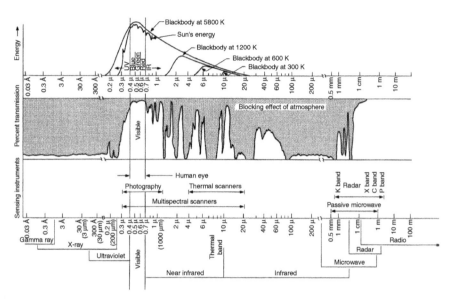

FIGURE 1.4
The electromagnetic spectrum illustrating atmospheric attenuation and general sensor categories. (From Way, D.S. *Terrain Analysis*, 2nd ed., Dowden, Hutchinson & Ross, Stroudsburg, Pennsylvania, 1978.)

The panchromatic mode has a spatial resolution of 5 m. Swath width is 60 × 60 km. SPOT is capable of acquiring stereocoptic (overlapping) imagery.

- IKONOS, Quick Bird (DigitalGlobe), and OrbView-3 (Orbimage) are recently launched U.S. satellites that are reported to have a multispectral spatial resolution of 4 m and a panchromatic spatial resolution of 1 m. Swath width is of the order of 8 × 8 km.
- Other countries including India, Russia, Japan, and China have also launched satellites.

Satellite Imagery (Radar Sensors)

- SAR (*synthetic aperture radar*) images have been obtained by the European Space Agency with satellites ERS-1 and ERS-2 since 1992. Since SAR is based on reflected signals from radio waves, images can be obtained during day or night, and through cloud cover.
- InSAR (*interferometric SAR*), is also known as *DifSAR (differential synthetic aperture radar interferometry)*. Using two different satellite passes and applying a complex and computer-intensive process, an interferogram is created. Because of precise, specific geometry, the sources combine to create bright and dark patterns of interference or "fringes," which can be converted to ground heights with a precision reported down to 0.7 cm possible. The images are color-enhanced using filters to produce color-coded interferograms. They have been used to record ground subsidence over large areas (Professional Surveyor, 1999).

Airborne Imagery (Digital Sensors)

- *Airborne digital imagery* is obtained from aircraft flying at less than 3000 m. Data procurement is normally on a site-, time-, and weather-specific basis and sequential

FIGURE 1.5
False-color satellite image of northern New Jersey (ERTS-1, 1972). (From EROS Data Center, 1972.)

missions are practical. One imaging system available is composed of a digital camera, onboard GPS, and inertial measurement unit. The digital camera collects imagery in three bands and can provide both true color and color infrared. Images are obtained with resolutions of 0.3 to 1 m, and stereo-images are possible (Professional Surveyor, 2002).

Airborne Imagery (Radar Sensors)

- SLAR (*side-looking airborne radar, real-aperture system*) has been in use for many years. It penetrates cloud cover and to some degree vegetation, providing low-resolution images normally at scales of 1:125,000 to 1:100,000, and occasionally smaller (see Figure 1.6). Worldwide coverage, including the United States, is spotty.
- LiDAR (*light detection and ranging, or laser radar; GPS based*): This airborne system sends out pulses of laser energy to the ground surface that reflect the energy, providing a measure of the distance. The beams rebound to sensitive detectors in the aircraft where the resulting data are analyzed with a computer program that ignores trees and other ground cover. A topographic map or an orthophoto is created without the use of ground control. Precision is reported to be as fine as 10 cm (DeLoach and Leonard, 2000; Gillbeaut, 2003).

FIGURE 1.6
SLAR image of northern New Jersey. (Courtesy of USGS, 1984.)

Airborne Imagery (Aerial Photography)

- *High-altitude stereo-aerial photographs* provide the smallest scale images for stereo-viewing, ranging in scale from 1:125,000 to 1:100,000, yielding substantial detail on terrain features (Figure 1.7). Worldwide coverage, including the United States, is spotty.
- *Stereo pairs of aerial photographs* provide the basis for detailed engineering geologic mapping. They can be obtained in panchromatic, true color, or color infrared (CIR). Single photos in true color and CIR are given in Figure 1.8. Detailed studies of large to small areas should be based on stereoscopic interpretation. Because aerial photographs are the basis for modern topographic mapping they are available on a worldwide basis, at least at scales of 1:50,000.

Hyperspectral Imagery

Multispectral scanners discussed above obtain images over a small number of broad spectral bands of the electromagnetic spectrum. They lack sufficient spectral resolution for precise surface studies. Hyperspectral imaging (imagery spectrometry) has been developed in recent years to acquire spectral data over hundreds of narrow, descrete, contiguous spectral bands. Various systems include SIS, AIS, AVIRIS, HIRIS and HYDICE, and others. Some are carried by satellite (HIRIS), while others, such as AVIRIS (Airborne Visible/Infrared Imaging Spectrometer), are flown on aircraft platforms at heights of up to 100 km above sea level.

The main objective of AVIRIS is to identify, measure, and monitor constituents of the Earth's surface based on molecular absorption and particle scattering signatures. Some applications include mineral identification for the mining industry for new sites and for mine waste studies (Henderson III, 2000).

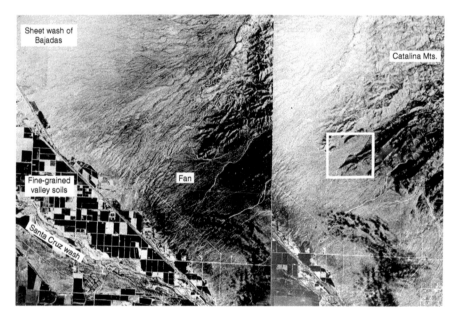

FIGURE 1.7
NASA high-altitude stereo-pair of an area northwest of Tucson, Arizona (scale 1:125,000). Apparent are sheet wash and sheet erosion of the "bajadas" alluvial fans of granular soils, valley fill of fine-grained soils, and the "dry wash" of the Santa Cruz River, all typical depositional forms in valleys adjacent to mountains in arid to semiarid climates. (Original image by NASA reproduction by US Geological Survey. EROS Data Center.)

(a)

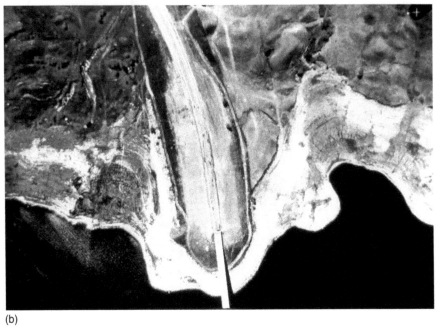

(b)

FIGURE 1.8
Aerial photos of a bridge abutment illustrating the advantage of infrared over true color. On the left is vegetation growing over shallow, poor-draining marine shales. On the right are relatively free-draining granular glacial soils with sparce vegetation. (a) True color photo; (b) color infrared photo (CIR). (Courtesy of Woodward–Clyde Consultants.)

Seaborne Imagery

- *Side-scan sonar* provides images of the seafloor or other water bodies, which often have features indicative of significant geologic conditions. The images given as Figure 1.9 were obtained for a landslide study where the failure surface passed

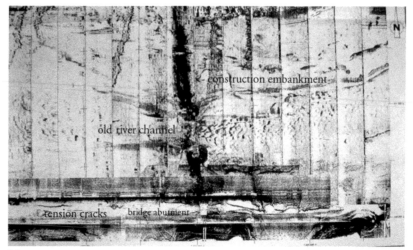

(a)

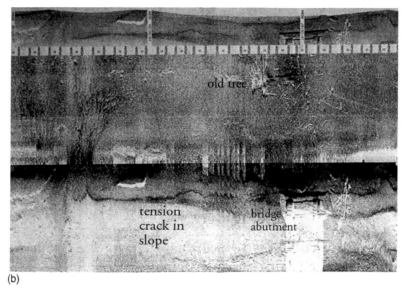

(b)

FIGURE 1.9
Side-scan sonar images of a reservoir bottom. (a) Side-scan sonar mosaic of reservoir bottom, water depth = 180 ft; (b) detailed side-scan sonar image. A portion of the mosaic in (a). (Courtesy of Woodward–Clyde Consultants.)

into a reservoir. Figure 1.9a is a mosaic of a number of passes where water is 180 ft deep. Figure 1.9b is a portion of one image showing the escarpment or tension crack along the reservoir margin. These images usually are obtained on a project-based need.

1.2.4   Terrain Analysis

General

Significance

Terrain analysis is often the most important part of any geotechnical investigation. Landforms (topographic expression) and other surface characteristics are strong indicators of geologic conditions. Characteristic terrain features reveal rock type and structural forms, where the rock is relatively shallow and subject to weathering and erosion, or represent typical soil formations in terms of their origin and mode of deposition where deposits are sufficiently thick.

Objectives

The delineation and mapping of the significant aspects of the geologic environment are the objectives of terrain analysis. Information is provided on rock types and structures, soil types and formations, groundwater conditions, and floodplains; and, on the locations of such hazards as landslides and other slope failures, sinkholes, and other instances of ground collapse and subsidence.

Methodology

Terrain analysis is based on the interpretation of features evident on topographic maps and remotely sensed imagery. Imagery interpretation as applied to engineering geologic mapping is summarized in Table 1.3, and as applied to environmental and natural resource studies in Table 1.4. The elements of imagery interpretation are summarized in Tables 1.5 and 1.6. Figures of USGS quadrangle sheets and stereo-pairs of aerial photos included in this book are summarized in Appendix B.

Topographic maps, such as USGS quadrangle sheets, show landforms, drainage patterns, stream shape, and surface water conditions, all indicators of geologic conditions. Because of their availability and usefulness they should be procured as a first step in almost any study.

There are many forms of remote-sensing imagery presenting the features evident on the topographic maps. They are useful for environmental as well as geological studies.

TABLE 1.3

Uses of Remote Sensing for Engineering Geologic Mapping

Information Desired	Applicable Imagery
Regional geologic mapping and delineation of major structural features	Satellite and SLAR imagery
(a) Global coverage, moderate resolution	(a) LANDSAT
(b) High resolution, but incomplete global coverage	(b) SPOT, IKONOS, etc.
(c) Useful for areas of perennial cloud cover and heavy vegetation; low resolution	(c) SLAR
Detailed mapping of rock type, structure, soil formations, drainage, groundwater, slope failures, sinkholes, etc.	Stereo-pairs of aerial photos B&W, true color, CIR
(a) Moderately large areas	(a) Scale 1:100,000
(b) Large areas, general mapping	(b) Scale 1:60,000–1:40,000
(c) Small areas, detailed mapping	(c) Scale 1:20,000–1:8,000
Improved definition of surface and groundwater conditions on large- to local- area basis, such as land–water interface, seepage, ground moisture (important for sinkhole and fault identification)	Stereo-pairs of color infrared (CIR)
Seafloor and other underwater conditions (rock outcrops, soils, sunken vessels, pipelines, etc.)	Side-scan sonar

Notes: (1) Normal studies of large land areas, such as for highways, airports, industrial zones, new communities should be based on the interpretation of aerial photos of at least two scale ranges (1:60,000–1:40,000 and 1:20,000–1:8,000).

(2) Studies of areas where seismicity is of concern should always begin with interpretations of ERTS/LAND-SAT imagery, then be supplemented by interpretation of normal study imagery scales.

TABLE 1.4

Uses of Remote Sensing for Environmental and Natural Resource Studies

Information Desired	Applicable Imagery
Regional environmental studies of air, water and vegetation quality, flooding	Satellite imagery
(a) On a changing or seasonal basis	(a) LANDSAT
(b) High resolution but incomplete global coverage	(b) SPOT, IKONOS, etc.
Surface and groundwater studies (*large to local areas*)	CIR and Satellite (MSS) imagery
(a) General	(a) CIR and MSS imagery
(b) Thermal gradients indicative of pollution or saltwater intrusion of surface water	(b) Thermal IR scanner (ETM)
(c) Subsurface seepage	(c) Thermal scanner (ETM)
Vegetation: forestry and crop studies (large to local areas) identify types, differentiate healthy from diseased vegetation	CIR and Satellite (MSS) imagery
Mineral resource studies	MSS and hyperspectral imagery
(a) Based on landform analysis	
(b) Based on plant indicators	

The selection of imagery depends upon availability, the study purpose, and the land area involved. Stereoscopic examination and interpretation of aerial photographs is the basic analytical method.

Remote-Sensing Imagery and Interpretation

Regional Geologic Studies

On a regional basis, landform is the most important element of interpretation for geotechnical studies. In general terms, landform reflects the relative resistance of geologic materials to erosion. Some relationships among landform, rock type, and structure are apparent on the portion of the physiographic diagram of northern New Jersey in Figure 1.10. The geology of the area is shown in Figure 1.2. Landform is also evident in Figure 1.5, a false-color satellite image, and in Figure 1.6, a SLAR image.

Millions of years ago the area illustrated in the figures was a peneplain. Modern physiography is the result of differential erosion between strong and weak rocks. Much of the area has been subject to glaciation, and the limit of glaciation (the terminal moraine) is given on the Geologic Map. The ridges and uplands, apparent on the physiographic diagram and the figures, are underlain by hard rocks resistant to erosion, such as the conglomerate of the Delaware River Water Gap, the crystalline rocks of the Reading Prong (primarily gneiss), and the basalt dikes of the Watchung Mountains. Also apparent is the scarp of the Ramapo Fault, along the contact of the upland gneiss and the Triassic sandstones and shales. The folded sedimentary rocks in the northwest portion of the figures include the conglomerate ridge, but are mostly relatively soft shales, and soluble dolomites and limestones. The shales and soluble rocks are much less resistant to erosion than the conglomerate, gneiss, and basalt. They erode more quickly and, therefore, underlie the valleys. The Great Valley is mostly underlain by the soluble rocks, i.e., dolomite and limestone. Farther to the northwest are the essentially horizontal beds of sandstones and shales of the Pocono Mountains. Their apparently irregular surface has been gouged by the glacier.

Satellite imagery is most important for terrain analysis where detailed geologic maps are generally not available, such as in parts of Africa, Asia, and South America. Digital sensors may not provide adequate coverage for areas with frequent cloud cover; radar, which penetrates clouds, is an option.

TABLE 1.5

Elements of Imagery Interpretation

Imagery Feature	Imagery Type	Interpretation
Topography	Satellite images; SLAR: stereo-pairs of aerial photos and topographic maps	• Rock masses as formed or subsequently deformed have characteristic land forms as do soil formations classified by mode of deposition or occurrence, which in all cases depend strongly on climate • Slope inclinations and heights are related to material types in terms of strength and structure • Slope failures, sinkholes, erosion gullies, etc. have characteristic forms
Drainage patterns and stream forms	Satellite images; SLAR: stereo-pairs of aerial photos and topographic maps	• Drainage patterns on a regional and local bases reflect rock type and variations, rock structure, and where soil cover adequately thick, the soil type • Stream form is also related to its geologic environment • Streams, lakes, and swamps are indicators of the ground water table, which usually follows the surface at depressed contours
Gully characteristics	Stereo-pairs of aerial photos (large scale)	Various soil types have characteristic gulley shapes
Photo tone	B&W aerial photos	Tone shows relative ground moisture and texture. Some general relationships are • White — concrete, or free-draining soils above the water table • Light gray — primarily coarse soils with some fines; acid rocks • Dull gray — slow-draining soils; basic rocks Dark gray to black — poor draining soils, organic soils groundwater near the surface
Vegetation	B&W aerial photos	Vegetation varies with climate, geologic material, and land use Tree lines often delineate floodplain limits and fault traces
Land use	B&W aerial photos	Most significant are the locations of man-made fills, cut for roadways, borrow pits, open-pit mines, and other man-made features. Development is usually related to landform
Color-enhanced (false-color) imagery	Satellite images Red normally used for near infrared. Can be presented in various colors to enhance specific features	Filtered through red: color significance • Vegetation — the brighter the red the healthier is the vegetation • Water bodies — water absorbs sun's rays, clear water shows black. Silt reflects sun's rays, sedimentation shows light blue • Urban areas — bluish-gray hues
	Multispectral photos or color IR	Various filters are used to emphasize the desired feature (vegetation type, water-body pollution, thickness of snow field, etc.)
	Thermal IR	Various filters are used to emphasize a particular feature. Can delineate water gradients to 1°F

Note: In all color-enhanced imagery, ground truth is required to identify the feature related to a specific color.

TABLE 1.6

Interpretation of Color Infrared Photos (CIR)

Color	Interpretation
Red	Healthy vegetation
Bright	Pasture, winter wheat
Darker	Evergreens: pine, conifers
Dark	Cypress
Pink	Damaged or stressed vegetation
Light	Dead or unhealthy vegetation
Light blue green	Dead or unhealthy vegetation
Bluish-gray	Dormant vegetation
Dark green-black	Wetlands
Greenish white	Fallow fields
White	Bare fields, dry soil
	Sandy beaches, gravel roads, snow
Gray	Bare fields, wet soil; urban areas
Blue	Water bodies; lakes, rivers; land fills
Light	Heavy sediment load
Blue	Moderate sediment load
Dark	Very little sediment load
Black	Clear water; or sediment

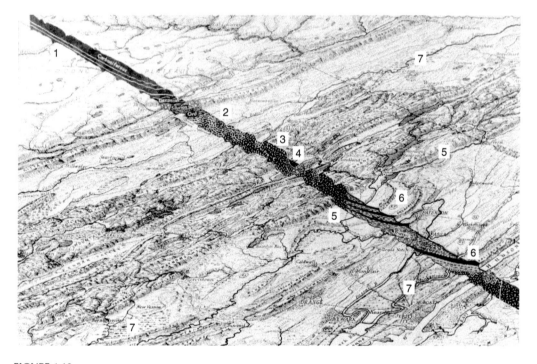

FIGURE 1.10

Physiographic diagram of northern New Jersey illustrating relationships between rock type, structure, and landform. 1, horizontally bedded sedimentary rocks; 2, folded sedimentary rocks; 3, batholith of recambrian gneiss; 4, graben formed by fault blocks; 5, scrap of the Ramapo fault; 6, basalt dikes and diabase sill; 7, glacial lake beds. See also Geology Map (Figure 1.2), satellite image (Figure 1.5), and SLAR image (Figure 1.6). (Figure drawn by E.J. Raisz, courtesy of Geographical Press, a division of Hammond World Atlas Corp. 12643.)

Environmental Studies

An important advantage of satellite imagery is the recording of changing conditions with time for a given area, such as deforestation, detection of degradation of vegetation from pollution or other causes, and the extent of river-basin flooding.

Some applications for water bodies as shown by color variations include: varying water depths in shallow water, different concentrations of sediment at mouths of rivers, and thermal gradients indicative of pollution or saltwater intrusion.

Aerial Photographs

Stereoscopic examination of stereo-pairs of aerial photographs is an extremely useful method in the determination of geologic conditions. For large areas it is preferable to obtain photos at two scale ranges 1:60,000 to 1:40,000 (Figure 1.11) and 1:20,000 to 1:8,000 (Figure 1.12), with the smaller scales providing an overview and the larger scales providing details. A stereo-pair

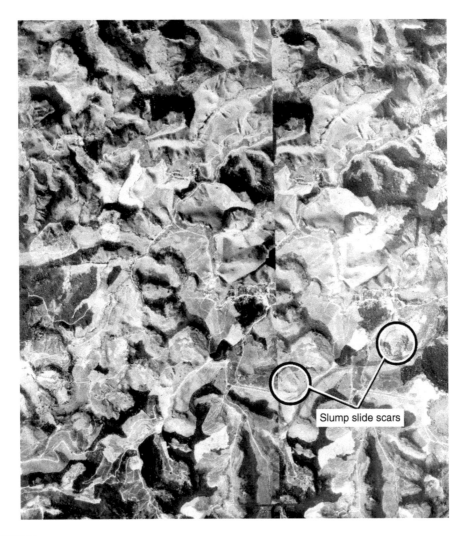

FIGURE 1.11
Stereo-pair of serial photos (scale 1:40,000) shows landform developing in metamorphic rocks from subtropical weathering. Severe erosion and a number of landslides are apparent, including the rotational slides shown on the larger scales as in Figure 1.12.

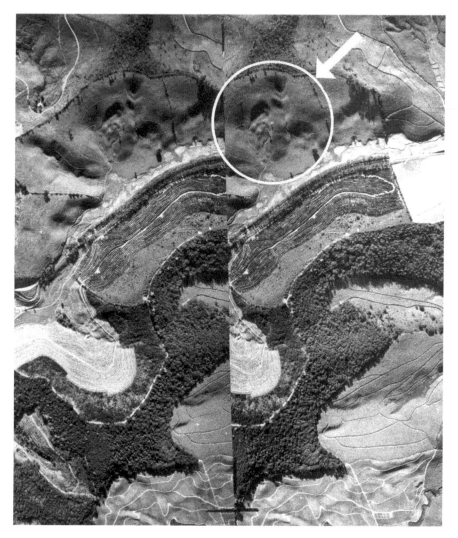

FIGURE 1.12
Stereo-pair of aerial photos (scale 1:8,000) of a portion of Figure 1.11 provides substantially more detailed information on the soil conditions and their distribution. The spoon-shaped slide that occurred in residual soils is clearly apparent. Its rounded forms, resulting from erosion, indicate that the slide is relatively old.

at a scale of 1:40,000 is given in Figure 1.11. It shows the landform developing in metamorphic rocks from subtropical weathering. Slump slide scars are observed on the photos. Figure 1.12 is a stereo-pair at a scale of 1:8,000 of a portion of the area in Figure 1.11. A geologist interpreting stereo-pairs with a stereoscope is shown in Figure 1.13.

The elements of imagery interpretation are summarized in Table 1.5. CIR (Figure 1.8) depicts landscapes in colors very different from those in true color photos. Healthy vegetation, for example, appears in shades ranging from bright pink to deep-reddish brown. Table 1.6 lists different colors of some typical surface materials as shown in CIR photographs.

Samples of stereo-pairs of aerial photographs included in this book are summarized in Appendix B. Techniques of terrain analysis and air photo interpretation are described by Way (1978), Lueder (1959), ASP (1960), Belcher (1948), and Avery and Berlin (1992), among others.

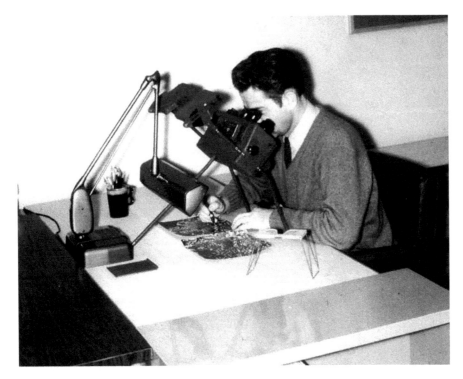

FIGURE 1.13
Stereoscopic interpretation of aerial photographs. Two types of stereo-viewers are shown.

Interpretive Features of Topographic Maps

Topographic maps, such as the USGS Quadrangle sheets, provide the simplest and least costly data for terrain analysis. Scales of 1:50,000 to 1:250,000 show regional landforms and drainage patterns and can indicate rock type and regional structural features such as folds and lineations, the latter often representing faults. Older maps often show the locations of mines. Scales of 1:10,000 to 1:24,000 provide more local detail on features such as slopes, soil formations, and sinkholes.

Interpretation of geologic conditions is based primarily on landforms as disclosed by contour lines and drainage patterns. A list of the USGS quad sheets included in this book are given in Appendix B. They illustrate many relationships between landform and geology.

The topographic map (Figure 1.3) illustrates the relationship between landforms as disclosed by contour lines and geologic conditions. Landforms reflect the differential resistance to erosion of the various rock types, as shown on the 3-D diagram inset in Figure 1.3. The area has been glaciated and is underlain with hard granitic gneiss and relatively weak shales, and soluble dolomite and limestone in the valleys. The closely spaced contours in Figure 1.3 in the gneiss indicate the steep slopes characteristic of hard rock. The marshy lands and the Wallkill River are indicative of the groundwater table in the valleys; the marshy area in the gneiss uplands, high above the river, indicate a perched watertable condition. Topographic maps are very useful in estimating groundwater conditions, although seasonal variations must be considered.

The landforms evident in Figure 1.14, a copy of the USGS Quad of Wallingford, Connecticut, are indicative of several geologic formations; the steep-sloped, irregularly shaped form in the upper left is an area of very resistant granite gneiss at shallow depths.

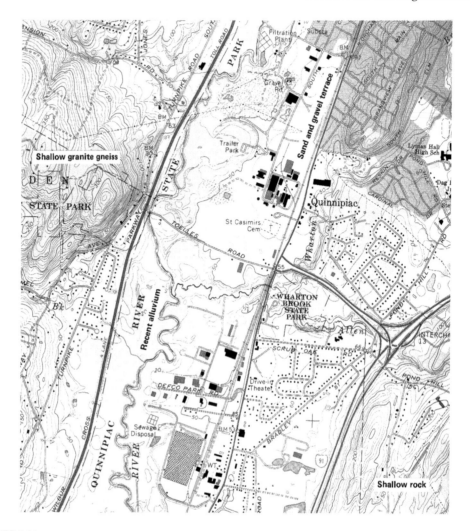

FIGURE 1.14
USGS quadrangle map, Wallington, Connecticut (scale 1:24, 000). Map provides detailed information on terrain features. (Courtesy of USGS.)

The Quinnipiac River is in its middle stage or floodway zone, flowing over the remnants of a glacial lake bed and depositing fine-grained soils during flood stages (recent alluvium). The very flat areas, generally between the railroad and the river extending through the middle portion of the map to about Elev 50, constitute a sand and gravel terrace formation, which at one time was the valley floor. A gravel pit is noted in the terrace formation, and areas of poor drainage (swamps and ponds) are apparent in the right-hand portion of the map, along Pond Hill Road. These perched water conditions above the valley floor result from the poor internal drainage of the underlying clayey glacial till.

Engineering Geology Maps

General

Data obtained from terrain analysis can be plotted to prepare an engineering geology map that provides information on geologic conditions over an entire study area.

When interpreted with experience, significant knowledge regarding the engineering characteristics of the formations and materials becomes available.

Preliminary map assessment permits conclusions regarding:

1. Abandonment of the site to avoid extremely hazardous conditions
2. Location of structures to avoid unfavorable conditions
3. General requirements for foundations and excavation
4. Formulation of the program of subsurface investigation

Map Preparation

Using a topographic map as a base map, select a scale convenient to the study area and purpose. Plot the data obtained from terrain analysis including boundaries of the various soil and rock types, major structural features, areas of shallow groundwater and rock, and locations of hazards such as sinkholes and landslides. A suggested nomenclature for the identification of various soil and rock formations is given in Table 1.7.

TABLE 1.7

Suggested Map Symbols for Engineering Geology Maps

Classification	Symbol	Modifiers Based on Coarseness or Occurrence
Residual soil	R	Rm — massive Rs — saprolite Rc — coarse-grained (granular) Rf — fine-grained (clayey or cohesive)
Colluvial soil	C	Cr — originally residual soil Cm — originally glacio-marine soils Cl — originally glacio-lacustrine soils T — talus
Alluvail soil	A	Ao — oxbow lake At — terrace Ar — recent alluvium; usually silt–sand mixtures with organic matter Ac — coarse grained; sand and gravel mixtures Am — medium grained; sand–silt mixtures Af — fine grained; silt–clay mixtures
Eolian soil	E	El — loess Ed — dunes
Glacial soils	G	Gm — moraine Gt — till Gs — stratified drift (outwash plains) Gk — kane Ge — esker Gl — lakebed
Organic soils	O	Om — marsh Os — swamp
Man-made fill	F	
High watertable	Hg	

Rock symbols		
Igneous	**Sedimentary**	**Metamorphic**
gr — granite	sg — conglomerate	qz — quartzite
ry — rhyolite	ss — sandstone	ma — marble
sy — syenite	si — siltstone	hr — hornfeld
mo — monzonite	sh — shale	gn — gneiss
di — diorite	ls — limestone	sc — schist
ga — gabbro	ak — arkose	ph — phyllite
ba — basalt	do — dolomite	sl — slate

Preparation is first on a preliminary basis; a final map is prepared after site reconnaissance and, preferably, after at least some subsurface investigation. To aid in reconnaissance, all significant cuts and other surface exposures should be noted on the map, as well as areas of questionable conditions.

Assessment of Mapped Conditions

Soil formations: Soils may be geologically classified by their origin and mode of occurrence, the engineering significance of which lies in the characteristic properties common to the various classes. Therefore, if a soil formation is classed in terms of origin and mode of occurrence, preliminary judgments can be made regarding their influence on construction.

Rock formations: The various rock types have characteristic engineering properties and structural features, either as originally formed or as deformed by tectonic or other geologic activity. The identification of rock-mass features allows the formulation of preliminary judgments regarding their influence on construction.

Three examples of engineering geology maps illustrate the approach:

1. A proposed new community in a region of glacial soils is illustrated in Figure 1.15. Conditions may be generally interpreted for engineering purposes solely on the basis of the soil types, classified geologically as follows:

    (a) *Foundation conditions*: RX, GT — good support all loads; GK — good support moderate loads; GL, GT/GL-possible suitable support for light to moderate

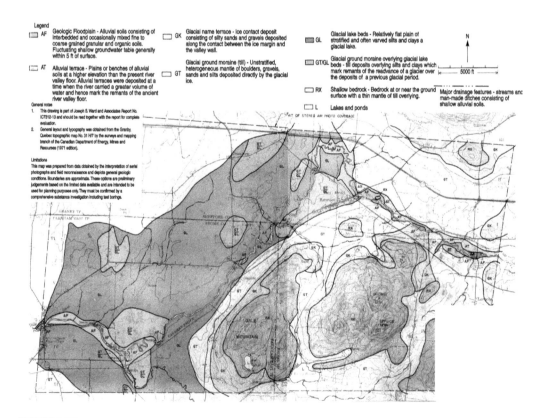

FIGURE 1.15
Preliminary engineering geology map of Bromont, Quebec, Canada. Conditions may be generally interpreted for engineering evaluations as described in the text. (Courtesy of Joseph S. Ward and Assoc.)

loads; AF-probable poor support. Areas of GK, GL, GT/GL, and AR in particular require detailed investigation. (NOTE: Terms such as "moderate loads" or "poor support" require definition in the report accompanying the map.)

(b) *Excavation conditions*: RX areas will require blasting. High groundwater can be expected in areas of AF and GL.

(c) *Borrow materials*: Coarse-grained granular soils are found in GK.

(d) *Septic tanks*: High groundwater, clayey soils, or shallow rock over much of the area imposes substantial constraint to their use.

(e) *Groundwater for potable water supply*: Most feasible locations for wells are at the base of slopes in the GT and GK materials. Buried channel aquifers may exist in the valley under the GL deposits.

2. An interstate highway planned for an area with potential slope stability problems is illustrated in Figure 1.16; slope failures in the area are shown on the stereo-pair (Figure 1.17). The area is characteristic of many glaciated valleys in the northeast United States, which were once the locations of glacial lakes.

3. A community to receive a new sanitary sewer system is illustrated in Figure 1.18a. An aerial photo of the area is given in Figure 1.18b and stereo-pairs in Figure 1.19.

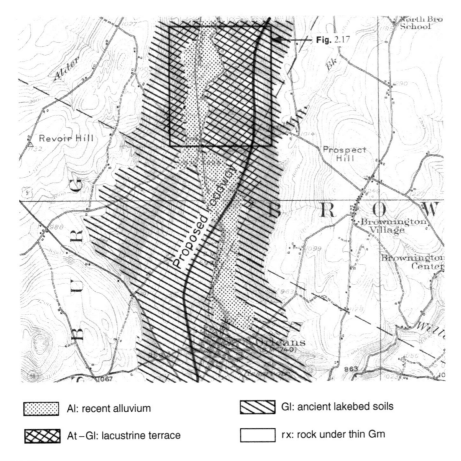

Al: recent alluvium   Gl: ancient lakebed soils

At–Gl: lacustrine terrace   rx: rock under thin Gm

FIGURE 1.16
General engineering geology map prepared for interstate highway through a valley with glacial lacustrine soils on the slopes (Barton River Valley, Orleans, Vermont). The lower slopes in the At-GL material are subjected to active movements (see stereo-pair, Figure 1.17); the upper slopes in overconsolidated GL soils (stiff to hard varved clays) will tend to be unstable in cut. In many areas, the Al soils will be highly compressible under embankment fills.

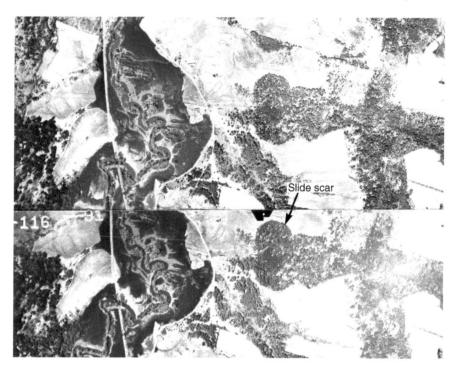

FIGURE 1.17
Stereo-pair of aerial photos showing old slide scars in glacial lakebed terrace soils (Barton River Valley, Orleans, Vermont).

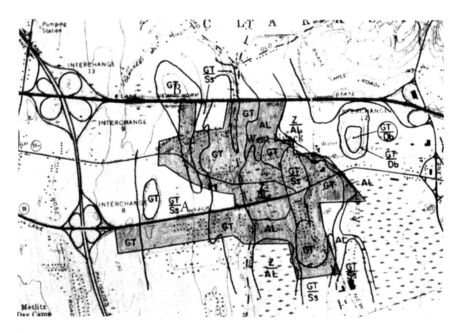

FIGURE 1.18 (A)
Preliminary engineering geology map. Sanitary sewer study, West Nyack, New York. *Note:* GT=glacial till; Ss=Sandstone; GT/Ss < 10′ GT over Ss; GT/Db < 10′ GT over diabase; AL=recent alluvium (lakebed soils); Z/AL=organic soils over alluvium. (Courtesy of Joseph S. Ward and Assoc.)

FIGURE 1.18(B)
Aerial photo of West Nyack area. Soil cut at (B); rock cut at (A).

The conditions shown on the preliminary map were investigated with seismic refraction surveys and a few test borings and pits. For sanitary sewer studies, the major concerns are depth of rock and groundwater as they affect excavations, and foundation problems caused by weak soils such as organics and lakebed soils. The large cuts along the highways expose conditions and serve as very large test pits. A rock cut in sandstones is indicated as A, and a soil cut in glacial till is indicated as B on the stereo-pairs of Figure 1.19.

Other map forms prepared for engineering studies can include:

1. *Geologic hazard,* or *risk maps,* which delineate geologic conditions in terms of various degrees of hazard or risk such as terrain where soil liquefaction or slope failures are of concern.
2. *Geologic constraint maps* form the basis for the preparation of land-use maps.

1.2.5   Site Reconnaissance

General

All sites should be visited by an experienced professional to collect firsthand information on geology, terrain and exploration equipment access, existing structures and their condition, existing utilities, and potentially hazardous conditions. Prior examination of aerial photos will identify many of the points to be examined.

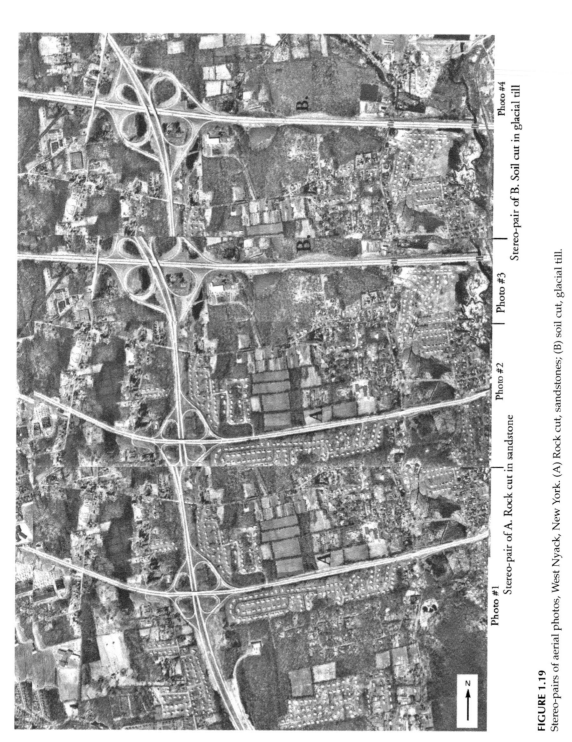

Photo #1

Stereo-pair of A. Rock cut in sandstone

Photo #2

Photo #3

Stereo-pair of B. Soil cut in glacial till

Photo #4

N

FIGURE 1.19
Stereo-pairs of aerial photos, *West Nyack, New York.* (A) Rock cut, sandstones; (B) soil cut, glacial till.

Reconnaissance Checklist

1. Examine exposures of soils and rocks in cuts (highways, rail-roads, building excavations, gravel pits, quarries, stream banks, and terraces), and on the surface, and note effluent groundwater seepage.
2. Examine slopes for signs of instability (creep ridges, tilted and bent trees, tilted poles, and slope seepage).
3. Examine existing structures and pavements for signs of distress.
4. Note evidence of flood levels along streams.
5. Contact local architects and engineers for information on foundations and local soil conditions.
6. Contact local well drillers for information on groundwater conditions.
7. Contact local public officials for building code data and information on foundations, soil conditions, and on-site utilities.
8. Note site conditions imposing constraints on access for exploration equipment.
9. Note present land use.

Revise Engineering Geology Map

The information gathered during site reconnaissance is used to revise the preliminary engineering geology map where necessary.

1.2.6 Preparation of Subsurface Exploration Program

Prepare the subsurface exploration program, considering the necessity of:

- Confirming the boundaries of the various geologic formations as mapped.
- Obtaining data for the preparation of geologic sections.
- Obtaining samples for identification, classification, and laboratory testing.
- Obtaining *in situ* measurements of the engineering properties of the materials.

1.3 Subsurface Exploration

1.3.1 General

Objectives

- To confirm or supplement the engineering geology map showing shallow and surficial distributions of the various formations.
- To determine the subsurface distribution of the geologic materials and groundwater conditions.
- To obtain samples of the geologic materials for identification and laboratory testing (Section 1.4).
- To obtain *in situ* measurements of engineering properties (Chapter 2).

Exploration Method Categories

General Categories

- *Direct methods* allow the examination of materials, usually with the recovery of samples; examples are excavations and test borings.

- *Indirect methods* provide a measure of material properties; examples are geophysical methods and the use of the cone penetrometer, which through correlations allow an estimation of material type.

Specific Categories

- Geophysical methods provide indirect data (Section 1.3.2).
- Reconnaissance methods provide direct and indirect data (Section 1.3.3).
- Continuous cone penetrometer (CPT) (Section 1.3.4).
- Test and core borings provide direct data (Section 1.3.5).
- Remote borehole sensing and logging provide direct and indirect data (Section 1.3.6).

Method Selection

Basic Factors

Selection is based on consideration of the study objectives and phase, the size of the study area, project type and design elements, geologic conditions, surface conditions and accessibility, and the limitations of budget and time.

The various methods in terms of their applicability to general geologic conditions are listed in Table 1.1.

Key Methods

Geophysical methods, particularly seismic refraction surveys, provide the quickest and often the most economical method of obtaining general information over large land areas, or in areas with difficult access, such as mountainous regions or large water bodies. They are particularly useful in investigating shallow rock conditions.

Test pits and trenches are rapid and economical reconnaissance methods for obtaining information on shallow soil, groundwater conditions, and depth and rippability of rock, and for investigating landfills of miscellaneous materials.

Test borings are necessary in almost all investigations for the procurement of soil and rock samples below depths reachable by test pits.

Other methods can be generally considered to provide information supplemental to that obtained by key methods.

1.3.2   Geophysical methods

The more common geophysical methods are summarized in Table 1.8.

Seismic Methods: General

Theoretical Basis

Elastic waves, initiated by some energy source, travel through geologic media at characteristic velocities and are refracted and reflected by material changes or travel directly through the material, finally arriving at the surface where they are detected and recorded by instruments (Figure 1.20). There are several types of elastic waves.

Compression or primary (P) waves are body waves that may propagate along the surface and into the subsurface, returning to the surface by reflection and refraction, or that may travel through the materials as direct waves. P waves have the highest velocities V_p and arrive first at the recording instrument.

Shear (S) waves are also body waves propagating and traveling in a manner similar to P waves. S waves travel at velocities V_s, from approximately $0.58V_p$ for well-consolidated

TABLE 1.8

Geophysical Methods of Exploration Summarized

Method	Applications	Comments
Seismic refraction from surface	Obtain stratum depths and velocities, land or water. Geologic sections interpreted	Most suitable if velocities increase with depth and rock surface regular
Seismic direct (crosshole, uphole, downhole)	Obtain velocities for particular strata; dynamic properties; rock mass quality	Requires drill holes. Crosshole yields best results. Costly
Seismic reflection	General subsurface section depicted. Water bodies yield clearest sections	Land results difficult to interpret. Velocities not obtained. Stratum depths comps require other data
Electrical resistivity	Locate saltwater boundaries, clean granular and clay strata, rock depth, and underground mines by measured anomalies	Difficult to interpret. Subject to wide variations. No engineering properties obtained. Probe configurations vary
Electrical conductivity	Obtain subsurface sections by data interpretation. Identify contaminant plumes by measured anomalies	Difficult to interpret. Subject to wide variations. No engineering properties obtained
Gravimeter	Detect faults, domes, intrusions, cavities, buried valleys by measured anomalies	Precise surface elevations needed. Not commonly used. Measures density differences
Magnetometer	Mineral prospecting, location of large igneous masses	Normally not used in engineering or groundwater studies
Ground-probing radar	General subsurface section depicted. Most useful to show buried pipe, bedrock, voids boulders	Interpretation difficult. Limited to shallow depths. No engineering properties
Thermography	Shallow subsurface section depicted. Useful for water pipeline leaks	Interpretation difficult. Limited to shallow depths. No engineering properties

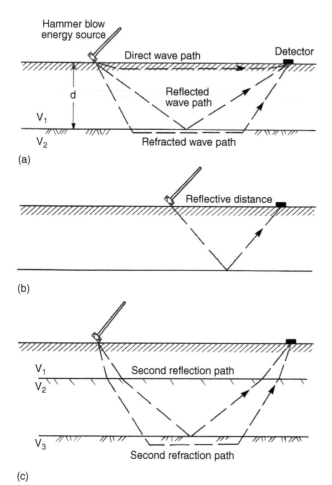

(a)

(b)

(c)

FIGURE 1.20
Transmission paths of (a) direct,
(b) reflected, and (c) refracted seismic
waves through shallow subsurface.

materials to $0.45V_p$ for poorly consolidated soils. They are not transmitted through water or across air gaps.

Rayleigh (R) waves propagate only near the surface as a disturbance whose amplitude attenuates rapidly with depth, traveling at a velocity approximately $0.9V_p$. The recorded velocity may be less because R waves travel near the surface where lower-velocity material normally occurs, and usually consist of a trail of low-frequency waves spread out over a long time interval.

Transmission Characteristics

In a given material, the arrival time of each wave at the recording instrument depends on the travel distance between the energy source and the detector, which is in turn a function of the depth of the stratum. In a sequence of strata with successively higher velocities, there is a distance between the energy source and the detector at which the refracted wave is transmitted through a higher velocity material and arrives at the detector before the direct or reflected wave. Even though the direct and reflected waves travel shorter distances, they are transmitted at lower velocities.

For land explorations to depths of less than about 1000 ft (300 m), seismic refraction techniques traditionally have been used rather than reflection because the direct and refracted waves arrive first and tend to mask the reflected waves. Reflection seismology is normally used for deep exploration and for marine studies for profiling, but does not directly yield velocity data as do refraction and direct techniques.

Seismic Exploration Techniques

Refraction techniques are used to measure compression (P) wave velocities in each geologic stratum, which are indicative of type of material and location of the groundwater table, to estimate the depths of various substrata, and to indicate the locations of faults and large caverns.

Direct techniques provide information on rock-mass characteristics, such as fracture density and degree of decomposition, and on dynamic soil and rock properties including Young's modulus, Poisson's ratio, shear modulus, and bulk modulus (Sections 2.5.3 and 2.5.5).

Reflection techniques have been used primarily in marine investigations. They provide a pictorial record of the sea-bottom profile showing changes in strata, salt domes, faults, and marine slides. Since velocities are not directly measured, material types and depths of strata can only be or cannot be inferred unless inferred when correlations are made with other data.

Energy Sources for Wave Propagation

Impact source (hammer or weight drop), used for shallow explorations on land, tends to generate disproportionately large Rayleigh surface waves, but also produces large P waves, helpful for engineering studies. *Explosives*, used for land and subaqueous studies, convert a smaller portion of their energy into surface waves, especially when placed at substantial depths below the surface. *High-energy spark* is used for subaqueous studies. See also Griffiths and King (1969) and Mooney (1973).

Seismic Refraction Method

General

Seismic refraction techniques are used to measure material velocities, from which depths of changes in strata are computed. Material types are judged from correlations with velocities.

Basic equipment includes an energy source (hammer or explosives); elastic-wave detectors (seismometers), which are geophones (electromechanical transducers) for land exploration or hydrophones (pressure-sensitive transducers) for aqueous exploration; and a recording seismograph that contains a power source, amplifiers, timing devices, and a recorder. Equipment may provide single or multiple-recording channels.

The recorded elastic waveforms are presented as seismograms.

Operational Procedures

Single-channel seismograph operation employs a single geophone set into the ground, a short distance from the instrument. A metal plate, located on the ground about 10 ft (3 m) from the instrument, is struck with a sledge hammer (Figure 1.21). The instant of impact is recorded through a wire connecting the hammer with the instrument. The shock waves travel through the soil media and their arrival times are recorded as seismograms or as digital readouts. The plate is placed alternately at intervals of about 10 ft (3 m) from the geophone and struck at each location with the hammer. Single-channel units are used for shallow exploration under simple geologic conditions.

Multi-channel seismographs employ 6 to 24 or more geophones set out in an array to detect the seismic waves, which are transmitted and recorded simultaneously and continuously. The older seismographs recorded data on photographic film or magnetic tape; and modern seismographs record digital data in discrete time units. The energy source is usually some form of an explosive charge set on the surface or in an auger hole at a shallow depth. The desired depth of energy penetration is a function of the spread length (distance between the

FIGURE 1.21
Single-channel refraction seismograph. Man in upper right with hammer striking metal plate causes seismic waves.

shot point and the farthest geophone), which should be, in general, 3 to 4 times the desired penetration depth. A normal spread would be about 300 ft (100 m) to investigate depths to 100 ft (30 m) with geophones spaced at 30 ft (10 m) intervals to define the velocity curves. The geophysicist determines the spread length and the geophone spacing to suit the anticipated geological conditions. In practice, a shot is usually set off at one end of the spread, and then another at the opposite end (reverse profiling) to detect stratum changes and sloping rock surfaces. At times, charges are set off in the middle of the spread or at other locations. Multi-channel units are used for deep exploration and all geologic conditions.

Seismograms

Seismic waveforms are usually recorded on photographic paper as seismograms. In Figure 1.22, the P wave, traveling at the highest velocity, is the first arrival to be recorded and is easily recognized. It is used to determine the depths to the various strata on the basis of their characteristic transmission velocities.

The S wave appears later in the wave train as a large pulse and is often difficult to recognize. In the figure, it is observed crossing the spread at an intermediate angle from the first arrivals, indicating a lower velocity. S wave velocities are used in conjunction with P wave velocities to compute the dynamic properties of the transmitting media (Section 2.5.3).

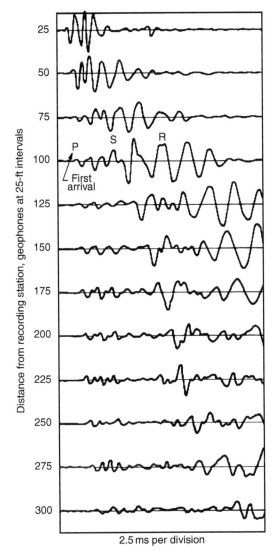

2.5 ms per division

FIGURE 1.22
Seismograms of waveforms recorded at 25-ft intervals as caused by 300-lb weight drop.

The Rayleigh wave (R) appears as a large-amplitude, low-frequency signal arriving late on the train. In the figure, it crosses the spread at an angle larger than the S wave and leaves a record at about the 200 ft spread. Although easy to recognize, the beginning is essentially indeterminate and does not provide much information for engineering studies.

Time–Distance Charts and Analysis

A typical time–distance chart of the first arrivals obtained with a single-unit seismograph is given in Figure 1.23. As the distance from the geophone to the shot point is increased, eventually the shock waves will have sufficient time to reach the interface between media of lower and higher velocities, to be refracted and travel along the interface at the higher velocity, and to arrive before the direct and reflected waves traveling through the shallower, lower-velocity material.

The travel times of the first arrivals are plotted against the distance from the geophone and the velocity of the various media determined from the slopes of the lines connecting the plotted points as shown in the figure.

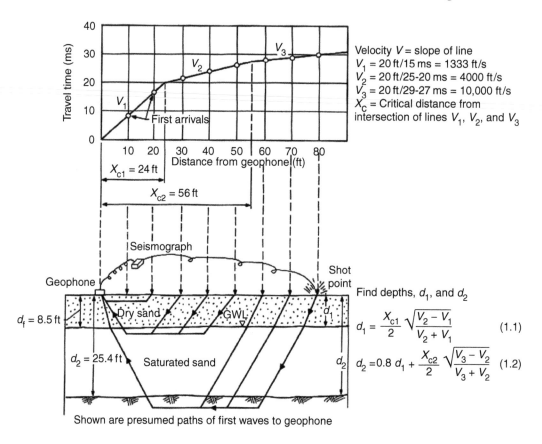

FIGURE 1.23
Time–distance graph and the solution to a three-layer problem.

Various formulas are available for computing the depth of the interfaces of the various layers, varying from simple to complex depending on the number of layers involved and the dip of the beds. The formulas for computing the depths of the relatively simple three-layer problem are given in Figure 1.23 (see equations (1.1) and (1.2)). For interpretation some information on topography must be available.

Actual seismograms for three shots along the same spread (each end and the middle) are given in Figure 1.24, and the time–distance plots in Figure 1.25. The example, a three-layer problem in a residual soil profile, is from a continuous profiling study for a railroad. The resulting subsurface section is given in Figure 1.26.

Limitations

- Softer, lower-velocity material will be masked by overlying denser, higher-velocity material and cannot be directly disclosed.
- A stratum with a thickness of less than about one fourth the depth from the ground surface to the top of the stratum cannot be distinguished.
- Erratic or "average" results are obtained in boulder formations, areas of irregular bedrock surfaces, or rock with thin, hard layers dipping in softer rock.
- Well-defined stratum interfaces are not obtained where velocity increases gradually with depth, as in residual soil grading to weathered to sound rock.

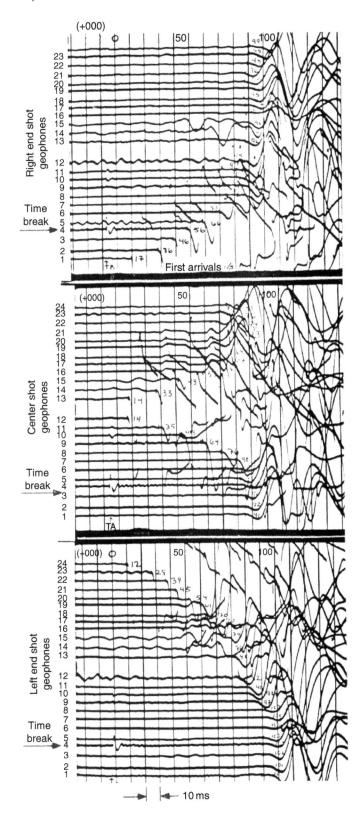

FIGURE 1.24
Seismograms for one 24-channel spread with three shot points. (Courtesy of Technosolo.)

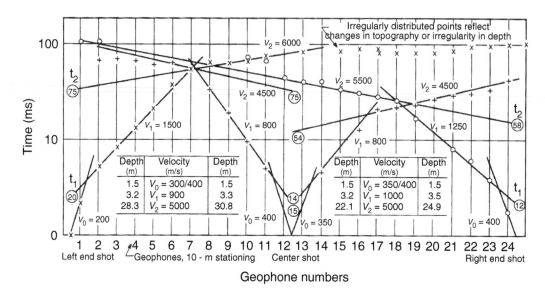

FIGURE 1.25
Time–distance graphs for the seismograms of Figure 1.24. (Courtesy of Technosolo.)

- In frozen ground, the shot point and geophones must be below the frozen zone because the shock waves travel much faster through frost than through the underlying layers.
- Application in urban areas is limited because of utility lines, pavements, foundations, and extraneous noise sources.

Applications

The method is most suitable as an exploration tool where there are media with densities that increase distinctly with depth, and fairly planar interface surfaces. In such instances, it can economically and efficiently provide a general profile of geologic conditions (Figure 1.26).

Material types are estimated from computed P wave velocities. Typical velocities for many types of materials are given in Table 1.9, and for weathered and fractured igneous and metamorphic rocks in Table 1.10.

Velocity data are also used to estimate rock rippability (see Table 2.7).

Seismic Direct Methods

Applications

Seismic direct methods are used to obtain data on the dynamic properties of rocks and soils (Chapter 2), and to evaluate rock-mass quality. See also Auld (1977), Ballard Jr. (1976), and Dobecki (1979).

Techniques (Figure 1.27)

Uphole survey: The geophones are laid out on the surface in an array, and the energy source is set off in an uncased mud-filled borehole at successively decreasing depths starting at the bottom of the hole. The energy source is usually either an explosive or a mechanical pulse instrument composed of a stationary part and a hammer.

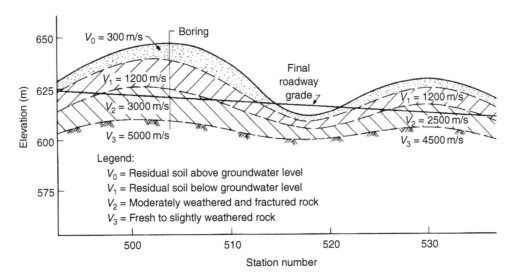

FIGURE 1.26
Example of subsurface section prepared from a continuous refraction profile. Velocities are correlated with rippability as shown in Table 2.7.

Downshole survey: The energy source is located on the surface and the detectors incorporated in a sonde which is raised or lowered in the borehole, to give either a continuous or intermittent log of adjacent materials.

Crosshole survey: The energy source is located in a center test boring and the detectors are placed at the same depth as the energy source in a number of surrounding boreholes.

Advantages over Refraction Surveys

In the uphole and downhole methods, the influence of reflection and refraction from the layers surrounding the layer of interest is substantially reduced.

In the crosshole method, the influence of surrounding layers is eliminated (unless they are dipping steeply) and the seismic velocities are measured directly for a particular stratum. *Crosshole* is the dominant and most useful technique.

Seismic Reflection Method

Application

Seismic reflection surveys obtain a schematic representation of the subsurface in terms of time and, because of the very rapid accumulation of data over large areas are used in engineering studies, primarily offshore. In recent years, as technology has improved, the method is being used with increasing frequency to pictorialize stratigraphy in land areas. The results for a landslide study are shown in Figure 1.28. The sliding mass was 3500 ft in length and the survey was to identify wedge boundaries in the failure zone.

For marine surveys, reflection methods are much more rapid than refraction surveys and, when obtained from a moving vessel, provide a continuous image of subseafloor conditions.

Operational Procedures

Continuous marine profiling is usually performed with an electric–sonic energy source generating continuous short-duration pulses, while towed behind a vessel, in conjunction with a hydrophone that detects the original pulses and reflected echo signals. The output

TABLE 1.9

Typical Compression-Wave Velocities in Soils and Rocks

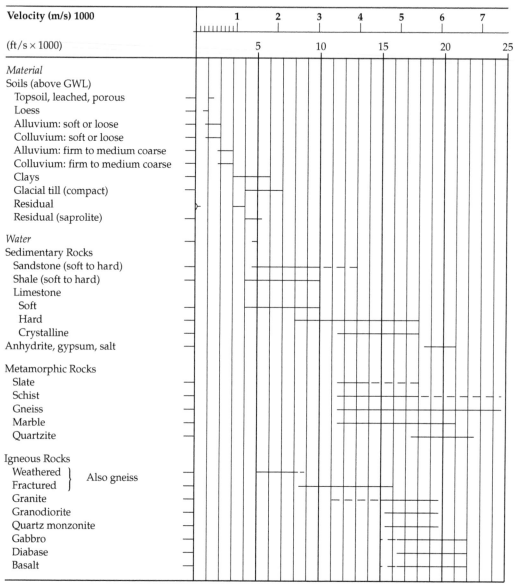

Notes: (1) Velocities given are for dry soil conditions. As the percent saturation increases so do velocities: therefore, a velocity change may indicate the water table rather than a material change. (2) The wide range of velocities given for rocks reflect the degree of weathering and fracturing: sound, massive rock yields the highest velocities.

of the hydrophone is amplified and passed to a recorder, which transcribes each spark event to sensitized paper, directly resulting in a pictorial section beneath the water surface. Equipment in use is generally of two types:

- "Subbottom Profilers," "Boomers," and "Bubble Pulsers" provide penetration depths below the water-body floor in the range of 50 to 100 m. In general, Subbottom Profilers and Boomers provide higher resolution and Bubble Pulsers provide deeper penetration. They are often used together.

TABLE 1.10

Typical *P*-Wave Velocities of Weathered and Fractured Igneous and Metamorphic Rocks

Material	Grade	V_p (m/sec)
Fresh, sound rock	F	5000+
Slightly weathered or widely spaced fractures	WS	5000–4000
Moderately weathered or moderately close fractures	WM	4000–3000
Strongly weathered or close fractures	WH	3000–2000
Very strongly weathered (saprolite) or crushed	WC	2000–1200[a]
Residual soil (unstructured saprolite), strong	RS	1200–600[a]
Residual soil, weak, dry	RS	600–300[a]

[a] V_p (water) $\approx$ 1500 m/sec. V_p (min) of saturated soil $\approx$ 900 m/sec.

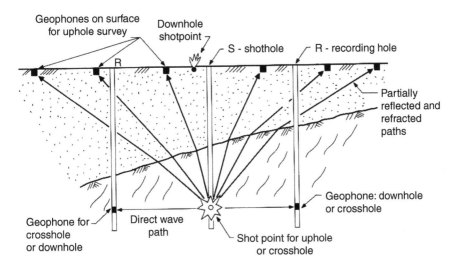

FIGURE 1.27
Direct seismic methods to measure dynamic properties of soils and rocks, and assess rock mass quality. A single borehole is used in the uphole or downhole survey; array of usually four borings is used in the crosshole survey. In uphole surveys, the geophones should be set on rock. If possible, to obtain measurements of rock quality.

- "Sparkers" and "airguns" provide penetrations to depths of 50 to 1000 m or more. They generally operate in water depths from 10 to 600 m with resolution capabilities of about 15 to 25 m.

Interpretive Information Obtained

A pictorial section beneath the seafloor, or other large water body, is obtained, showing the general stratigraphy, and such features as slumps and gas pockets when the resolution is high. Figure 1.29 is a Bubble Pulser image below a reservoir bottom in about 100 ft of water. The study purpose was to locate possible rupture zones at the toe of a large land-slide. Figure 1.9 is a side-scan sonar image of the area.

Deeper penetration methods, such as sparker surveys, can indicate major geologic structures such as faults and salt domes (Figure 1.30) and massive submarine slides.

Velocities cannot be calculated with reliability since distances are not accurately known, and therefore material types and stratum depths cannot be evaluated as in refraction

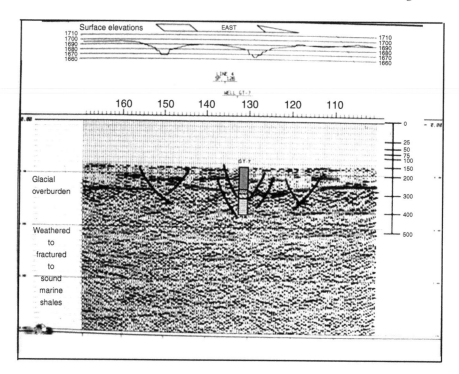

FIGURE 1.28
Seismic reflection profile for landslide study. Shown are interpreted slope failure surfaces. (Courtesy of Woodword–Clyde Consultants.)

methods. Depths are estimated by assuming a water velocity of 2500 m/sec, but variations in strata impedance affect the thickness scale. Test borings or refraction studies are necessary for depth- and material-type determinations.

Electrical Resistivity Methods

Applications

The resistivity of soil or rock is controlled primarily by pore water conditions that vary widely for any material. Therefore, resistivity values cannot be directly interpreted in terms of soil type and lithology. Some applications are:

- Differentiation between clean granular materials and clay layers for borrow-material location.
- Measurement of the thickness of organic deposits in areas difficult to access.
- Measurement of the depth to a potential failure surface in "quick" clays in which the salt content, and therefore the resistivity, is characteristically different near the potential failure surface.
- Location of subsurface saltwater boundaries.
- Identification of variations in groundwater quality in homogeneous granular deposits, as may be caused by chemical wastes leaking from a storage basin.
- Measurement of depth to bedrock and delineation of varying rock quality.
- Location of solution cavities in limestone and underground mines (not always successful).

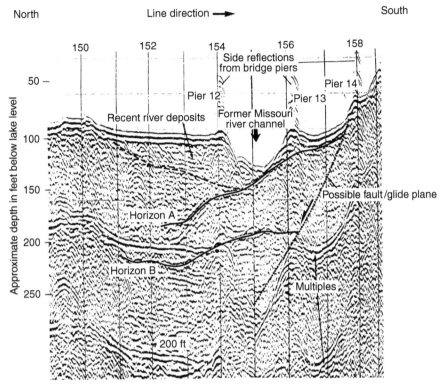

FIGURE 1.29
Subbottom profiling with bubble pulser to locate possible failure surface at toe of landslide. Section is in area of side-scan sonar image (see Figure 1.9). (Courtesy of Woodward–Clyde Consultants.)

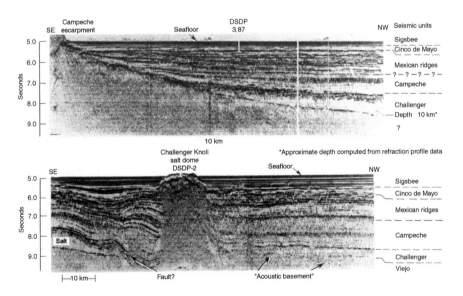

FIGURE 1.30
Part of multichannel seismic reflection section, abyssal western Gulf of Mexico. (From Ladd, J. W. et al., Geology, GSA, 1976. With permission.)

Theoretical Basis

Various subsurface materials have characteristic conductances for direct currents of electricity. Electrolytic action, made possible by the presence of moisture and dissolved salts within the soil and rock formations, permits the passage of current between electrodes placed in the surface soils. In general, conductance is good in materials such as moist clays and silts, and poor in materials such as dry loose sands, gravels, and sound rocks.

Resistivity refers to the resistance to current flow developed in geologic materials and is expressed as Ω cm^2/cm, or simply as Ω cm or Ω ft. Some typical values of resistivity for various geologic materials are given in Table 1.11.

Apparatus

The electrical resistivity apparatus consists of a battery as energy source, a milliammeter, a potentiometer, and electrodes (Figure 1.31). There are three basic electrode configurations as shown in Figure 1.32.

Wenner array: Commonly used in the United States, it employs four equally spaced electrodes. All four electrodes are moved between successive operations.

Schlumberger array: Commonly used in Europe, it is similar to the Wenner, except that the spacing between the two center electrodes is made smaller than that between the other two. In operation, the inner electrodes remain fixed and the outer electrodes are moved. The test is repeated changing the spacing between the inner electrodes.

Dipole–dipole array: The source dipole is separated from the receiving dipole and the distances between the two dipoles are varied.

Pole–dipole array (not shown in Figure 1.32): One of the current electrodes is placed at a distance from the survey area approximately 5 to 10 times the desired survey depth. The receiving dipoles are stepped away from the current electrode, providing resistivity measurements that are reasonably representative of media encountered at a specific depth and distance from the current electrode.

Operational Procedures

With a battery as a direct-current source, a current flow is established between the two outer electrodes. The current drop is detected by the two inner electrodes and recorded on

TABLE 1.11

Typical Resistivity Values for Geologic Materials[a]

Materials	Resistivity	
	Ω ft	Ω m
Clayey soils: wet to moist	5–10	1.5–3.0
Silty clay and silty soils: wet to moist	10–50	3–15
Silty and sandy soils: moist to dry	50–500	15–150
Bedrock: well fractured to slightly fractured with moist soil-filled cracks	500–1000	150–300
Sand and gravel with silt	About 1000	About 300
Sand and gravel with silt layers	1000–8000	300–2400
Bedrock: slightly fractured with dry soil-filled cracks	100–8000	300–2400
Sand and gravel deposits: coarse and dry	>8000	>2400
Bedrock: massive and hard	>8000	>2400
Freshwater	67–200	20–60
Seawater	0.6–0.8	0.18–0.24

[a] From Soiltest, Inc.

Note: (1) In soils, resistivity is controlled more by water content than by soil minerals. (2) The resistivity of the pore or cleft water is related to the number and type of dissolved ions and the water temperature.

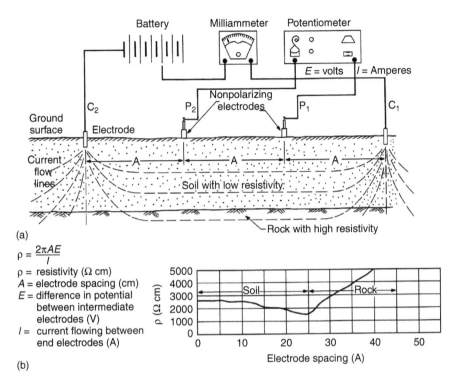

FIGURE 1.31
(a) Components of the electrical resistivity apparatus and the common four- electrode configuration of the Wenner array. (b) Typical resistivity curve. (From ASTM, Symposium on Surface and Subsurface Reconnaissance, Philadelphia, 1951. Copyright ASTM International, reprinted with permission.)

the potentiometer, and the "apparent" resistivity (Wenner array) is computed from the expression

$$\rho = (2\pi AE)/I \qquad (1.3)$$

where ρ is the soil resistivity, A the distance between electrodes (in cm), E the differential potential between intermediate electrodes (in V), and I the current flowing between end electrodes (in A).

"Apparent" resistivity signifies an average value resulting from layering effects.

In *vertical profiling*, the electrode spacing is increased while the resistivity changes are recorded, and a curve of resistivity vs. electrode spacing is drawn as shown in Figure 1.31. As the value of resistivity obtained is largely dependent upon material resistivities to a depth equal to the electrode spacing A, material changes can be inferred from the change in the slope of the curve. For any given depth, in terms of electrode spacing, the *lateral variations* in resistivity are measured by moving the rear electrode to a front position and marching the array laterally. Subsurface conditions are inferred from the variations in vertical and lateral values. In multilayered systems, interpretations must be correlated with test borings.

To interpret vertical profiling, a set of empirical curves (Wetzel–Mooney curves) is used to estimate the depth to an interface and the resistivity. The curves, log–log plots of apparent resistivity vs. electrode separation, are matched to the curves drawn from the field data.

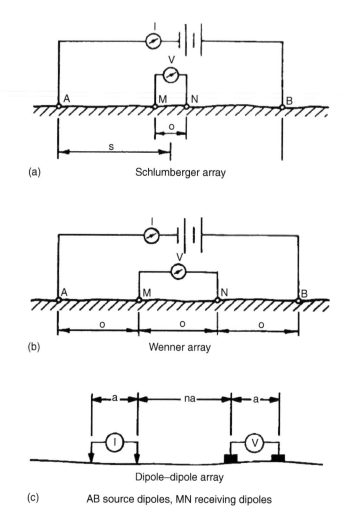

(a)   Schlumberger array

(b)   Wenner array

(c)   AB source dipoles, MN receiving dipoles

FIGURE 1.32
Electrode (array) configurations for resistivity measurements. (From ASCE *Technical Engineering and Design Guides*, adapted from the U.S. Army Corps of Engineers, 1998.)

Underground Coal Mine Study

A colleague advised the author of a study of an underground coal mine, where an expert in electrical resistivity used a "pole–dipole" method. Three different signatures were obtained, as shown in Figure 1.33: intact rock, caved rock, and voids. The signatures, initially considered as anomalies, were confirmed by core borings and comparisons with old mine maps.

Railway Tunnel Through Poor Quality Rock

Dahlin et al. (1996) report on a resistivity investigation for a railway tunnel in Sweden. The imaging system consisted of a resistivity meter, a relay-matrix switching unit, four electrode cables, a computer, steel electrodes, and various connectors. The total length of investigation was of the order of 8200 m to depths of 120 m. Color-coded results showed variation in rock quality along the entire proposed alignment. Low resistivities indicated very poor rock. Core borings confirmed the interpretations.

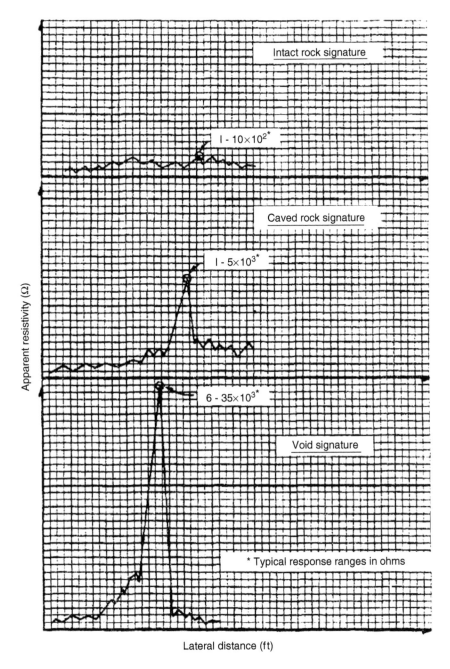

Lateral distance (ft)

FIGURE 1.33
Results of electrical resistivity study for coal mine using pole–dipole method. (Courtesy of Woodward–Clyde Consultants.)

Limitations

Since resistivity is a function of water content and soluble salts, materials with widely differing engineering properties can have the same resistivity. Therefore, correlations from one location to another may not be possible. Differentiation between strata may not be possible where the overlying material has an extremely high resistance.

Water-table location often limits the depth for practical study because conductivity rises sharply in saturated materials and makes differentiation between horizons impossible.

Because of the difficulties of relating measured resistivity values to specific soil or rock types, subsurface conditions are usually inferred from the vertical and lateral variations in the measured values. In multilayered systems, interpretations must be confirmed by correlations with test boring data and, in general, electrical resistivity should be considered as a preliminary exploration method.

Electrical Conductivity (EM) Surveys

Applications

Terrain conductivity meters read directly the apparent conductivity, and interpretation of profiles is usually qualitative, based on showing anomalies that are then investigated by other methods. Some applications are:

- Mapping nonorganic contamination of groundwater, which usually results in an increase in conductivity over "clean" groundwater. An example is acid mine drainage.
- Mapping soil and groundwater salinity.
- Mapping depth to basement rock.
- Locating buried metal tanks and drums.
- Locating buried mine adits.

Operational Procedures

Conductivity is compared with resistivity in Figure 1.34. The electrical conductivity meter, the Geonics EM 34, uses two dipoles which can be used in either the vertical or horizontal mode. A single dipole, in the horizontal mode, is shown in Figure 1.35. Each mode gives a significantly different response. One person carries a small transmitter coil, while a second person carries a second coil that receives the data from the transmitter coil. Electrical contact with the ground is not required and rapid exploration to depths of 60 m are possible.

The electrical conductivity meter, EM 31 (Figure 1.36), is operated by one person. Exploration is rapid, but the effective depth of exploration is about 6 m.

Underground Mine Study

Old maps indicated that an adit for a lead–zinc mine was located somewhere in an area to be developed for expensive home construction. The general area had suffered from

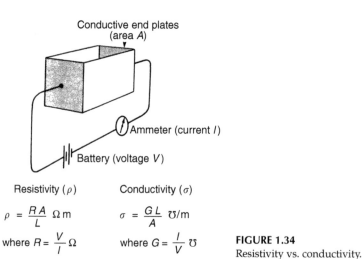

Conductive end plates
(area *A*)

Ammeter (current *I*)

Battery (voltage *V*)

Resistivity (ρ)

$$\rho = \frac{RA}{L} \; \Omega\,m$$

where $R = \frac{V}{I} \; \Omega$

Conductivity (σ)

$$\sigma = \frac{GL}{A} \; \mho/m$$

where $G = \frac{I}{V} \; \mho$

FIGURE 1.34
Resistivity vs. conductivity.

FIGURE 1.35
Terrain conductivity meter (EM34). (Courtesy of Geonics.)

occasional ground collapse in the old mines. The area was surveyed using an EM34 and the results identified an anomaly with 0 Ω (Figure 1.37). Continuous airtrack probes and core borings confirmed that the mine was at a depth of about 40 ft.

Ground-Probing Radar (GPR)

Applications

Ground-probing (ground-penetrating) radars are used as a rapid method of subsurface profiling. They are designed to probe solids relatively opaque to radar waves, such as pavement-reinforcing rods and bases, as well as subpavement voids, buried pipes, bedrock surfaces and overlying boulders, caverns, tunnels, clay zones, faults, and ore bodies. Images consist essentially of wavy lines. They are difficult to interpret and primarily locate anomalies that require additional investigation.

Theoretical Basis

Energy is emitted in the radio portion of the electromagnetic spectrum (Figure 1.4), of which some portion is reflected back to the radar equipment. Various materials have differing degrees of transparency to radar penetration. New GPRs are digital with an improved signal-to-noise ratio using fiber optics cables and improved antenna design.

FIGURE 1.36
Terrain conductivity meter (EM31). (Courtesy of Geonics.)

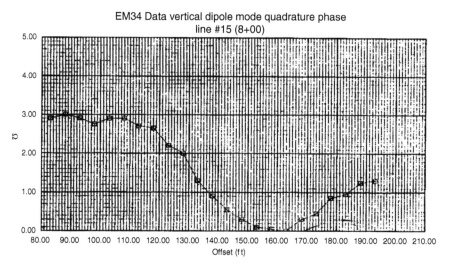

FIGURE 1.37
EM34 survey profile to locate a lead–zinc mine ca. 1860: 0U was found to be the location of the mine. (Courtesy of Woodward–Clyde Consultants.)

Techniques

Continuous subsurface profiling by impulse radar, also termed electromagnetic subsurface profiling (ESP), provides a registration of a continuously reflected radar pulse similar to seismic reflection images. A sled-mounted antenna is towed behind a small vehicle or boat containing the ESP system. It has been used since 1970 to locate buried sewer lines and cables, evaluate pavement conditions, detect voids, and profile the bottoms of rivers and lakes. The ESP technique provides clear images in low-conductivity materials such as sand, freshwater, or rock, and poor results in high-conductivity materials such as wet clay because the penetration depth is limited by the strong attenuation of the signal.

Magnetometer Surveys

Applications

Magnetometer surveys are used for the detection of magnetic ore bodies or rocks that are strongly magnetic, such as the crystalline types as differentiated from sedimentary types. They are seldom used for engineering studies.

Theoretical Basis

Many rocks contain small but significant quantities of ferromagnetic materials which vary with rock type. The weak magnetization modifies the Earth's magnetic field to an extent that can be detected by sensitive instruments.

Operational Procedure

Magnetometers provide the measurements and, when towed behind aircraft, they can cover large areas and provide appreciable data in a relatively short time.

Data Presentation

Contour maps are prepared showing lines of equal value that are qualitatively evaluated to locate anomalies indicative of ore bodies or rock-type changes.

Gravimeter Surveys

Applications

In their normal geologic application, gravimetric or microgravity surveys, are used for the detection of major subsurface structures such as faults, domes, anticlines, and intrusions. Gravimetric surveys have been used in engineering studies to detect cavities in limestone and the location of old mine shafts (Ghatge, 1993). Modern instruments are extremely sensitive, however, and the requirement for the precise determination of surface elevations may cause the application of the method to be relatively costly.

Theoretical Basis

Major geologic structures impose a disturbance on the Earth's gravitational field. The part of the difference between the measured gravity and theoretical gravity, which is purely a result of lateral variations in material density, is known as the Bouguer anomaly. Other factors affecting gravity are latitude, altitude, and topography, and have to be require considered during gravitational measurements to obtain the quantity representing the Bouguer anomaly.

Gravimeters consist of spring-supported pendulums similar in design to a long-period seismograph.

Data Presentation

Isogal maps are prepared showing contours of similar values given in milligals (mgal) to illustrate the gravity anomalies. (Note: 1 mgal = 0.001 gal; 1 gal = acceleration due to gravity = 1 cm/s^2.)

Cavity Exploration in Soluble Rock

The bulk density of limestones is about 160 pcf (2.6 g/cm^3), and that of soils generally ranges from 100 to 125 pcf (1.6 to 2.0 g/cm^3). In karst regions, a gravity-low anomaly may indicate an empty cavity, a cavity filled with low-density material, or a change in soil or groundwater conditions. Microgravimetric instruments have been developed in recent years which permit a precision of 0.01 mgal (10 μgal) or better (Greenfield, 1979), equivalent to a change in soil thickness of about 10 in. (24 cm) for a density contrast between soil and rock of 62.4 pcf (1.0 g/cm^3). A detected anomaly is then explored with test borings.

Infrared Thermography

An infrared scanner, which can detect very small variations in temperature, is mounted on a sled that can be towed by a person or a vehicle. Multicolor images are displayed and recorded. Cooler areas are depicted in green or blue and the warmer areas in orange or red. In engineering studies, infrared scanners are most useful in detecting water leaks in sewers or water lines. In limestone areas, such leaks are often precursors to sinkhole development.

1.3.3   Reconnaissance Methods

General

Reconnaissance methods of exploration are divided into two general groups as follows:

- *Large excavations* allow close examination of geologic materials and include test pits, test trenches, and large-diameter holes, which can be made relatively rapidly and cheaply. Adits and tunnels, although costly to excavate, are valuable for investigating rock mass conditions.
- *Hand tools and soundings* provide low-cost and rapid means of performing preliminary explorations. Samples are recovered with the hand auger and 1in. retractable plug sampler, and probings with bars provide indications of penetration resistance to shallow depths. The cone penetrometer test (CPT) and the standard penetration test (SPT) are used also as reconnaissance methods for preliminary explorations.

Methods

The various methods are summarized in Table 1.12, including:

- Test pit or trench excavation
- Large-diameter holes
- Adits and tunnels
- Bar soundings
- Hand auger or posthole digger
- One-inch retractable plug sampler
- Continuous penetrometer test (see Sections 1.34 and 2.4.5)
- Standard penetration test (see Section 2.4.5)

1.3.4   Continuous CPT(ASTM D5778)

History

In 1932, the Dutch developed a simple device to measure continuously soil properties *in situ*. At the end of steel rods was a penetrometer consisting of a cone tip. Over the years, the rods were advanced by driving them with a hammer (dynamic force), applying dead

TABLE 1.12

Reconnaissance Methods of Exploration

Method	Applications	Procedure	Limitations
Test pit or trench excavation	• Detailed examination of soil strata • Observation of groundwater seepage • Identification of GWL • Recovery of disturbed or undisturbed samples above GWL and *in situ* density tests • Examination of fault zones • Examination of miscellaneous and rubble fills • Identification of rock surface and evaluation of rippability • Borrow material investigations	• Excavation by backhoe, bulldozer, or by hand • Can be extended below GWL by sheeting and pumping if soils have at least some cohesion	• Usually limited in depth by water table (GWL), rock depth, or reach of equipment • Can be dangerous if left unsheeted and depths are above 4 to 5 ft
Large-diameter holes	• Detailed examination of strong cohesive soil strata and location of slickensides and other details affecting stability and seepage	• Holes 60 to 100 cm in diameter excavated by rotating large auger bucket (Figure 1.52), or excavated by hand	• Strong cohesive soils with no danger of collapse • Rock penetration limited except by calyx drilling (Section 1.4.5) • Very costly
Adits and tunnels	• Used in rock masses for preparation of detailed geological sections and *in situ* testing; primarily for large dams and tunnels	• Excavation by rock tunneling methods	• Rock masses that do not require lining for small-diameter tunnels are left open for relatively short time intervals
Bar soundings	• To determine thickness of shallow stratum of soft soils	• A metal bar is driven or pushed into ground	• No samples obtained • Penetration limited to relatively weak soils such as organics or soft clays
Hand auger or posthole digger	• Recovery of disturbed samples and determination of soil profile to shallow depths • Locate GWL (hole usually collapses in soils with little to slight cohesion)	• Rotation of a small-diameter auger into the ground by hand	• Above GWL in clay or granular soils with at least apparent cohesion • Below GWL in cohesive soils with adequate strength to prevent collapse • Penetration in dense sands and gravels or slightly plastic clays can be very difficult
One-inch retractable-plug sampler	• Blows from driving give qualitative measure of penetration resistance to depths of 30 m in soft clays • One-inch diameter samples can be retrieved up to 1 m in length	• Small-diameter casing is driven into ground by 30 lb slip hammer dropped 12 in. • Samples are obtained by retracting driving plug and driving or pressing the casing forward	• The entire rod string must be removed to recover sample • Penetration depth in strong soils limited • Small-diameter samples
Continuous cone penetrometer (CPT)(see Sections 1.3.4 and 2.4.5)	• Continuous penetration resistance including side friction and point resistance for all but very strong soils on land or water	• Probe is jacked against a reaction for continuous penetration	• No samples are recovered
Standard penetration test (SPT)(see Section 2.4.5)	• Recovery of disturbed samples and determination of soil profile • Locate GWL	• Split-barrel sampler driven into ground by 140 lb hammer dropped 30 in.	• Penetration limited to soils and soft rocks. Not suitable for boulders and hard rocks

weights (static force), or pushing hydraulically. Initially, only penetration resistance at the cone tip was measured (q_c or q_t). Later, a cone was developed with a sleeve to measure shaft (side) friction f_s (Begemann cone) in addition to tip resistance. The Begemann cone was termed a subtraction cone. It measures the total sleeve plus tip force on the cone and the tip resistance when pushed into the ground. Sleeve friction is calculated by subtracting the tip resistance from the total resistance.

Fugro, ca. 1965, developed an electric cone (the compression cone) that measured and recorded both tip resistance and shaft friction separately. Some electric cones have a maximum value for sleeve friction of the order of 20 tons. The subtraction cone has no sleeve friction limit; the only limit is the total penetrometer force. Subtraction cones can be used where sleeve friction is high, such as in very stiff clay, and the limit of the electric cone is exceeded.

CPT Operations

Modern cones are pushed continuously into the ground by a hydraulic-force apparatus reacting against a machine. The apparatus can be mounted on a variety of platforms, including truck or track mounts, small portable units, and barges or drill ships. The interior of a truck-mounted CPT is shown in Figure 1.38a. Large modern rigs have capacities of up to 30 metric tons. CPT rigs are often mounted with test boring drill rigs, but the reaction force is limited.

Advanced by hydraulic thrust, the electric cones employ load cells and strain gages that measure electronically both tip resistance and local sleeve friction simultaneously. The results are recorded digitally at the surface with an accuracy of measurement of usually better than 1%. Readings are usually taken at 5 cm intervals. Cones vary in size with areas of 10 and 15 cm^2 the most common because ASTM criteria apply. The 15 cm^2 cones can push well in loose gravels, cemented sands, and very stiff fine-grained soils and weathered rock. Various cone sizes are shown in Figure 1.38b.

The CPT method permits rapid and economical exploration of thick deposits of weak to moderately strong soils and provides detailed information on soil stratification. There have been many modifications to cone penetrometers in the past 20 years. The test can measure *in situ* many important soil properties applicable to geotechnical and environmental studies as summarized below. The interpretation of strength and compressibility properties are covered in Section 2.4.5.

Although soil sampling is possible with a special tool, soil samples are normally not obtained. CPT data are usually confirmed with test borings and soil sampling, but the number of borings is significantly reduced.

See also ASTM D5778, Sanglerat (1972), Schmertmann (1977), and Robertson et al. (1998).

Engineering Applications

Standard CPT. The common application of the CPT is to obtain measurements of engineering strength properties. In relatively permeable soils, such as fine and coarser sands, pore pressure effects during penetration at standard rates often have negligible influence, and the CPT measures approximately fully drained behavior. In homogeneous, plastic clays, the CPT measures approximately fully undrained behavior. Mixed soils produce in-between behavior.

Piezocones are currently in common use (CPTU). They have a porous element near the tip and a built-in electric transducer to measure pore water pressure in addition to tip resistance and shaft friction. Information on stratification and soil type is more reliable than the standard CPT. The interpretation of material strength properties is improved, and data are obtained on deformation characteristics. To obtain pore-pressure data, penetration is

FIGURE 1.38 (a)
Cone penetrometer test equipment: (a) interior of CPT truck showing hydraulic force apparatus. (Courtesy of Fugro.)

stopped at the desired depth and readings are taken until the pore pressure generated by the penetration has dissipated. The coefficient of consolidation (c_h) and the coefficient for horizontal hydraulic conductivity (k_h) are determined. Hole inclination is also recorded with the piezocone illustrated as in Figure 1.39a. A plot of a CPTU log showing point and shaft friction, friction ratio, pore pressures, and a soil log is given in Figure 1.40. CPT plots now normally include the friction ratio and, with the CPTU, pore pressure measurements.

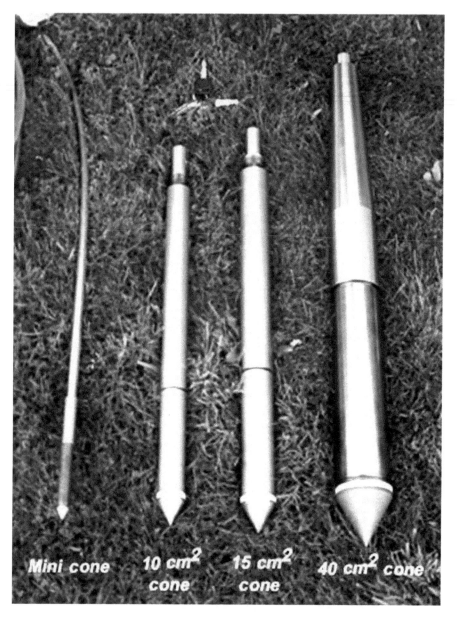

FIGURE 1.38 (b)
Cone penetrometer test equipment: (b) various cone sizes. (Courtesy of ConeTec.)

The *seismic cone penetration test* (SCPT) (Figure 1.39a) combines the piezocone with the measurement of small strain shear wave velocities (P and S waves). A small geophone or accelerometer is placed inside a standard cone, and seismic wave velocities are measured during pauses in cone penetration. A hammer blow to a static load on the surface can provide the shear wave source. Explosives can be used offshore or the "downhole seismic test" onshore. The results have been used for evaluating liquefaction potential (Robertson, 1990).

The *active gamma penetrometer* (GCPT) (Figure 1.39b) measures *in situ* soil density. The test is particularly important in clean sands, that are difficult to sample in the undisturbed state.

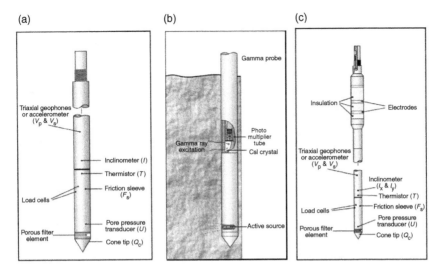

FIGURE 1.39
Various types of cone penetrometers. (a) Piezo cone penetrometer (CPTU): measures tip resistance, shaft friction, pore pressures, temperature, inclination, and shear wave velocities. (b) Active gamma penetrometer: GCPT measures *in situ* soil density, particularly useful in sands which are difficult to sample undisturbed. (c) Electrical resistivity cone (RCPT): provides measures of relative soil resistivity. (Courtesy of ConeTec.)

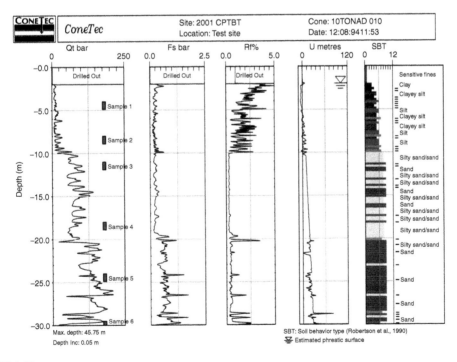

FIGURE 1.40
Sample of CPTU log. Data plots are provided in real time. (Courtesy of ConeTec.)

A *vision cone penetrometer* (VisCPT) has been developed recently to overcome the problem of no recovered samples (Hryciw et al., 2002). Miniature cameras are installed in the CPT probe, and continuous images of the soil's stratigraphy are obtained through synthetic sapphire windows mounted in the side of the probe.

Environmental Applications

Cone penetrometers have been modified for environmental studies and the identification of contaminants (Robertson et al., 1998). Sensors have been developed to measure temperature, pH, radioactivity (gamma), resistivity, ultraviolet-induced fluorescence, and total petroleum hydrocarbons and other contaminants. SCAPS (site characterization and analysis penetrometer system) is a program in use by the government agencies in which the cone penetrometer is used for hazardous waste site characterization. An electrical resistivity cone (RCPT) is shown in Figure 1.39c.

Classification of Materials

Correlations

CPT values are influenced by soil type and gradation, compactness, and consistency, which also affect the relationship of q_c with f_s. Correlations have been developed between cone-tip resistance (q_c is also referred to as cone-bearing capacity) and the friction ratio $R_f (= f_s/q_c)$ to provide a guide to soil classification as given in Figure 1.41. These charts do not provide a guide to soil classification based on grain size, but rather on soil behavior type. Figure 1.41 is based on data obtained at predominantly less than 30 m, and overlap between zones should be expected (Robertson, 1990).

It has been found that cones of slightly different designs will give slightly different values for q_c and f_s, especially in soft clays and silts. This apparently is due to the effect that water pressures have on measured penetration resistance and sleeve friction because of unequal end areas. It has also been recognized that overburden pressure increases with depth also affecting strength values, as it does with the Standard Penetration Test results.

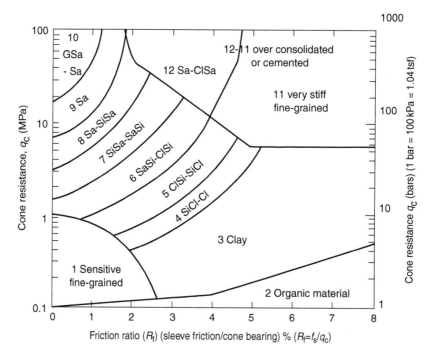

FIGURE 1.41
Simplified soil behavior type classification for standard electric friction cone. (Adapted from Robertson, P.K., et al., *Proceedings, In situ '86*, Blacksburg, VA, 1986.)

For these reasons, correction factors have been proposed for the strength parameters and cone geometry. For equal end area cones only q_t is normalized. It is noted that cone resistance q_c is corrected to total cone resistance q_t as follows (Robertson, 1990):

$$q_t = q_c + (1-a)u \qquad (1.4)$$

where u is the pore pressure measured between the cone tip and the friction sleeve and a is the net area ratio.

New classification charts have been proposed based on normalized data (Robertson, 1990).

Operations Offshore

General

Offshore shallow-water investigations, such as for ports and harbors, normally involve water depths of 3 to 30 m. Jack-up rigs or spud barges are used and the cone is pushed from the vessel by conventional methods. A drill casing is lowered to the seafloor to provide lateral support for the CPT rods.

For offshore deep-water exploration, such as for oil-production platforms, the CPT is usually operated in conjunction with wire-line drilling techniques (Section 1.3.5), with equipment mounted on large vessels such as shown in Figure 1.42. The major problem, maintaining adequate thrust reaction from a vessel subjected to sea swells, can be overcome by a motion compensator and the drill string. Thrust reaction can be provided by weighted frames set on the seafloor as shown in Figure 1.42.

Seafloor Reaction Systems

Underwater cone penetrometer rigs that operate from the seafloor have been developed by several firms. The Fugro-McClelland system, called "Seaclam," operates in water depths up to 300 m. A hydraulic jacking system, mounted in a ballasted frame with a reentry funnel, is lowered to the seabed (Figure 1.43). Drilling proceeds through the Seaclam and when sampling or testing is desired, a hydraulic pipe clamp grips the drill string to provide a reaction force of up to 20 tons. A string of steel rods, on which the electric friction cone is mounted, is pushed hydraulically at a constant rate of penetration. Data are transmitted digitally to the drill ship.

ConeTec have developed an underwater CPT that can operate in water depths up to 2500 ft that can penetrate to 30 ft below the mudline. It is lowered over the side of a steel vessel and set on the seafloor. Fugro also has an underwater ground surface CPT which presently has a penetration of about 6 m.

Sampling and In Situ Testing

The various underwater sampling and *in situ* tools using the Fugro Seaclam are illustrated in Figure 1.43. Some sampling and testing is obtained by free-falling down the drill pipe. Fugro have developed the "Dolphin" system for piston sampling, *in situ* CPT, and vane shear testing to water depths of at least 3000 m. Tools that require controlled thrust for operation, such as the cone penetrometer and piston sampler, employ a mud-powered thruster assembly at the base of the drill string. Vane testing and push sampling do not require the use of the thruster.

The Dolphin cone penetrometer is illustrated in Figure 1.44, and a plot of cone resistance data is shown in Figure 1.45. Penetration distances are 3 m or may refuse penetration at less than 3 m, and tip resistance and side friction are recorded. Drilling advances the hole to the next test depth and the CPT repeated. The remote vane shear device is shown in

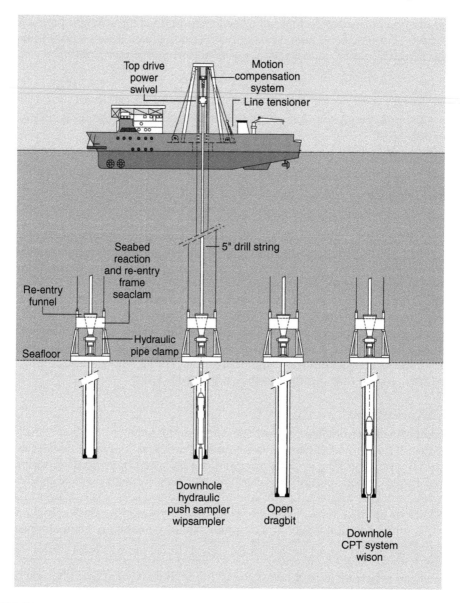

FIGURE 1.42
Schematic diagram of the operational procedures from a drill ship. Shown are the Fugro Seaclam for deep water drilling, push sampling, and *in situ* testing using a cone penetrometer. (Courtesy of Fugro.)

Figure 1.46 and a data plot is shown in Figure 1.47. The vane data, compared with undrained laboratory test results from piston samples, are shown in the figure.

1.3.5 Test and Core Borings

Purpose

Test and core borings are made to:

- Obtain samples of geologic materials for examination, classification, and laboratory testing

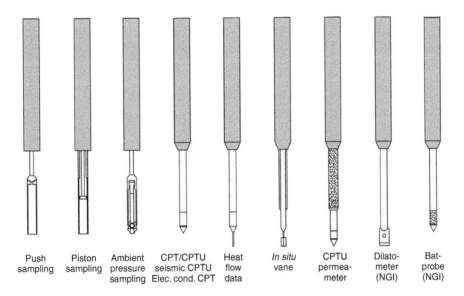

Push sampling	Piston sampling	Ambient pressure sampling	CPT/CPTU seismic CPTU Elec. cond. CPT	Heat flow data	*In situ* vane	CPTU permea- meter	Dilato- meter (NGI)	Bat- probe (NGI)

FIGURE 1.43
Various sampling and *in situ* testing tools used with the Fugro Seaclam (see Figure 1.42). BAT probe is an electrical resistance piezometer. (Courtesy of Fugro.)

- Permit *in situ* measurements of the physical and engineering properties of the materials
- Obtain information on groundwater conditions

Boring Types

They may be classified according to sampling operations:

- *Wash sample borings* are made to recover completely disturbed samples for general classification only.
- *Sample borings* are made to recover partially disturbed samples (SPT) or undisturbed samples (UD).
- *Core borings* are made to recover rock cores.
- *Rotary probes* recover only rock cuttings and are made to provide a rapid determination of the bedrock depth.
- *Air track probes* result in rock cuttings at the surface and are made to provide a rapid determination of rock quality.

Operational Elements

The execution of a boring requires fragmentation of materials, removal of the materials from the hole, and stabilization of the hole walls to prevent collapse.

Fragmentation

Materials in the hole are fragmented for removal by:

- Circulating water in loose sands or soft clays and organic soils
- Chopping while twisting a bit by hand (wash boring), or rotary drilling or augering in moderately strong soils

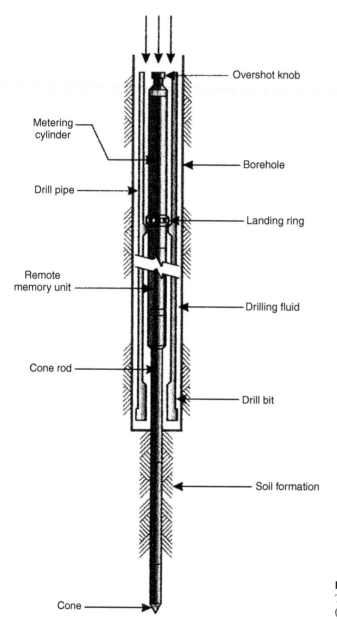

Overshot knob

Metering cylinder

Drill pipe

Borehole

Landing ring

Remote memory unit

Drilling fluid

Cone rod

Drill bit

Soil formation

Cone

FIGURE 1.44
The Dolphin Cone Penetrometer.
(Courtesy of Fugro.)

- Blasting or rotary drilling with a rock bit for "floating" boulders
- Rotary or percussion drilling in rock

Material Removal

Materials are removed to form the hole by:

- *Dry methods* used in cohesive soils, employing continuous-flight augers above the water table and the hollow-stem auger above and below the groundwater level (GWL).
- *Circulating fluids* from a point in the hole bottom, which is the more common procedure, accomplished by using clean water with casing, mud slurry formed

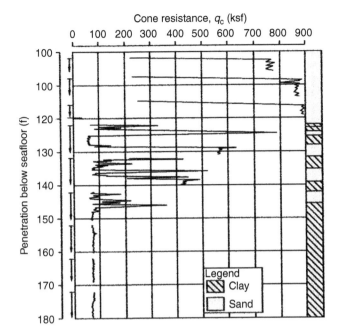

FIGURE 1.45
Plot of cone resistance data. (Courtesy of Fugro.)

either naturally or with additives used without casing, or air pressure in highly fractured or cavernous rock where circulating water is lost and does not return to the surface.

Hole Stabilization

Some form of stabilization is often needed to prevent hole collapse. None is required in strong cohesive soils above GWL.

Casing, used in sands and gravels above the water table, in most soils below GWL, and normally in very soft soils, is usually installed by driving with a 300 lb hammer, although 140 lb hammers are often used with lighter tripod rigs. The *hollow-stem auger* serves as casing and *wireline* and *Odex* methods install casing by drilling.

Driven casing has a number of disadvantages:

- Installation is slow in strong soils and casing recovery is often difficult.
- Sampling at stratum changes is prevented unless they occur at the end of a driven section, and *in situ* testing is limited.
- Obstacles such as boulders cannot be penetrated and require removal by blasting or drilling. The latter results in a reduced hole diameter which restricts sampling methods unless larger-diameter casing is installed before drilling.
- Loose granular soils below GWL tend to rise in casing during soil removal, resulting in plugged casing and loosened soils below.
- Removal of gravel particles is difficult and requires chopping to reduce particle sizes.
- Casing plugged with sand or gravel prevents sampler penetration adequate for recovery of undisturbed samples and representative SPT values. (The drill rod length and sampling tool must be measured carefully to ensure that the sampler rests on the bottom slightly below the casing.)

Mud slurry, formed naturally by the mixing of clayey soils during drilling, or by the addition of bentonite, is a fast and efficient method suitable for most forms of sampling and *in situ* testing.

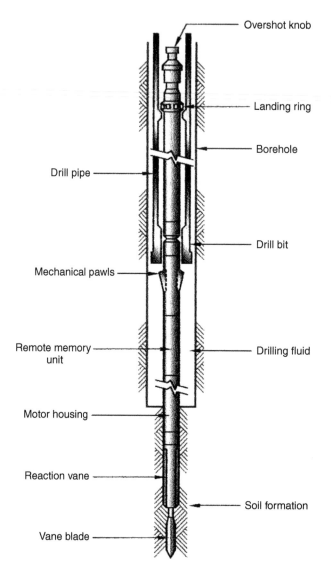

Overshot knob

Landing ring

Borehole

Drill pipe

Drill bit

Mechanical pawls

Remote memory
unit

Drilling fluid

Motor housing

Reaction vane

Soil formation

Vane blade

FIGURE 1.46
The Dolphin Remote Vane. (Courtesy of
Fugro.)

There are several disadvantages to mud slurry:

- Hole closure may occur in soft soils or crushed, porous materials such as shell beds.
- Relatively large pumps are required to circulate the slurry, particularly when boring depths exceed 30 ft (10 m).
- Mud-cased holes do not permit accurate water level readings, unless environmentally friendly biodegradable muds are used. Such muds incorporate an organic substance that degrades in a period of 24 to 48 h, allowing GWL measurements.
- Excessive wear on pumps and other circulating equipment occurs unless sand particles are removed in settling pits.
- Mud may penetrate some soils and contaminate samples.
- Mud loss is high in cavity-prone and highly fractured rock.

Grouting is used where closure or hole collapse occurs in fractured or seamy zones in rock masses. Cement grout is injected into the hole in the collapsing zone and then the hole is redrilled.

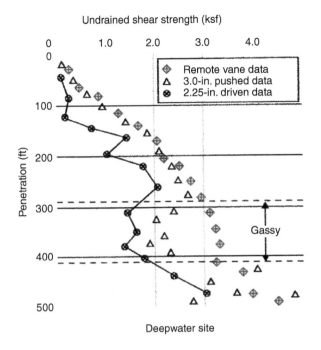

Deepwater site

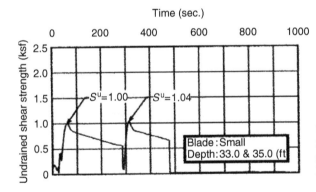

FIGURE 1.47
Remote vane results compared with lab test results from push samples. (Courtesy of Fugro.)

Boring Inclination

- *Vertical borings* are normal in soil formations and most rock conditions during investigation.
- *Angle borings* are often drilled in rock masses to explore for joints, faults, or solution cavities, or for the installation of anchors in soil or rock.
- *Horizontal borings* are drilled to explore for tunnels or the installation of rock bolts, instrumentation, or horizontal drains. Maintaining a straight horizontal boring is extremely difficult. At the start of the boring, gravity tends to pull the drill bit downward; then as penetration increases, gravity acts on the heavy drill string and the bit may tend to drift upward. Rock quality variations will also cause inclination changes. New technology employs directional drilling (Civil Engineering, 1998).

Standard Drilling Machines and Tools

There are a large number of hole-making methods and drilling machines. A summary is given in Table 1.13 in terms of application, method, advantages, and limitations. The basic

TABLE 1.13

Comparison of Various Drilling Machines and Methods

Machine	Application	Drilling Methods	Advantages	Limitations
Tripod with block and tackle (Figure 1.48) or motor-driven winch	• Procure soil samples • Exploratory borings for preliminary studies • Holes for some types of *in situ* testing	• Hole advanced by chopping while twisting rods and washing with pump-circulated water • Commonly called wash-boring method	• Requires only minumum-skill labor • Almost any location accessible to the light, portable equipment	• Slow operation, especially below 10 m • Penetration difficult in strong soils and impossible in rock • Difficult to remove gravel from casing; leads to poor samples • UD sampling difficult except in very soft soils because of lack of reaction
Rotary Drills Skid-mounted (Figure 1.49) Truck-mounted (Figure 1.50) Trailer-mounted Track-mounted	• Procure all types of soil and rock samples • Make hole for many types of *in situ* testing • Drilling inclined holes in soil or rock for horizontal drains or anchors	• Hole advanced by cutting bit on end of power-driven rotating drill rod to which pressure is applied hydraulically • Hole normally retained by mud slurry	• Relatively rapid • Can penetrate all types of materials • Suitable for all types of samples	• Equipment access in swampy or rugged terrain difficult • Requires trail or road • Requires level platform for drilling • Efficiency of drilling varies with rig size
Continuous-Flight Auger (Figure 1.54)	• Drill small- to moderate-size hole for continuous but disturbed samples • Other samples possible • Normally used in cohesive soils with adequate strength to prevent open hole collapse	• Rotating continuous flights of helical augers • Removal of all flights allows soil examination	• Rapid procedure for exploratory boring in strong cohesive soils and soft rock • SPT sampling possible when hole remains open after auger removal	• Hole collapses when auger is withdrawn from weak cohesive or cohesionless granular soils, thereby limiting depth, usually to near water table • Auger samples disturbed • Sampling methods limited • Requires rig samples and soil stratum examination modification
Hollow-Stem Auger (Figure 1.51)	• Drill small- to moderate-size holes for soil sampling	• Similar to continuous-flight auger, except hollow stem is screwed into ground to act as casing	• Rapid method in weak to moderately strong soils • SPT and UD sampling possible	• Penetration in strong soils to significant depths or through gravel layers difficult, and not possible through boulders and rock • Considerable disturbance may occur from auger bit

Method	Procedure	Applications/Advantages	Limitations
Large-Diameter Augers (Figure 1.52) Bucket auger, Disk auger, Helical auger	• Rotating large-diameter auger cuts soil to form hole	• Drill large-diameter holes (to 4 ft) for disturbed samples and soil strata examination in cohesive soils where hole remains open	• Rapid method • Enables close examination of subsurface soil conditions	• Depth limited by groundwater and rock conditions • Large machine requires easy access • Not suitable in cohesionless soils, soft clays, or organic soils • Samples disturbed
Percussion drills (cable-tool or churn-drilling)	• Heavy bits are raised and dropped to break up materials and form a slurry which is removed by bailers or sand pumps. Casing retains the hole	• Commonly used to drill water wells • Recovers "wash" samples in bailers • Define rock depth	• Relatively economical method of making large-diameter holes through any material (up to 2 ft [60 cm])	• Equipment large and cumbersome • Slow progress in strong soils and rock • Disturbance around bit from high-energy impact seriously affects SPT values • Rock coring and UD sampling not possible
Hammer drills	• Similar to percussion • Diesel pile-driving hammer used to drive double-wall casing while circulating air through annulus to blow cuttings from inner barrel	• Water wells • Exploratory holes through cobbles and boulders	• Relatively rapid penetration through cobbles and boulders	• Similar to percussion drills, except progress is much more rapid
Pneumatic percussion drill (Air track probes, Figure 1.52)	• Percussion rock bit chips and crushes rock with hammer blows as bit rotates. Chips removed by air pressure	• Drilling Holes for Rock anchors • Blasting • Rock characterization	• Rapid procedure for making small-diameter holes in hard rock	• Samples are only small chips. Not used for sampling. • Possible to lose entire drill stem in loose, fractured rock, clay seams, wet shale, etc. • Best use is hard massive rock
Impact drill	• Pneumatically energized tungsten carbide bit hammers hard rock at as high a rate as 700 blows/min	• Rapid drilling of exploratory hole in rock. (One case: 640-ft. hole, 6.5-in. diameter, drilled in 24 h)	• Very rapid penetration in rock masses. Could be used to drill pilot holes to substantial depths for tunnel studies, and core rock at critical depths with rotary methods	• Limited to rock masses • No sample recovery • Danger of hole closure in loose fractured or seamy rock zones. (Could be corrected by cement injection and redrilling)

components required for test or core borings include a drilling machine, casing, drill rods, drilling bits, and sampling tools.

Drilling machines consist of a power source, a mast for lifting apparatus, and a pump for circulating water or mud (or a compressor for air drilling) to lower, rotate, and raise the drilling tools to advance the hole and obtain samples. Test borings for obtaining representative or undisturbed samples under all conditions are normally made by rotary drills, and under certain conditions, with the tripod, block, and tackle. Some of the more common machines are illustrated in Figures 1.48 through 1.51.

Exploratory holes in which only disturbed samples are obtained are made with continuous-flight augers, large-diameter augers in clays or by percussion or hammer drilling in all types of materials. Large-diameter augers (Figure 1.52) can provide holes of diameters adequately large to permit visual examination of the borehole sides for detailed logging, if the inspector is provided with protection against caving. Pneumatic percussion and impact drills advance holes rapidly in rock, without core recovery.

Casing is used to retain the hole in the normal test boring operation, with tripods or rotary machines, at the beginning of the hole and for the cases described under Hole Stabilization. Boring cost is related directly to casing and hole size. Standard sizes and

FIGURE 1.48
The wash boring method. The hole is advanced by hand by twisting a bladed bit into the soil as water under pressure removes cuttings from the hole. In the photo, a 140 lb hammer is being positioned before driving an SPT sample.

FIGURE 1.49
A skid-mounted rotary drilling machine which advances the hole in soil or rock with a cutting bit on the end of a power-driven rotating drill rod to which pressure is applied by a hydraulic ram. In the photo, the drill rods are being lowered into the hole before the hole is advanced.

dimensions are given in Table 1.14; casing is designed for telescoping. There are several types of casing as follows:

- *Standard drive pipe* has couplings larger than the outside pipe diameter and is used for heavy-duty driving.
- *Flush-coupled casing* has couplings with the same diameter as the outside pipe diameter. For a given diameter, it is lighter and easier to drive because of its smooth outside surface than a standard pipe, although more costly to purchase.
- *Flush-jointed casing* has no couplings and is even lighter in weight than flush-coupled casing, but it is not as rugged as the other types and should not be driven.

Drill rods connect the drilling machine to the drill bits or sampler during the normal test or core boring operation with rotary machines (or tripods for soil borings). Standard sizes are given in Table 1.14. Selection is a function of anticipated boring depth, sampler types, and rock-core diameter, and must be related to machine capacity. The more common diameters are as follows:

- "A" rod is normally used in wash boring or shallow-depth rotary drilling to take SPT samples. In developing countries, a standard 1 in. pipe is often substituted for wash borings because of its low weight and ready availability.
- "B" rod is often used for shallow rotary core drilling, especially with light drilling machines.
- "N" rod is the normal rod size for use with large machines for all sampling and coring operations; it is especially necessary for deep core drilling (above 60 ft or 20 m).
- "H" rod is used in deep core borings in fractured rock since it is heavier and stiffer than "N" rod and will permit better core recoveries.

FIGURE 1.50
Rotary drilling with a truck-mounted Damco drill rig using mud slurry to prevent hole collapse. Rope on the cathead is used to lift rods to drive the SPT sampler, which in the photo is being removed from the hole. A hydraulic piston applies pressure to rods during rock coring, or during pressing of undisturbed samples. The table supporting the hydraulic works is retractable to allow driving of SPT samples or casing.

Drilling bits are used to cut soil or rock; some common types are shown in Figure 1.53. Chopping bits (others such as fishtail or offset are not shown) are used for wash borings. Drag bits (fishtail or bladed bits) are used for rotary soil boring. They are provided with passages or jets through which is pumped the drilling fluid that serves to clean the cutting blades. The jets must be designed to prevent the fluid stream from directly impinging on the hole walls and creating cavities, or from directing the stream straight downward and disturbing the soil at sampling depth. Low pump pressure is always required at sampling depth to avoid cavities and soil disturbance. Rock bits (tricone, roller bit, or "Hughes" bit) are used for rock drilling. Core bits (tungsten carbide teeth or diamonds) are used for rock coring while advancing the hole. Sizes are given in Table 1.14.

Sampling tools are described in Section 1.4.

Standard Boring Procedures

1. Take surface sample. (In some cases, samples are taken continuously from the surface to some depth. Sampling procedures are described in Section 1.4.2.)

FIGURE 1.51
Rotary drill rig advancing boring with the continuous hollow-stem flight Auger.

2. Drive "starter" casing or drill hollow-stem auger to 5 ft penetration (1 m penetration in metric countries).
3. Fragment and remove soil from casing to a depth of about 4 in. (10 cm) below the casing to remove material at sampling depth, which will be disturbed by the force of the plugged casing during driving. Completely disturbed "wash" samples (cuttings) may be collected at the casing head for approximate material classification.

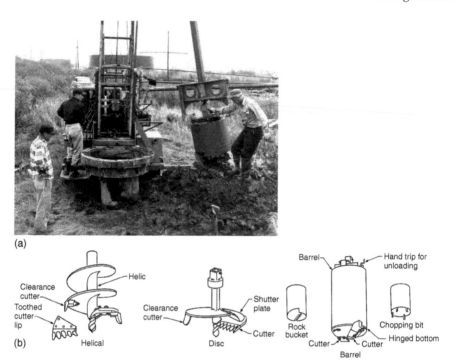

(a)

(b)

FIGURE 1.52
(a) Heavy-duty auger machine excavating with a large-diameter barrel bucket. Generally suitable only in soils with cohesion where the hole remains open without support. (b) Types of large augers. (From USBR, *Earth Manual*, U.S. Bureau of Reclamation, Denver, Colorado, 1974.)

TABLE 1.14

Standard Sizes of Drill Tools[a]

Size	O.D.		I.D.		Weight		Coupling O.D.	
	in.	mm	in.	mm	lb/ft	kg/m	in.	mm
Drill Rods — Flush Coupled								
E[b]	1 5/16	33.3	7/8	22.2	2.7	4.0	7/16	11.1
A[b]	1 5/8	41.3	1 1/4	28.5	3.7	5.7	9/16	14.3
B[b]	1 7/8	47.6	1 1/4	31.7	5.0	7.0	5/8	15.9
N[b]	2 3/8	60.3	2	50.8	5.2	7.5	1	29.4
EW[c]	1 3/8	34.9	15/16	23.8	3.1	4.7	7/16	11.1
AW[c]	1 3/4	44.4	1 1/4	31.8	4.2	6.5	5/8	15.9
BW[c]	2 1/8	54.0	1 3/4	44.5	4.3	6.7	3/4	19.3
NW[c]	2 5/8	66.7	2 1/4	57.1	5.5	8.4	1 3/8	34.9
HW[c]	3 1/2	88.9	3 1/16	77.8	7.7	11.5	2 3/8	60.3
Casing — Flush Jointed								
EW	1 13/16	43.0	1 1/2	38.1	2.76	4.2		
AW	2 1/4	57.2	1 29/32	48.4	3.80	5.8		
BW	2 7/8	73.9	2 3/8	60.3	7.00	10.6		
NW	3 1/2	88.9	3	76.2	8.69	13.2		
HW	4 1/2	114.3	4	101.6	11.35	16.9		
PW	5 1/2	139.7	4 7/8	127.0	15.35	22.8		
SW	6 7/8	168.3	6 1/32	152.4	19.49	29.0		
UW	7 5/8	193.7	7	177.8	23.47	34.9		
ZW	8 5/8	219.1	8 3/32	203.2	27.80	41.4		

(Continued)

TABLE 1.14

(*Continued*)

Size	O.D. in.	O.D. mm	I.D. in.	I.D. mm	Weight lb/ft	Weight kg/m	Coupling O.D. in.	Coupling O.D. mm
			Casing — Flush Coupled					
EX	1 13/16	46.0	1 5/8	41.3	1.80	2.7	1 1/2	33.1
AX	2 1/4	57.2	2	50.8	2.90	4.4	1 29/32	48.4
BX	2 7/8	73.0	2 9/16	65.1	5.90	8.8	2 3/8	69.3
NX	3 1/2	83.9	5 3/16	81.0	7.80	11.8	3	76.2
HX	4 1/2	114.3	4 1/8	104.8	8.65	13.6	3 15/16	100.0

Size (in.)	O.D. in.	O.D. mm	I.D. in.	I.D. mm	Weight lb/ft	Weight kg/m	Coupling O.D. in.	Coupling O.D. mm
			Casing — Standard Drive Pipe					
2	2 3/8	60.3	2 1/16	52.4	5.5	8.3	2 7/8	73.0
2 1/2	2 7/8	73.0	2 15/32	62.7	9.0	13.6	3 3/8	85.7
3	3 1/2	88.9	3 1/16	77.8	11.5	17.4	4	101.6
3 1/2	4	101.6	3 9/16	90.5	15.5	23.4	4 5/8	117.3
4	4 1/2	114.3	4 1/32	102.4	18.0	27.2	5 3/16	131.8
			Casing — Extra Heavy Drive Pipe					
2	2 3/8	60.3	1 15/16	49.2	5.0	7.6	2 7/32	56.4
2 1/2	2 7/8	73.0	2 21/64	59.1	7.7	11.6	2 5/8	66.7
3	3 1/2	88.9	2 29/32	73.8	10.2	15.4	3 1/4	82.5
3 1/2	4	101.6	3 23/64	85.3	12.5	18.9	3 3/4	95.3
4	4 1/2	114.3	3 53/64	97.2	15.0	22.7	4 1/4	107.8

Diamond Core Bits

DCDMA Standards Size	Core Diam. (Bit I.D.) in.	Core Diam. (Bit I.D.) mm	Hole Diam. (Reaming Shell O.D.) in.	Hole Diam. (Reaming Shell O.D.) mm
EWX and EWM	0.845	21.5	1.485	37.7
AWX and AWM	1.185	30.0	1.890	48.0
BWX and BWM	1.655	42.0	2.360	59.9
NWX and NWM	2.155	54.7	2.930	75.7
2 3/4 in., 3 7/8 in.	2.690	68.3	3.875	98.4
4 in., 5 1/2 in.	3.970	100.8	5.495	139.6
8 in., 7 3/4 in.	5.970	151.6	7.755	196.8
Wireline Size				
AQ	1 1/16	27.0	1 57/64	48.0
BQ	1 7/16	36.5	2 23/64	60.0
NQ	1 7/8	47.6	2 63/64	75.8
HQ	2 1/2	63.5	3 25/32	96.0
PQ	3 11/32	85.0	4 53/64	122.6

[a] From Diamond Core Drill Manufacturers Association (DCDMA).

[b] Original diamond core drill tool designations.

[c] Current DCDMA standards.

FIGURE 1.53
Several types of drilling bits. From left to right: three-bladed soil-cutting bits, tricone roller rock-cutting bits, carbolog-tooth bits for cutting soft rock, and diamond rock-coring bit.

4. Take sample.
5. Advance hole through the next interval of 5 ft (or 1 m) by either driving casing and removing the soils, as in step 3, or using a mud slurry or the hollow-stem auger to retain the hole, and take sample.
6. Continue sequences until prescribed final boring depth is reached. (In deep borings, or supplemental boring programs, sampling intervals are often increased to 10 or even 20 ft [3–6 m].)
7. When rock is encountered, set casing to the rock surface to permit coring with clean water to keep the bit cool and clean and prevent clogging.
8. Record casing lengths; measure and record drill rod and bit length and drill rod and sample tool length each time that the hole is entered to ensure that the hole bottom is reached, that the hole is not collapsing if uncased and that the sampler is at the required depth below the casing or final boring depth prior to sampling.

Other Drilling Machines and Methods

Continuous Hollow-Stem Auger (ASTM 5784-95)

The hollow-stem auger has been used with increasing frequency in recent years to avoid the use of drilling muds to retain the hole (Figure 1.51). Its use, therefore, is common for environmental investigations. During advance, the auger flights remove the soil from the hole and the hollow stem serves as casing. A bottom bit cuts the soil and a removable plug on a rod prevents soil from rising in the hollow portion of the stem (Figure 1.54). At sampling depth, the inner rods and plug are removed and either disturbed (SPT) samples or undisturbed tube samples can be obtained. In "clean" sands and silts below the water table, pressure equalization by filling the stem with water is usually necessary to prevent the saturated soils from entering the auger. Drilling is difficult or impossible in very hard

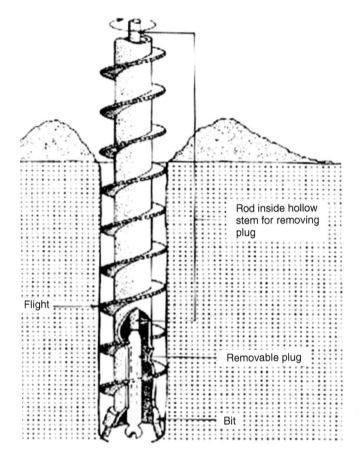

Rod inside hollow
stem for removing
plug

Flight

Removable plug

FIGURE 1.54
Continuous flight, hollow-stem auger.
Plug at tip prevents soil from entering
the stem during hole advance.

Bit

soils, gravels, cobbles, or boulders. To save time, drillers may advance the stem without the plug at the end. Below the water table, soils will rise in the hollow stem and sampling results will be affected. Boring inspectors should always insist that the plug is in place during hole advance.

Wireline Drilling

Wireline drilling eliminates the necessity of removing a string of drill rods for sampling and coring and is therefore a very efficient method for deep core drilling on land or offshore. The coring device is integral with the drill rods, which also serve as casing. Normally, it is not necessary to remove the casing except when making bit changes. The drill string is a 4 to 6 in. pipe with a bit at the end. The drill string is rotated as the drilling fluid is pumped down through it. (In offshore drilling, the mud, mixed onboard, is normally not recirculated, but rather flows up through the hole onto the seafloor.)

Soil sampling, rock coring, and *in situ* testing are carried out from the inner barrel assembly (Figure 1.82). Core samples are retrieved by removal of the inner barrel assembly from the core barrel portion of the casing drill rods. An "overshot" or retriever is lowered by the wireline through the drill rod to release a locking mechanism in the inner barrel head. The inner barrel with the core is then lifted with the wireline to the surface, the core removed, and the barrel returned to the bottom. In deep holes, it is necessary to pump the inner barrel into place with fluid pressure. Wireline core diameters are given in Table 1.14.

Odex Drilling

Odex drilling is used to set casing in formations where hole collapse occurs and drilling with mud is inefficient due to high mud loss. Such conditions include coarse and porous soils, highly fractured rock, or rock with frequent cavities. It is also used to install mini-piles socketed into rock.

The Odex drill includes casing with a bit, and an inner core barrel. When drilling a reamer on the Odex bit swings out and drills a hole larger than the external diameter of the casing. Cuttings are flushed from the casing with water or air. When the required depth has been reached, the drill is reversed and the reamer swings to its minimum diameter, allowing the bit to be lifted up through the casing, which remains in the hole. Rock drilling continues with a drill bit or a core barrel. Data on rock quality can be monitored electronically, providing measures on penetration rates, thrust on the bit, rotation torque, rotation speed, fluid pressure, and cross section of the hole.

Pneumatic Percussion Drills (Air Track Rigs)

Air tracks (Figure 1.55) provide a rapid and efficient method of characterizing rock masses in terms of quality. Rates of penetration per second are recorded; the harder the rock the lower are the penetration rates as shown on the logs given in Figure 1.56. Rock chips, removed by air pressure, can be examined at the hole entrance. Rock core borings should be drilled for correlations.

Subaqueous Drilling

Various types of platforms for the drilling equipment and applications for subaqueous test borings are given in Section 1.4.4.

Planning and Executing a Test Boring Program

Equipment Selection

The study phase, terrain features and accessibility, geologic conditions, boring depths, and the sample types required are considered when a test boring program is planned.

Boring Types

- *Exploratory borings* are normally performed first to determine general subsurface conditions. Only disturbed samples, and at times rock cores, are obtained.
- *Undisturbed sample borings* follow to obtain UD and perform *in situ* tests, usually in cohesive soils.
- *Core borings* are programmed to obtain rock cores.

Boring Spacing

In the feasibility and preliminary studies, borings are located to explore surface boundaries and stratigraphy as depicted on an engineering geology map. Additional borings may be required for increased definition. Grid systems may be appropriate in uniform conditions and, depending on the study area, size may range in spacing from 100 to 300 ft (30 to 100 m).

Final study programs depend upon the project type as follows:

- Structures (buildings, industrial plants, etc.) in urban areas usually are required by code to be investigated by borings at spacings that provide at least one boring for a given building area. In other than code-controlled areas, boring layout depends on building configuration, and spacing is generally about 50 to 100 ft

FIGURE 1.55
Air track drilling in limestone for deep foundation investigation. Figure 1.56 gives airtrack logs from an investigation in limestone.

    (15 to 30 m) depending on the uniformity of geologic conditions, the importance of the structure, and the foundation type.

- Dams are usually investigated on a grid of about 50 to 100 ft (15 to 30 m) spacing.
- Highway and railroad study programs depend on the adequacy of data obtained during the geologic mapping phase as supplemented by geophysical and reconnaissance studies (excavations, augers, probes, etc.), unless specified otherwise by a highway department or other owner. A minimum program requires borings in major cuts and fills, and at tunnel portals and all structure locations.
- In all cases, flexibility must be maintained and the program closely observed to permit the investigation of irregular or unforeseen conditions as they appear.

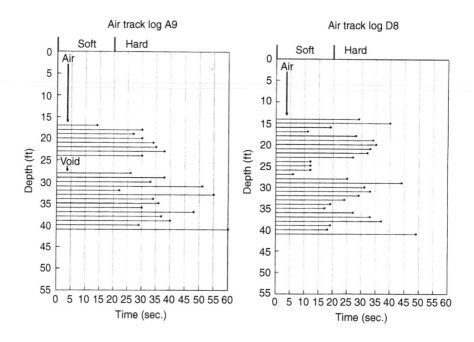

FIGURE 1.56
Air track logs from drilling in limestone. "Air" is soil drilling. Note voids on logs. Hard rock was defined as advance ≥20 sec. (Courtesy of Woodward–Clyde Consultants.)

Boring Depth

Excavations (open cuts for buildings, highways, subways, etc. and closed excavations for tunnels, caverns, mines) require borings adequately deep to explore to at least a short depth below final grade, or deeper if conditions are unfavorable, and to determine piezometric levels which may be artesian. The latter condition may require borings substantially below the final grade.

Foundations for buildings and other structures require boring depths programed and controlled to satisfy several conditions. As a general rule, borings must explore the entire zone of significant stress (about 1½ to 2 times the minimum width of the loaded area) in which deformable material exists. (Note that significant stress can refer to that imposed by a controlled fill and the floor it supports rather than the foundations bearing the fill, or can refer to stresses imposed at some depth along a pile group rather than at basement or floor level.)

The primary objective is to locate suitable bearing for some type of foundation and to have some knowledge of the materials beneath the bearing stratum. Drilling to "refusal" is never a satisfactory procedure unless the materials providing refusal to penetration (often inadequately defined by an SPT value) and those underlying have been previously explored and adequately defined. In rock, penetration must be adequate to differentiate boulders from bedrock.

Specifications

Boring type (exploratory, undisturbed, and core boring), spacing and depth, sample type and intervals, sample preservation and shipment, groundwater depths, and often drilling procedures are covered in specifications provided to the boring contractor. Standard boring specifications require modification to suit a particular project.

Boring Surveillance (Inspection)

During all phases of the boring program experienced surveillance should be provided to ensure that the intent of the specifications is properly interpreted and executed, and that the desired results are achieved.

Functions of the person performing the surveillance are generally as follows:

1. Enforce the specifications.
2. Maintain liaison with the structural engineer–architect and modify the program as necessary (add or delete borings, change types, depths and intervals of sampling, etc.).
3. Ensure complete and reliable drilling information (accurate reporting of depths, proper drilling, and sampling techniques).
4. Identify accurately all geologic conditions encountered and prepare reports and field logs that include all pertinent information (Section 1.4.7).

Some conditions where experienced geologic interpretations are necessary include:

- Differentiating a fill from a natural deposit. Some granular fills may appear to be natural, when in reality they overlie a thin organic stratum, or contain zones of trash and rubbish. Mistakenly identifying a deposit as fill can result in unnecessarily costly foundations when local building codes prohibit foundations on fill.
- Judging the recovered sample material to be wash remaining in the casing rather than undisturbed soil.
- Determining groundwater conditions.
- Differentiating boulders from bedrock on the basis of rock identification (boulders may be of a rock type different from the underlying bedrock, especially in glaciated terrain).

Supplemental Information

Many types of devices are available for use in boreholes to remotely sense and log various subsurface conditions, which should be considered in the planning of any subsurface exploration program, as described in Section 1.3.6.

1.3.6 Borehole Remote-Sensing and Logging

A number of devices and instruments can be lowered into boreholes to obtain a variety of information. They are particularly useful for investigating geologic conditions in materials from which it is difficult or impossible to obtain UD, such as cohesionless granular soils and badly fractured rock masses. Applications, equipment, operation, and limitations of various devices for remotely sensing and logging boreholes are summarized in Table 1.15.

Borehole cameras (TV and photographic) furnish images of borehole walls in fractured rock masses.

Seisviewer, an ultrasonic acoustical device, is also used to obtain images of borehole walls in fractured rock masses.

The 3-D velocity probe (Figure 1.57) ranges up to 15 ft in length, 3 in. in diameter, and about 150 lb in weight. It is lowered into the borehole to measure compression and shear-wave velocities, which provide information on fracture patterns in rock masses and, with the gamma–gamma probe for density measurements, provide the basis for computing dynamic properties. The 3-D sonic logger was so named because the records show the amplitude and arrival times of the sonic energy for a given travel distance (between the

TABLE 1.15

Borehole Remote-Sensing and Logging Methods

Device	Applications	Equipment	Operation	Limitations
Borehole film camera	Provides continuous undistorted record, with depth and orientation controlled, of all geologic planes in proper polar coordinates in rock masses. Records fractures to 0.01 in.	Camera with conical or rotating reflecting mirror, transmitter, illumination device, and compass contained in assembly with length of 33 in. and diameter of 2.75 in.	Camera lowered into dry or water-filled hole of NX diameter	In NX hole, depth limited to about 150 m. Images affected by water quality. Lens has limited depth of focus and cannot "see" depth of openings beyond more than a few centimeters
Borehole TV camera	Examination of boreholes in rock. Some types allow examination of voids such as caverns	TV camera. Some can be fitted with zoom lens and powerful miniature floodlight. Record can be taped	Camera lowered in to dry, or water-filled hole	Resolution less than photographic image
Seisviewer	To procure image of borehole walls showing fractures and discontinuities in rock masses	Ultrasonic acoustical device mounted in probe produces images of the entire hole	Probe lowered into dry or water-filled hole	Rock masses
3-D velocity probe (3-D sonic logger) (Figure 1.57)	To procure images of sonic compression and shear-wave amplitudes and rock fracture patterns. V_p and V_s used to compute dynamic properties	A sonde from which a sonic pulse is generated to travel into the rock mass (see the text)	Sonde is lowered into borehole. Arrival times of compression and shear waves are transmitted to surface, amplified and recorded	Fracture patterns in rock masses revealed a depth of about 1 m
Mechanical calipers	Continuous measurement of borehole diameter to differentiate soft from hard rock and swelling zones. Useful for underground excavations in rock	Mechanical caliper connected to surface recorder	Lowered into hole and spread mechanically as it is raised. Soft rock gives large diameters from drilling operation	Measures to about 32 in. maximum in rock massures
Recording thermometers	To procure information on groundwater flow and evaluate grouting conditions	Recording thermometer	Lowered into borehole filled after fluid temperature stabilizes	Rock masses and water-filled holes

Method	Purpose	Operation	Procedure	Remarks
Electric well log	To obtain continuous record of relative material characteristics and ground-water conditions from ground surface in all materials	Devices to record parent resistivities and spontaneous potential (natural) generated in borehole. Similar to electrical resistivity measurements from surface	Instrument lowered into mud or water-filled hole. Dry-hole devices available	Data are essentially qualitative for correlations with sample or core borings
Gamma–gamma probe	To measure material density *in situ*. Particularly useful in cohesionless soils and fractured rock masses[a]	Nuclear probe measures back-scatter of gamma rays emitted from a source in the probe	Probe lowered into cased or mud-filled hole. Back-scatter must be calibrated for the casing or mud	Measurements of *in situ* density
Neutron–gamma probe	To measure material moisture content *in situ*[a]	Nuclear probe measures backscatter of gamma and neutron rays (resulting from bombardment by fast neutrons), which gives a measure of hydrogen contents of materials	Probe lowered into cased or mud-filled hole. Back-scatter must be calibrated for the casing or mud. Hydrogen content is correlated with the water or hydrocarbon content	Measurements of *in situ* water content
Scintillometer	Locate shales or clay zones. Used primarily for petroleum exploration	Nuclear probe measures gamma rays emitted naturally from the mass	Probe lowered into borehole. Shales and clays have a high emission intensity compared with sands	Qualitative assessments of shale or clay formations
Rock detector (acoustic sounding technique) (Figure 1.58)	To differentiate boulders from bedrock	Microphone transmitter set into bedrock at various locations to produce noise which is monitored with earphones and observed and recorded by oscilloscope	One observer listens to and series of holes drilled drilling sounds while the other records drilling depth for each significant change Characteristic sounds enable differentiation between soil, boulders, and rock	Area about 60–100m from each geophone can be investigated. In overburden, drill bits tend to clog in clays and hole collapses below GWL. Best conditions are shallow deposits of dry, are shallow deposits of dry, slightly cohesive granular soils
Tro-pari	Measure borehole inclination and direction (see text)			

[a] Nuclear probes (gamma and neutron) have been used to monitor changes in moving slopes (Cotecchia, 1978) and even to locate the failure zone in a uniform deposit that was evidenced by a sudden change in change in density and moisture on a relatively uniform log.

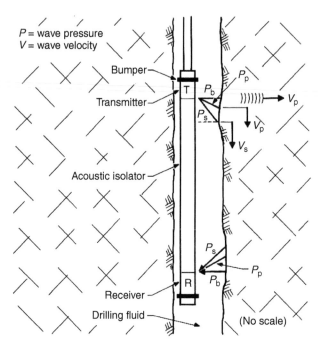

FIGURE 1.57
Schematic of the principles of the three–dimensional velocity probe. (From Myung, J.T. and Baltosser, R.W., *Stability of Rock Slopes*, ASCE, New York, 1972, pp. 31–56.)

transmitter and receiver). For velocity computations, the hole diameter must be known and is measured with mechanical calipers.

- *Mechanical calipers* are used to measure hole diameter in rock masses.
- *Recording thermometers* are used to measure fluid temperatures in rock masses.
- The *electric well logger* is used in soil and rock masses for continuous measurements of resistivity.
- The *gamma–gamma probe* is used in soil and rock masses to obtain continuous measurements of *in situ* densities.
- The *neutron–gamma probe* is used in soil and rock masses to obtain continuous measurements of *in situ* moisture contents.
- The *scintillometer* is used to locate shales and clay zones in soil and rock masses.
- The *rock detector* is an acoustical sounding device used to differentiate boulders and other obstructions from bedrock. A geophone is set into bedrock and connected to an amplifier, headphone, and oscilloscope. A series of holes is drilled with a wagon drill, or another drilling machine (Figure 1.58), while the observer listens to the volume and nature of the generated sounds.
- The *Tro-pari surveying instrument* is used to measure borehole inclination and direction in relatively deep borings in good-quality rock. The instrument works with a clockwork mechanism that simultaneously locks a plumb device and a magnetic compass when set in the borehole. The inclination and direction of the borehole, respectively, are read when the instrument is retrieved at the surface.

1.3.7   Groundwater and Seepage Detection

General Groundwater Conditions

Figure 1.59 illustrates various groundwater conditions, which are summarized as follows:

- *Static water table or level* (GWL) is located at a depth below which the ground is continuously saturated and the water is at atmospheric pressure.

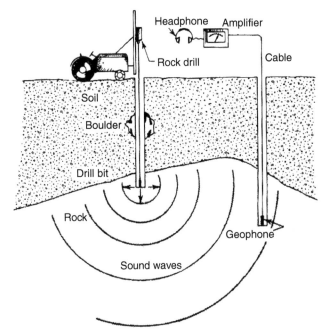

FIGURE 1.58
The elements of the rock indicator, or acoustical sounding technique, to differentiate boulders from bedrock.

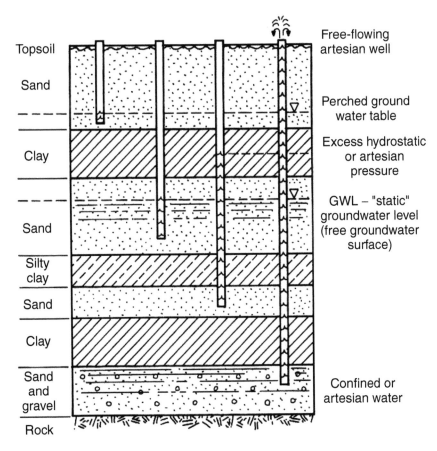

FIGURE 1.59
Various groundwater conditions.

- *Perched water table* can also be a measured GWL, and represents a saturated zone overlying an impervious stratum below which the ground is not saturated.
- *Artesian conditions* result from groundwater under a head in a confined stratum which is greater than the static water-table head and can result in free flow at the surface when the confined stratum is penetrated.
- Variations in conditions occur with time, and are affected by seasonal conditions, tidal fluctuations, flooding, and pumping. Variations also occur with physical conditions in terms of soil type and density, ground contours, surface drainage, rock-mass discontinuities, etc.

Determining Conditions

Terrain analysis (Section 1.2.4, Interpretation of Topographic Maps) provides general information on watertable location.

Geophysical methods (seismic refraction and electrical resistivity) provide indirect measures of the approximate depth to groundwater.

Reconnaissance methods employing excavations locate the GWL in a positive manner but within a short time interval. Probable variations with time must always be considered during site analysis. Test pits are the best method for measuring short-term water levels and seepage rates. Auger borings may also clearly reveal the water table, especially in slightly cohesive soils. In cohesionless soils, the hole will tend to collapse within a few centimeters of the GWL but direct measurement is not usually possible.

Test Borings

The moisture condition of drive samples (SPT) may provide an indication of GWL. Samples will range from dry to moist to wet (saturated) as the GWL is approached. The condition should be noted on boring logs.

In cased borings, the water level is generally determined by pumping or bailing water from the hole and permitting stabilization for 24 h. This method is reliable in uniform sand strata or other pervious materials, but differentiating perched from static conditions is difficult. Site stratigraphy provides some clues for judgment. The method is unreliable when the casing terminates in an impervious stratum which blocks the entrance of water. The casing should be raised until it terminates in a permeable stratum.

If the casing ends in an aquifer with an artesian head, water may flow from the casing when pumping ceases during drilling operations, or it may rise to a point above the estimated GWL.

Casing water levels should be noted periodically during boring operations and for a period thereafter, and water loss as well as artesian and static conditions should be noted.

Boring with a mud slurry will not provide reliable water level readings unless the hole is flushed with clean water or unless biodegradable mud is used for boring. After flushing, most holes in permeable soils collapse near or slightly above the perched or static water level.

Borehole remote-sensing probes, such as the electric well logger and the neutron probe, provide good indications of perched or static conditions.

Piezometers (Section 3.4.2) provide the most accurate method of measuring groundwater conditions, and are especially useful for recording changes with time. Measurements made during pumping tests performed at various levels are useful in differentiating perched from static conditions.

Seepage Detection

Conditions of engineering significance include:

- Flow through, around, and beneath earth dam embankments.
- Flow beneath and around concrete dams.
- Groundwater pollution as caused by flow from sanitary landfills, mine tailings storage areas, chemical waste ponds, etc.
- Flows from slopes observable during site reconnaissance, depending on the season.

Detection Methods

Tracers placed in a water body may be useful in locating entrance points of seepage and detecting exit points. Nonradioactive tracers include fluorescent and nonfluorescent dyes. Radioactive tracers include ^{82}Br and ^{131}I that are readily detectable with a Geiger–Muller counter. Temperature can be sensed with a thermistor attached to the top of an insulated aluminum-tipped probe inserted into the ground.

Acoustical emission monitoring (Section 3.3.5) may detect large flows.

1.4   Recovery of Samples and Cores

1.4.1   General

Objectives

Samples of geologic materials are recovered to allow detailed examination for identification and classification, and to provide specimens for laboratory testing to obtain data on their physical and engineering properties.

Sample Classes Based on Quality

Totally disturbed samples are characterized by the complete destruction of fabric and structure and the mixing of materials such as that occurring in wash and auger samples.

Representative samples are partially deformed. The engineering properties (strength, compressibility, and permeability) are changed, but the original fabric and structure vary from unchanged to distorted, and are still apparent. Such distortion occurs with split-barrel samples.

Undisturbed samples may display slight deformations around their perimeter, but for the most part, the engineering properties are unchanged. Such results are obtained with tube or block samples.

Sampler Selection

A number of factors are considered in the selection of samplers, including:

- Sample use, which varies from general determination of material (wash sampler), to examination of material and fabric and *in situ* testing (split-barrel sampler), to performing laboratory index tests (split-barrel sampler), and to carrying out laboratory engineering-properties tests (UD).
- Soil type, since some samplers are suited only for particular conditions, such as soft to firm soils vs. hard soils.

- Rock conditions, since various combinations of rock bits and core barrels are used, depending on rock type and quality and the amount of recovery required.
- Surface conditions, which vary from land or quiet water to shallow or deep water with moderate to heavy swells.

Some common sampling tools and their applications to various subsurface conditions are illustrated in Figure 1.60. The various tools and methods and their applications and limitations are described in Table 1.16.

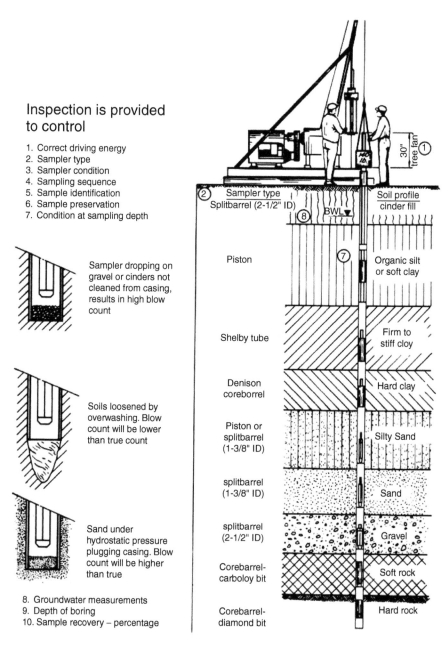

Inspection is provided to control

1. Correct driving energy
2. Sampler type
3. Sampler condition
4. Sampling sequence
5. Sample identification
6. Sample preservation
7. Condition at sampling depth

Sampler dropping on gravel or cinders not cleaned from casing, results in high blow count

Soils loosened by overwashing. Blow count will be lower than true count

Sand under hydrostatic pressure plugging casing. Blow count will be higher than true

8. Groundwater measurements
9. Depth of boring
10. Sample recovery – percentage

Sampler type
Splitbarrel (2-1/2" ID)

Piston

Shelby tube

Denison coreborrel

Piston or splitbarrel (1-3/8" ID)

splitbarrel (1-3/8" ID)

splitbarrel (2-1/2" ID)

Corebarrel-carboloy bit

Corebarrel-diamond bit

Soil profile
cinder fill

Organic silt or soft clay

Firm to stiff cloy

Hard clay

Silty Sand

Sand

Gravel

Soft rock

Hard rock

FIGURE 1.60
Common sampling tools for soil and rock and their application.

1.4.2   Test Boring Soil Sampling

Types Commonly Used

The split-barrel sampler (SS) is used in all soil types for *representative samples* used for identification and index tests.

Thin-wall tubes are used in soft to firm clays and *coring samplers* are used in stiff to hard cohesive soils for undisturbed samples used for engineering properties tests.

Required Boring Diameters

Wash or exploratory borings for split-barrel sampling are normally of 2½ in. diameter (casing I.D.). UC borings are normally of 4 in. diameter, but may be larger to improve sample quality. Core borings vary from 2 in. to larger, normally with NX core taken in a hole started with 4 in. diameter casing.

Sampling Interval

Samples are normally prescribed for 5 ft (or 1 m) intervals and a change in strata. Samples should also be taken at the surface to record the topsoil thickness, and continuously from the surface to below the depth of shallow foundations to assure information at footing depth (5 ft intervals often do not provide information at shallow footing elevations).

Continuous sampling is also important through miscellaneous fills that vary widely in materials and that often overlie a layer of organic soil, which may be thin but significant, and through formations with highly variable strata. In deep borings, sampling depths are often changed to 10 or 20 ft intervals after several normally sampled borings are completed and general subsurface conditions defined.

Factors Affecting Sample Quality

Sampler Wall Thickness

A large outside diameter relative to the inside diameter causes deformation by material displacement.

Sampler Conditions

- Dull, bent, or otherwise deformed cutting edges on the sampler cause sample deformation.
- Inside friction, increased by rust, dirt, or, in the case of tubes, omission of lacquer, causes distortions which are evidenced by a turning downward of layers, resulting in conical shapes under extreme cases.

Boring Operations

- Dynamic forces caused by driving casing can loosen dense granular soils or densify loose granular soils.
- Sands may rise in the casing when below the GWL (Figure 1.60).
- Overwashing, jetting, and high fluid pressures also loosen granular soils or soften cohesive materials (Figure 1.60).
- Coarse materials often remain in the hole after washing, particularly in cased borings (Figure 1.60). These "cuttings" should be removed by driving a split barrel sampler, by pushing a Shelby tube, or with a cleanout auger. Contamination is common after boring through gravel layers or miscellaneous fills containing cinders, etc.

TABLE 1.16

Sampling Tools and Methods

Category-Method and Tool	Application	Limitations
Reconnaissance		
Wash sample	Indication of material type only	Completely mixed, altered, segregated
Auger sample	Material identification	Completely disturbed
Retractable-plug sampler	Material identification	Slight disturbance, very small sample of soft soils
Black sample	Large undisturbed sample of cohesive materials	Taken from test pits, cohesive soils only
Test Boring Sampling (Soils)		
Split barrel (spoon)	Undisturbed samples in soils suitable for identification and lab index tests	Samples not suitable for engineering properties testing
Shelby tube	Undisturbed sample in firm to stiff cohesive soil. Can be driven into hard soils	Sampling impossible in very coarse granular soils Will not retrieve very or clean granular soils
Standard stationary piston	Undisturbed samples in soft to firm clays and silts	Will not penetrate compact sands, stiff clays, and other strong soils. Will not retrieve sands. Can be overpushed
Osterberg piston sampler	Undisturbed samples in all soils with cohesion except very strong. Less successful in clean sands	Usually cannot penetrate strong residual soil and glacial till. Some disturbance in sand and often loss of sample. User cannot observe amount of partial penetration
Shear-pin piston (Greer and McClelland)	Undisturbed samples in all soils with cohesion except very strong. Often recovers samples in sands and can be used to determine natural density	Usually cannot penetrate strong residual soil or glacial till. Disturbance in sands Cannot observe amount of partial penetration
Swedish foil sampler	Continuous undisturbed samples in soft to firm cohesive soils	Gravel and shells will rupture foil. Cannot penetrate strong soils
Denison sampler	Undisturbed samples in strong cohesive soils such as residual soils, glacial till, soft rock alternating	Not suitable in clean granular soils, and soft to firm clays
Pitcher sampler	Similar to Denison above. Superior in soft to hard layers. Can be used in firm clays	Similar to Denison above

Subaqueous Sampling Without Test Boring

Method	Description
Free-fall gravity coring tube	Samples firm to stiff clays, sand and fine gravel in water depths of 4000 m
Harpoon-type gravity sampler	Samples river bottom muds and silts to depths of about 3 m
Explosive coring tube	Small-diameter samples of stiff to hard ocean bottom soils to water depths of 6000 m
Gas-operated free-fall piston (NG)	Good-quality samples up to 10 m depth from seafloor
Vibracore	Undisturbed samples of soft to firm bottom sediment, 3 1/2 in. diameter to depths of 12 m
Free-fall gravity coring tube	Maximum length of penetration about 5 m in soft soils, 3 m in firm soils
Harpoon-type gravity sampler	Penetration limited to few meters in soft soils
Explosive coring tube (piggot tube)	Sample diameters only 1 7/8 in. Penetration only to 3 m below seafloor
Gas-operated free-fall piston (NG)	Penetration limited to 10 m below seafloor
Vibracore	Limited to soft to firm soils and maximum penetration of 12 m. Water depth limited to 60 m

Subaqueous Sampling with Test Boring

Method	Description
Wireline drive sample	Disturbed sample in soils
Wireling push samples	Relatively undisturbed samples may be obtained in cohesive materials
Wireline drive sample	Penetration length during driving not known
Wireling push samples	Often poor or no recovery in clean granular soils

Rock Coring

Method	Description
Single-tube core barrel	Coring hard homogeneous rock where high recovery is not necessary
Double-tube core barrel	Coring most rock types where high recovery is not necessary, and rock is not highly fractured or soft
Double-tube swivel-type core barrel	Superior to double-tube swivel-core barrel, above Particularly useful to obtain high recovery in friable, highly fractured rock
Wireline core barrel	Deep hole drilling in rock or offshore because of substantial reduction of in-out times for tools
Oriented core barrel	Determination of orientation of geologic structures
Integral coring method	Recover cores and determine orientation in poor-quality rock with cavities, numerous fractures, and shear zones
Calyx or shot coring	Obtain cores in medium- to good-quality rock up to 2-m diameter
Single-tube core barrel	Circulating water erodes soft, weathered, or fractured rock
Double-tube core barrel	Recovery often low in soft or fractured rocks
Double-tube swivel-type core barrel	Not needed in good-quality rock. Barrel is more costly and complicated than others mentioned above
Wireline core barrel	No more efficient than normal drilling to depths of about 30 m
Oriented core barrel	Procedure is slow and costly. Requires full recovery
Integral coring method	Slow and costly procedure
Calyx or shot coring	Slow and costly. Difficult in soft or seamy rock

- Hole squeezing may occur in soft clays if the drilling mud is too thin.
- Plastic clays may remain along the casing walls if cleaning is not thorough.
- Hollow stem augers can cause severe disturbances depending on the rate of advance and rotation, and the choice of teeth on the bit.
- Hollow-stem augers must be advanced with the plug in the auger prior to sampling to prevent soil from entering the auger.

Sampler Insertion

All ball check valves and other mechanisms should be working properly before the sampler is lowered into the hole. The sampler should be lowered to the bottom immediately after the hole is cleaned. Measurements of the total length of the sampler and rods should be made carefully to ensure that the sampler is resting on the bottom elevation to which the hole was cleaned and to avoid sampling cuttings. Since complete cleaning is usually not practical, split-barrel samplers are seated under the rod weight and often tapped lightly with the drive hammer; piston samples are forced gently through the zone of soft cuttings.

Soil Factors

- Soft to firm clays generally provide the best "undisturbed" samples, except for "quick" clays, which are easily disturbed.
- Air or gas dissolved in pore water and released during sampling and storage can reduce shear strength.
- Heavily overconsolidated clays may be subject to the opening of fissures from stress release during boring and sampling, thereby substantially reducing strength.
- Gravel particles in a clay matrix will cause disturbance.
- Cohesionless granular soils cannot be sampled "undisturbed" in the present state of the art.
- Disturbance in cohesive materials usually results in a decrease in shear strength and an increase in compressibility.

Sample disturbance and its effect on engineering properties are described in detail by Broms (1980). See Section 1.4.6 for sample preservation, shipment, storage, and extraction.

Split-Barrel Sampler (Split Spoon) (ASTM D1586-99)

Purpose

Split-barrel samplers are used to obtain representative samples suitable for field examination of soil texture and fabric and for laboratory tests, including measurements of grain-size distribution, specific gravity, and plasticity index, which require retaining the entire sample in a large jar.

Sampler Description

Split-barrel samplers are available with and without liners; the components are shown in Figure 1.61. O.D. ranges from 2 to 4½ in. A common O.D. is 2 in. with ¼ in. wall thickness (1½ in. sample). Larger diameters are used for sampling gravelly soils. Lengths are either 18 or 24 in.

A ball check valve prevents drill pipe fluid from pushing the sample out during retrieval. To prevent sample spillage during retrieval, flap valves can be installed in the shoe for loose sands, or a leaf-spring core retainer (basket) can be installed for very soft

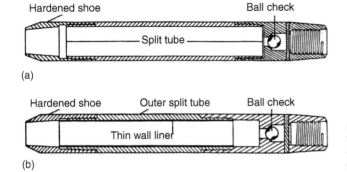

(a)

(b)

FIGURE 1.61
The split-barrel *sampler or split spoon*: (a) without liner; (b) with liner. (Courtesy of Sprague and Henwood, Inc.)

clays and fine cohesionless soils. Upon retrieval, the barrel between the head and the shoe is split open (Figure 1.62), the sample is examined and described, removed, and stored.

In some sampler types, brass liners are used for procuring drive samples of strong cohesive soils for laboratory direct-shear testing.

Sampling Procedure

The sampler is installed on the hole bottom, then driven into the soil with a hammer (normally 140 lb) falling on the drill rods. The number of blows required for a given weight and drop height, and a given penetration, are recorded to provide a measure of soil compactness or consistency as described in Section 2.4.5 (Standard Penetration Test).

Thin-Wall Tube Samplers

Purpose

Thin-wall tube samplers are used to obtain UD of soft to stiff cohesive soils for laboratory testing of strength, compressibility, and permeability.

Tube Materials

Cold-drawn, seamless steel tubing (trade name "Shelby tube") is used for most soil materials; brass tubes are used for organic soils where corrosion resistance is required. Wall thickness is usually 18 gage; heavier gages are available. Lacquer coating can provide corrosion protection and reduce internal frictional resistance and sample disturbance.

Tube diameters and lengths range from 2 to 6 in. in diameter, 24 to 30 in. in length. Tubes 2 in. in diameter are used in 2½ in. exploratory borings, but 2 in. diameter samples have a large ratio of perimeter disturbance to area and are considered too small for reliable laboratory engineering-property testing.

Tubes 3 in. (2.87″) in diameter are generally considered the standard type for laboratory test samples. The tube should be provided with a cutting edge drawn in to provide about 0.04 in. inside clearance (or 0.5 to 1.5% less than the tube I.D.), which permits the sample to expand slightly upon entering the tube, thereby relieving sample friction along the walls and reducing disturbance.

Tubes 4 to 6 in. in diameter reduce disturbance but require more costly borings. A 5 in. tube yields four samples of 1 in. diameter from the same depth for triaxial testing.

Operations

Thin-wall tubes are normally pressed into the soil by hydraulically applied force. After pressing, the sample is left to rest in the ground for 2 to 3 min to permit slight expansion

FIGURE 1.62
Split-barrel sample of sand.

and an increase in wall friction to aid in retrieval. The rods and sampler are rotated clockwise about two revolutions to free the sampler by shearing the soil at the sampler bottom. The sample is withdrawn slowly from the hole with an even pull and no jerking. In soft soils and loose granular soils, the sampler bottom is capped just before it emerges from the casing fluid to prevent the soil from falling from the tube.

Shelby Tube Sampling

A thin-wall tube is fitted to a head assembly (Figure 1.63) that is attached to drill rod. An "O ring" provides a seal between the head and the tube, and a ball check valve prevents water in the rods from flushing the sample out during retrieval. Application is most satisfactory in firm to hard cohesive soils. In firm to stiff soils, the tube is pushed into the soil by a steady thrust of the hydraulic system on a rotary drilling machine using the machine weight as a reaction. Care is required that the sampler is not pressed to a distance greater than its length.

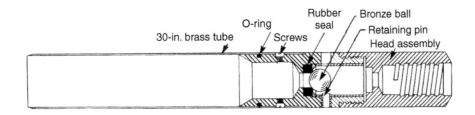

FIGURE 1.63
Thin-wall "Shelby tube" sampler. (Courtesy of Sprague and Henwood, Inc.)

Soft soils are difficult to sample and retain because they have insufficient strength to push the column of fluid in the tube past the ball check valve. In stiff to hard cohesive soils, samples are often taken by driving heavy-gage tubes.

Standard Stationary Piston Sampler

A thin-wall tube is attached to a head assembly. The tube contains a piston (Figure 1.64), which is connected to a rod passing through the drill rod to the surface. When at the bottom of the tube, the piston prevents soil from entering the tube as it is lowered into the hole and permits seating through soft cuttings. The rod connected to the piston is held fixed at the surface, while the hydraulic system on the drilling machine presses the tube past the piston into the soil. With light rigs, the reaction can be increased by using Earth anchors. In properly fitted piston samplers, a strong vacuum is created to hold the sample in the tubes during withdrawal from the hole. The stationary piston sampler is used to retrieve very soft to firm cohesive soils.

Osterberg Hydraulic Piston Sampler

A thin-wall tube contains an actuating and a fixed piston (Figure 1.65). An opening in the head assembly permits applying fluid pressure to the actuating piston at the top of the tube. Fluid pressure is applied to the actuating piston, which presses the tube past the fixed piston into the soil. The actuating piston eliminates the cumbersome rods of the standard stationary piston as well as the possibility of overpushing. The sampler is commonly used for very soft to firm cohesive soils.

Shear-Pin Piston

This device is similar to the Osterberg sampler except that the tube is attached to the piston with shear pins, which permit the fluid pressure to build to a high value before it "shoots" the piston when the pins shear. The sampler can be used in soft to stiff cohesive soils and loose sands. In loose sands disturbance is unavoidable. The "apparent" density, however, can be determined by measuring the weight of the total sample in the tube and assuming the volume calculated from the tube diameter and the stroke length. With the shear-pin piston, the sampling tube will almost always be fully extended because of the high thrust obtained.

Double-Tube Soil Core Barrels

Purpose

Double-tube soil core barrels are used to obtain UD in stiff to hard cohesive soils, saprolite, and soft rock.

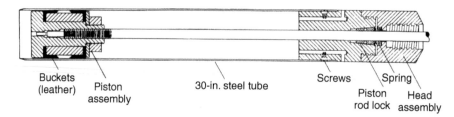

Buckets   Piston   30-in. steel tube   Screws   Spring
(leather)   assembly   Piston   Head
                                                                                   rod lock assembly

FIGURE 1.64
Stationary piston sampler. (Courtesy of Sprague and Henwood, Inc.)

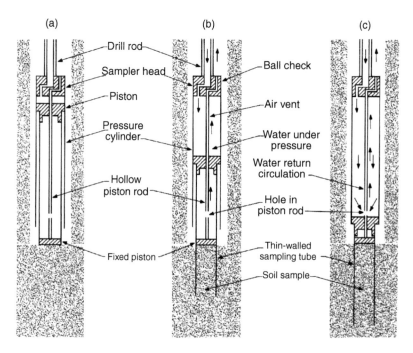

FIGURE 1.65
Operation of the Osterberg piston sampler. (a) Sampler is set on cleaned bottom of borehole. (b) Hydraulic pressure propels sampling tube into the soil. (c) Pressure is released to through hole in piston rod. (From ENR, *Engineering News Record*, 1952. Reprinted with permission of McGraw-Hill.)

Denison Core Barrel

The Denison core barrel is used in materials such as hard clays and cemented sands. It includes a rotating outer barrel and bit containing a fixed inner barrel with a liner, as shown in Figure 1.66. The cutting shoe on the inner barrel can extend below the cutting bit. Liners range from 28-gage galvanized steel to brass and other materials such as phenolic-resin-impregnated paper with a 1/16 in. wall. Various bits are available for cutting materials of varying hardness, to obtain samples ranging in diameter from 2 3/8 to 6 5/16 in. Sample tubes range from 2 to 5 ft in length. The extension of the cutting shoe below the cutting bit is adjustable. The maximum extension is used in relatively soft or loose materials, whereas in hard materials the shoe is maintained flush with the bit.

During drilling, pressure is applied by the hydraulic feed mechanism on the drill rig to the inner barrel, while the bit on the outer barrel cuts away the soil. The sampler is

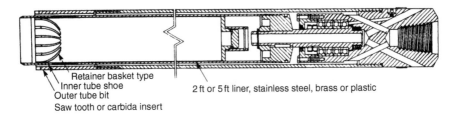

Retainer basket type
Inner tube shoe
Outer tube bit
Saw tooth or carbida insert

2 ft or 5 ft liner, stainless steel, brass or plastic

FIGURE 1.66
Denison core barrel for hard soils. (Courtesy of Sprague and Henwood, Inc.)

FIGURE 1.67
Removal of the cutting bit from the Denison core barrel after coring cemented silty sands.

retrieved from the hole, the cutting bit removed from the barrel (Figure 1.67), and the thin liners, retained in the inner tube by wall friction or a basket-type retainer, are removed.

Pitcher Sampler

The operation of the Pitcher sampler is similar to that of the Denison core barrel except that the inner barrel is spring-loaded and thus provides for the automatic adjustment of the distance by which the cutting edge of the barrel leads the coring bit (Figure 1.68). Because of adjusting spring pressure, the Pitcher sampler is particularly suited to sampling deposits consisting of alternating soft and hard layers.

1.4.3   Miscellaneous Soil-Sampling Methods

Wash Samples

Completely disturbed cuttings from the hole advance operation carried to the surface by the wash fluid and caught in small sieves or by hand are termed washed samples. Value is limited to providing only an indication of the type of material being penetrated.

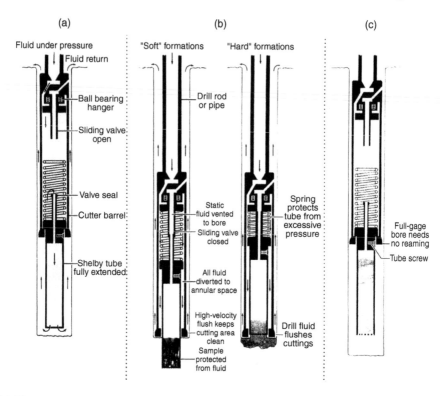

FIGURE 1.68
Operation of the Pitcher sampler: (a) down the hole; (b) sampling; (c) recovery. (Courtesy of Mobile Drilling Inc.)

Auger Samples

Completely disturbed cuttings from the penetration of posthole diggers (hand augers, Table 1.12), continuous-flight augers, or large-diameter augers are brought to the surface when the auger is removed from the hole. In cohesive soils they are useful for soil identification, moisture content, and plasticity index tests.

Retractable Plug Sampler

One-inch-diameter tubes containing slightly disturbed samples are obtained in soft to firm organic and cohesive soils suitable for soil identification and moisture content measurements (Table 1.12).

Test Pit Samples

High-quality undisturbed *block* or "chunk" samples or small cylinder samples in softer soils with some cohesion are taken from test pits. Block samples are particularly useful in soils difficult to sample UD such as residual soils. Strong cohesive soils are sampled by carefully hand-cutting a block from the pit walls. The sample is trimmed by knife and encased in paraffin on the exposed sides. The block is cut loose from the pit, overturned, and the remaining side is coated. It is then sealed in a box for shipment to the laboratory. Very large samples are possible.

Weaker soils with some cohesion are carefully hand-trimmed into a small cylinder and sealed. The method has been used to obtain samples for density tests in partially saturated silty or slightly clayey sands.

Swedish Foil Sampler

High-quality continuous samples in soft, sensitive cohesive soils, useful in locating the shear zone in a slope failure problem, are possible with the Swedish foil sampler. A sampling tube, usually 8 ft in length, is pushed into the soil by a special drill rig as a reaction. To eliminate friction between the sample and the tube walls, thin steel strips, or foils, unroll to follow along the sampler walls as the sampler penetrates the soil.

1.4.4   Subaqueous Sampling

Categories

Sample procurement under subaqueous conditions can be placed in one of four general categories on the basis of the sampling technique:

1. Normal cased-boring methods.
2. Wireline drilling techniques.
3. Sampling to shallow depths below the bottom without drill rigs and casing.
4. Sample recovery from deep borings in offshore sands. Borehole remote-sensing and logging methods (Section 1.3.6), such as the electric well logger and nuclear probes, should be considered since they provide important supplemental data.

Normal Cased-Boring Methods

General

Normal cased borings require a stable platform for mounting the boring equipment and procurement of samples. The up and down movements from swells severely affect drilling and sampling operations as bits and samplers are removed from contact with the hole bottom. Tidal effects require careful considerations in depth measurements.

Platforms

Floats (Figure 1.69) or barges (Figure 1.70) are used in shallow water, generally less than 50 ft (15 m) deep, with slight swells. Penetration depths are in moderate ranges, depending on drill rig capacity.

Large barges or jack-up platforms are used in water to depths of the order of 100 ft (30 m) with slight to moderate swells. Penetration depths below the bottom are moderate depending upon the drilling equipment.

Drill ships with wireline drilling techniques are used in deep water (Section 1.3.4 and Figure 1.42).

Wireline Drilling Methods

General

Wireline drilling techniques are used in deep water. Much deeper penetration depths are possible than are with normal cased borings, and operations can tolerate much more severe sea conditions than can cased borings.

Platforms

Large barges or moderately large ships are used in relatively calm water and water depths over 50 ft (15 m), where deep penetration of the seafloor is required. *Jack-up platforms* are used where heavy swells can occur. *Large drill ships* (Figure 1.42) are used in deep water where deep penetration below the seafloor is required.

FIGURE 1.69
Float-mounted tripod rig; casing is being driven prior to SPT exploratory sampling.

FIGURE 1.70
Barge-mounted rotary drill rig operating in the Hudson River for the third tube of the Lincoln Tunnel, New York City, a location with strong currents and heavy boat traffic.

Samplers

Drive samplers (Figure 1.71): Either split-barrel samplers or tubes are driven with a 176 lb (80 kg) hammer dropped 10 ft (3 m) by release of a wire-hoisting drum. Penetration is only approximated by measuring sample recovery since the sampler is attached to the wire, not to drill rods. Recovery is related to the blow count for a rough estimate of relative density. Tube samples recovered in deep water at substantial depths below the seafloor in stiff clays will undergo significant strength decrease from stress release upon extraction from the seabed and extrusion in the shipboard laboratory.

Pushed-tube and piston samplers: See Section 1.3.4 for operations offshore.

Sampling to Shallow Penetration without Drill Rigs and Casing

General

Various devices and methods are available for sampling shallow seafloor conditions without the necessity of mounting a drill rig on a platform and maintaining a fixed position for extended time intervals. Sampling procedure involves operating the equipment from the side of a vessel equipped with a crane. Sampling is generally not feasible in strong materials or to bottom penetration depths greater than about 40 ft (12 m), depending upon the device used.

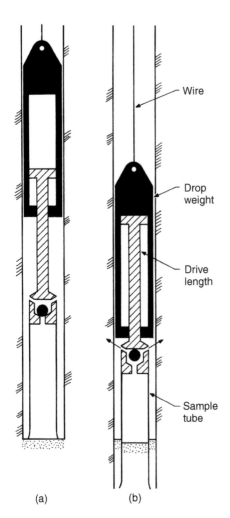

(a)   (b)

FIGURE 1.71
Wireline drive sampler: (a) before driving, (b) after driving.

Sampling Methods and Devices

The application, description, and penetration depths of the various devices and methods are summarized in Table 1.17. The devices include the following:

- Petersen dredge
- Harpoon-type gravity corer (Figure 1.72)
- Free-fall gravity corer
- Piston gravity corer (Figure 1.73)
- Piggot explosive coring tube
- Benthos Boomerang Corer (Figure 1.74)
- Vibracore (Figure 1.75) is the most practical method for obtaining long cores in soft soils in deep water. It can also be used on land to sample remotely contaminated lagoons where a soft crust makes access hazardous to people.

1.4.5   Rock Coring

Objectives

Rock coring is intended to obtain intact cores and a high percentage of core recovery.

Equipment

Rotary drilling machine, drill rods, a core barrel to receive the core, and a cutting bit are needed.

Operations

The core barrel is rotated under pressure from the drill rig applied directly to it, while water flows through the head, down the barrel, out through the waterways in the bit, and up through the rock hole and casing (in soil) to return to the surface. When the rock is first encountered in a borehole, the initial core runs are usually short because of the possibility that the upper rock will be soft and fractured. As rock quality improves, longer core runs are made.

The core barrel is generally rotated between 50 and 1750 r/min; rotation speed is a function of the bit diameter and rock quality. Slow speeds are used in soft or badly fractured rocks and high speeds are used in sound hard rocks. If large vibrations and "chatter" of the drill stem occurs, the speed should be reduced or core recovery and quality will be severely affected.

Bit pressure is also modified to suit conditions. Low bit pressure is used in soft rocks and high pressure is used in hard rocks. When vibrations and "chatter" occur, the pressure, which is imposed hydraulically, should be reduced.

Fluid pressure should be the minimum required to return the cuttings adequately to the surface to avoid erosion of borehole walls. If there is no fluid return, drilling should immediately stop and the core barrel returned to the surface to avoid overheating the bit, which would result in bit damage (loss of diamonds) and possible jamming in the hole.

Lack of fluid can result from:

- Blockage of the core barrel, which occurs in clayey zones. Continued drilling after a broken core has blocked entry into the core barrel results in core grinding. Indications of blockage may be heavy rod vibrations, a marked decrease in penetration rate accompanied by an increase in engine speed, return fluid more heavily laden with cuttings than normal, and a rise in circulation fluid pressure.

TABLE 1.17

Subaqueous Soil Sampling Without Drill Rigs and Casing

Device	Application	Description	Penetration Depth	Comments
Petersen dredge	Large, relatively intact "grab" samples of seafloor	Clam-shell type grab weighing about 100 lb with capacity of about 0.4 ft³	To about 4 in.	Effective in water depths to 200 ft. More with additional weight
Harpoon-type gravity corer (Figure 1.72)	Cores from 1.5 to 6 in. diameter in soft to firm soils	Vaned weight connected to coring tube dropped directly from boad Tube contains liners and core retainer	To about 30 ft	Maximum water depth depends only on weight. UD sampling possible with short, large-diameter barrels
Free-fall gravity corer (Figure 1.73)	Cores 1.5 to 6 in. diameter in soft to firm soils	Device suspended on wire rope over vessel side at height of above seafloor about 15 ft and then released	Soft soils to about 17 ft. Firm soils to about 10 ft	As above for harpoon type
Piston gravity corer (Ewing gravity corer)	2.5 in. sample in soft to firm soils	Similar to free-fall corer, except that coring tube contains a piston that remains stationary on the seafloor during sampling	Standard core barrel 10 ft; additional 10 ft sections can be added	Can obtain high-quality UD samples
Piggot explosive coring tube	Cores of soft to hard bottom sediments	Similar to gravity corer. Drive weight serves as gun barrel and coring tube as projectile. When tube meets resistance of seafloor, weighed gun barrel slides over trigger mechanism to fire a cartridge. The exploding gas drives tube into bottom sediments	Cores to 1 7/8 in. and to 10 ft length have been recovered in stiff to hard materials	Has been used successfully in 20,000 ft of water
Norwegian Geotechnical Institute gas-operated piston	Good-quality samples in soft clays	Similar to the Osterberg piston sampler, except that the piston on the sampling tube is activated by gas pressure	About 35 ft	
Benthos Boomerang corer (Figure 1.74)	High-quality representative samples in clays and sands	Weighed free-fall plastic core tube droped from a vessel penetrates the sea floor. Floats inflate and rise to surface with the core	Up to 80 in. At times, less in dense sands	Requires minimum water depth of 33 ft. Has been used to depth of 29, 000 ft
Vibracore (Figure 1.75)	High-quality sample in soft to firm sediments, diameter 3 1/2 in.	Apparatus is set on seafloor. Air pressure from the vessel activates an air-powered mechanical vibrator to cause penetration of the tube, which contains a plastic liner to retain the core	Length of 20 and 40 ft Rate of penetration varies with material strength. Samples a 20 ft core in soft soils in 2 min	Maximum water depth of about 200 ft

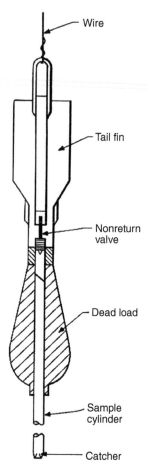

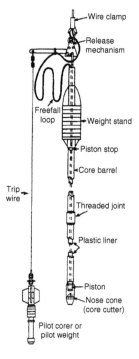

FIGURE 1.72
Harpoon-type gravity corer.

FIGURE 1.73
Schematic diagram of a typical piston-type gravity corer. (From USAEC 1996, Pub. EM 1110-1-1906. With permission.)

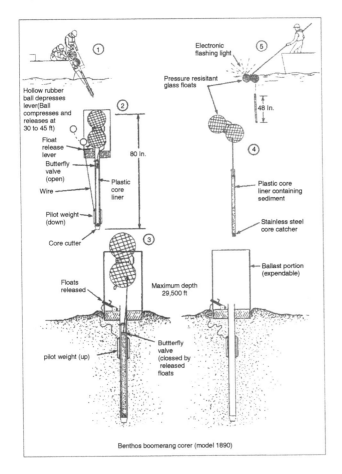

Benthos boomerang corer (model 1890)

FIGURE 1.74
The operating of sequence of the Boomerang Corer. (From USACE 1996, Pub. EM1110-1-1906. With permission.)

- Loss in caverns, large cavities, or highly fractured zones. In Figure 1.76, a light drilling mud is being used to minimize fluid loss (note the mud "pit"), while coring in limestone with highly fractured zones above the water table.

When the prescribed coring length is obtained the core barrel is retrieved from the ground. The core is removed from the barrel (Figure 1.77) and laid out in wooden boxes exactly as recovered (Figure 1.78). Wooden spacers are placed to divide each run. The depths are noted, the core is examined, and a detailed log is prepared.

Core Barrels

The selection of a core barrel is based on the condition of the rock to be cored and the amount and quality of core required. Core barrels vary in length from 2 to 20 ft, with 5 and 10 ft being the most common.

Table 1.18 provides summary descriptions of suitable rock conditions for optimum application, descriptions of barrel operation, and general comments. The types include:

- Single-tube core barrel (Figure 1.79).
- Double-tube rigid core barrel (Figure 1.80).
- Swivel-type double-tube core barrel, of two types: conventional and Series M (Figure 1.81). These types usually provide the best core recovery and are the most commonly specified for rock coring.
- Wireline core barrel (Figure 1.82).

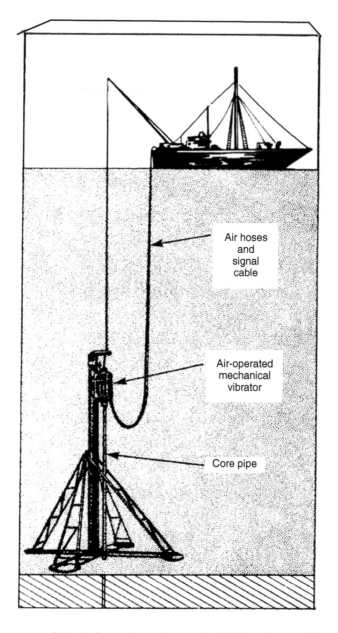

Air hoses
and
signal
cable

Air-operated
mechanical
vibrator

Core pipe

FIGURE 1.75
The Vibracore lowered to the seafloor.
(Courtesy of Alpine Ocean Seismic
Survey, Inc.)

- Oriented core barrel contains knives that scribe a groove on the rock core. The compass orientation of the groove is continuously recorded, which enables determination of the strike of joints and other fractures (Figure 1.83). During normal coring operations cores twist in the hole and accurate determination of joint strikes are not reliable.

Coring Bits

General

Types of coring bits are based on the cutting material, i.e., sawtooth, carbide inserts, and diamonds.

Waterways are required in the bits for cooling. Conventional waterways are passages cut into the bit face; they result in enlarged hole diameter in soft rock. Bottom-discharge bits should be

FIGURE 1.76
Core drilling with a Failing Holemaster. Light drilling mud is necessary in the fractured limestone above the water table to prevent drilling mud loss.

used for coring soft rock or rock with soil-filled fractures. Discharge occurs behind a metal skirt separating the core from the discharging fluid, providing protection from erosion.

Common bit sizes and core diameters are given in Table 1.14. The smaller diameters are used in exploratory borings for rock identification or in good-quality rock, but when maximum core recovery is required in all rock types, NX cores or larger are obtained. In seamy and fractured rock, core recovery improves with the larger diameters, and HX size is commonly used.

FIGURE 1.77
Removal of HX diameter limestone core from the inner barrel of a double-tube swivel-type core barrel.

Reaming shells, slightly larger than the core barrel diameter and set with diamonds or carbide insert strips, ream the hole, maintaining its gage and reducing bit wear.

Bit Types

Sawtooth bits are the lowest in cost and have a series of teeth cut in the bit which are faced with tungsten carbide. They are used primarily to core overburden and very soft rock.

Carbide insert bits (Figure 1.53) have tungsten carbide teeth set in a metal matrix and are used in soft to medium-hard rocks.

Diamond bits (Figure 1.53) are the most common type, producing high-quality cores in all rock types from soft to hard. Coring is more rapid, and smaller and longer cores are retrieved than with other bit types. The diamonds are either surface-set in a metal matrix, or the metal matrix is impregnated throughout with diamond chips. There are various designs for cutting various rock types, differing in quality, size, and spacing of the diamonds, matrix composition, face contours, and the number and locations of the waterways.

Core Recovery and RQD

Reporting Methods

Percent core recovery is the standard reporting method wherein core recovery is given as a percentage of total length cored. Rock Quality Designation (RQD) was proposed by Deere (1963) as a method for classifying core recovery to reflect the fracturing and alteration of rock masses. For RQD determination, the core should be at least 50 mm in diameter (NX) and recovery with double-tube swivel-type barrels is preferred.

FIGURE 1.78
Core recovery of 100% in hard, sound limestone: very poor recovery in shaley, clayey, and heavily fractured zones.

RQD is obtained by summing the total length of core recovered, but counting only those pieces of hard, sound core which are 10 cm (4 in.) in length or longer, and taking that total length as a percentage of the total length cored. If the core is broken by handling or drilling, as evidenced by fresh breaks in the core (often perpendicular to the core), the pieces are fitted together and counted as one piece.

TABLE 1.18

Types of Rock Core Barrels

Core Barrel	Suitable Rock Conditions	Operation	Comments
Single tube (Figure 1.79)	Hared homogeneous rock which resists erosion	Water flows directly around the core. Uses split-ring core catcher	Simple and rugged. Severe core loss in soft or fractured rock
Double tube, rigid type (Figure 1.80)	Medium to hard rock, sound to moderately fractured. Erosion-resistant to some extent	Inner barrel attached to head and rotates with outer barrel as water flows through annular space	Water makes contact with core only in reamer shell and bit area, reducing core erosion. Holes in inner tube may allow small flow around core
Double tube, swivel type (conventional series)	Fractured formations of average rock hardness not excessively susceptible to erosion	Inner barrel remains stationary, while outer barrel and bit rotate. Inner barrel terminates above core lifter	Torsional forces on core are eliminated minimizing breakage. Core lifter may tilt and block entrance to inner barrel, or may rotate with the bit causing grinding of the core
Double tube, rigid type (series M)(Figure 1.81)	Badly fractured. Soft, or friable rock easily eroded	Similar to conventional series, except that core lifter is attached to inner barrel and remains oriented. Inner barrel is extended to the bit face	Superior to the conventional series. Blocking and grinding minimized. Erosion minimized by extended inner barrel
Wireline core barrel (Figure 1.82)	Deep core drilling under all rock conditions	See Section 1.3.4	Retriever attached to wireline retrieves inner barrel and core without the necessity of removing core bit and drill tools from the hole
Oriented core barrel	Determine orientation of rock Oriented core (Figure 1.83)	Similar to conventional core barrels. Orienting barrel has three triangular hardened scribes mounted in the inner barrel shoe that cuts grooves in the core. A scribe is aligned with a lug on a survey instrument mounted in a nonmagnetic drill collar. The instrument contains a compass-angle device, multishot camera, and a clock mechanism. About 30 cm of coring the advance is stopped and a photograph of the compass clock, and lug is taken. Geologic orientation is obtained by correlation between photographs of the core grooves and the compass photograph	

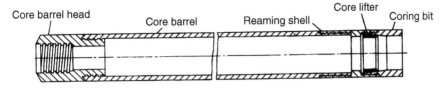

FIGURE 1.79
Single-tube core barrel. (Courtesy of Sprague and Henwood, Inc.)

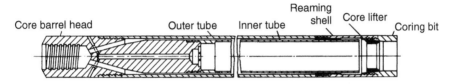

FIGURE 1.80
Rigid-type double-tube core barrel. (Courtesy of Sprague and Henwood, Inc.)

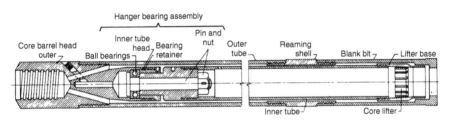

FIGURE 1.81
Swivel-type double-tube core barrel, series M. (Courtesy of Sprague and Henwood, Inc.)

Causes of Low Recovery

Rock conditions: Fractured or decomposed rock and soft clayey seams cause low recovery, for example, as shown in Figure 1.78. Rock quality can vary substantially for a given location and rock types as illustrated in Figuers 1.84 and 1.85, which show a formation of granite gneiss, varying from sound and massive to jointed and seamy. Core recovery in the heavily jointed zone is illustrated in Figure 1.86.

Coring equipment: Worn bits, improper rod sizes (too light), improper core barrel and bit, and inadequate drilling machine size all result in low recovery. In one case, in the author's experience, coring to depths of 30 to 50 m in a weathered to sound gneiss with light drill rigs, light "A" rods, and NX double-tube core barrels resulted in 40 to 70% recoveries and 20 to 30% RQD values. When the same drillers redrilled the holes within a 1m distance using heavier machines, "N" rod and HX core barrels, recovery increased to 90 to 100% and RQDs to 70 to 80%, even in highly decomposed rock zones, layers of hard clay, and seams of soft clay within the rock mass.

Coring procedure: Inadequate drilling fluid quantities, increased fluid pressure, improper drill rod pressure, or improper rotation speed all affect core recovery.

Integral Coring Method

Purpose

Integral coring is used to obtain representative cores in rock masses in which recovery is difficult with normal techniques, and to reveal defects and discontinuities such as joint openings and fillings, shear zones, and cavities. The method, developed by Dr. Manual Rocha of Laboratorio Nacional de Engenheira Civil (LNEC) of Lisbon, can produce cores

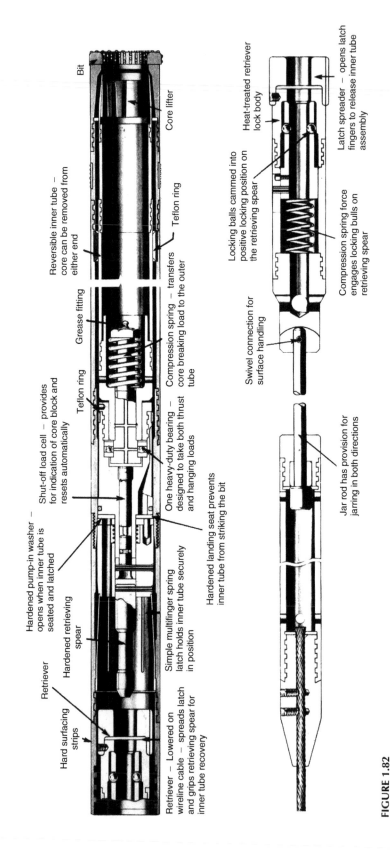

FIGURE 1.82

Wireline core barrel and retrieval assembly. (Courtesy of Sprague and Henwood, Inc.)

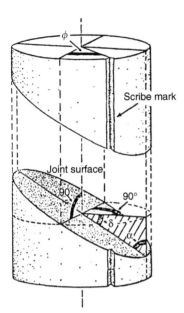

FIGURE 1.83
Oriented core with the joint surface intersecting the core wall at the joint dip angle. The boundaries of a horizontal line across the joint are located at angle ø providing the strike angle.

of 100% recovery with the orientation known. Defect orientation is an important factor in rock-mass stability analysis.

Technique
1. An NX-diameter hole or larger is drilled to where integral coring is to begin.
2. A second, smaller hole (nominally about 1 in. [26 mm] in diameter) is drilled coaxially with the first through the desired core depth, although usually not exceeding 1.5 m in depth.
3. A notched pipe is lowered into the hole and bonded to the rock mass with cement or epoxy resin grout, which leaves the pipe through perforations.
4. After the grout has set, a core is recovered by overcoring around the pipe and through the cemented mass.
5. During installation of the pipe, the notch positions are carefully controlled by a special adapter and recorded so that when the core is retrieved, the orientation of the fractures and shear zones in the rock mass are known.

Large-Diameter Cores by Calyx or Shot Drilling

Purpose
Calyx or shot drilling is intended to allow borehole inspection in rock masses in holes up to 6 ft (2 m) in diameter.

Method
Calyx drilling uses chilled shot as a cutting medium. The shot is fed with water and lodges around and partially embeds in a bit of soft steel. The flow of freshwater is regulated carefully to remove the cuttings but not the shot. The cores are recovered by a special core-lifter barrel, wedge pins, or mucking after removal of the core barrel.

Limitations
The method is limited to rock of adequate hardness to resist erosion by the wash water and to vertical or nearly vertical holes.

FIGURE 1.84
Massive, hard granite gneiss at the mouth of a water tunnel. Note diorite dike and seepage from joints. Core recovery in such materials should be high.

1.4.6   Sample and Core Treatment

Upon Retrieval

The sampler is dismantled carefully to avoid shocks and blows (in soils), obvious cuttings are removed, and the recovery is recorded (RQD is also recorded in rock masses).

FIGURE 1.85
Jointed granite gneiss and crushed rock zone at other end of tunnel of Figure 1.84 about 400 m distant. Core recovery is shown in Figure 1.86.

The sample is immediately described and logged. It is not allowed to dry out, since the consistency of cohesive soils changes and details of stratification become obscured. The sample is then preserved and protected from excessive heat and freezing.

FIGURE 1.86
Core recovery of about 90% in fractured diorite grading to gneiss; RQD about 40 to 70%. Coring with NX double-tube swivel-type barrel.

Preservation, Shipment and Storage

Split-Barrel Samples

Carefully place intact uncontaminated short cores in wide-mouth jars of sufficient size (16 oz) to store 12 in. of sample. The samples, which may be used for laboratory examination, should not be mashed or pushed into any container; such action would result in complete loss of fabric and structure. The jar caps should contain a rubber seal, be closed tightly, and be waxed to prevent moisture loss. Liner samples are preserved as thin-wall tube samples.

Thin-Wall Tube Samples

Remove all cuttings from the sample top with a small auger and fill the top with a mixture of paraffin and a microcrystalline wax such as Petrowax, applied at a temperature close to the congealing point. Normal paraffin is subjected to excessive shrinkage during cooling and should not be used or an ineffective moisture seal will result. The top is capped, taped, and waxed.

Invert the tube, remove a small amount of soil from the bottom and fill the tube with wax. Cap, tape, and wax the bottom.

Tubes should be shipped upright, if possible, in containers separating the tubes from each other and packed with straw.

In the laboratory, tubes that are not to be immediately tested are stored in rooms with controlled humidity to prevent long-term drying. Soil properties can change with time; therefore, for best results, samples should be tested as soon as they are received in the laboratory.

Rock Cores

Rock cores are stored in specially made boxes (Figure 1.86) in which wooden spacers are placed along the core to identify the depth of run.

Extrusion of UD Samples

Thin-Wall Pushed Samples

To obtain specimens for testing thin-wall tube, samples should be extruded from the tube in the laboratory in the same direction as the sample entered the tube, with the tube held

FIGURE 1.87
Vertical extrusion of Shelby tube sample in the same direction as taken in the field to minimize disturbance.
(Courtesy of Joseph S. Ward and Assoc.)

vertical, as shown in Figure 1.87. This procedure avoids the effects on sample quality of reverse wall friction and of the sample's passing the cutting edge of the tube.

Thin-Wall Cored Samples

Because they contain strong cohesive soils, wall friction in cored soil samples is usually too high to permit extrusion from the entire tube without causing severe disturbance. Removal normally requires cutting the tube into sections and then extruding the shorter lengths.

Field Extrusion

Some practitioners extrude the sample in the field, cutting off 6-in. sections, wrapping them in aluminum foil, and surrounding them with wax in a carton. The procedure simplifies transport, but leads to additional field and laboratory handling which may result in the disturbance of easily remolded soils.

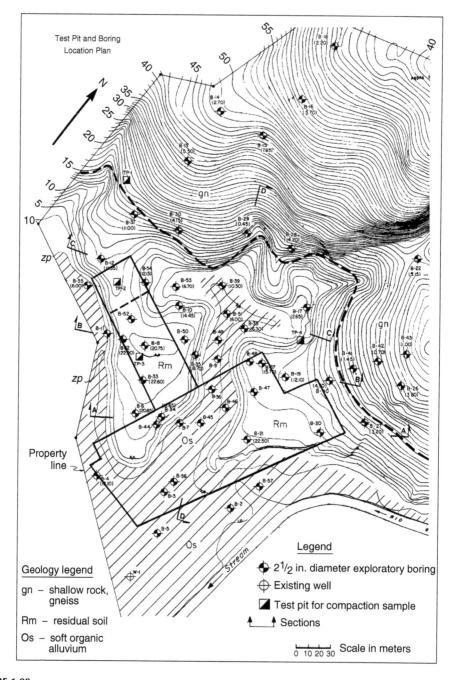

FIGURE 1.88

Example of test pit and boring location plan using topographic map and engineering geology map as base map. (Courtesy of Joseph S. Ward and Assoc.)

1.4.7   Data Presentation

Basic Elements

Location Plan

Locations of all explorations should be shown accurately on a plan. It is helpful to use a topographic map as a base map, which also shows the surficial geology as in Figure 1.88. A map providing the general site location is also useful, especially for future reference to local geologic conditions. Many reports lack an accurate description of the site location.

Geologic Sections

Data from the various exploration methods are used as a basis for typical geologic sections to illustrate the more significant geologic conditions, as in Figure 1.89. The objective is to illustrate clearly the problems of the geologic environment influencing design and construction.

For engineering evaluations, it is often useful to prepare large-scale sections on which are plotted all of the key engineering property data as measured in the field and in the laboratory.

Fence diagrams, or three-dimensional sections, are helpful for sites with complex geology. An example is given in Figure 1.90.

Logs

The results of test and core borings, test pits, and other reconnaissance methods are presented on logs which include all pertinent information.

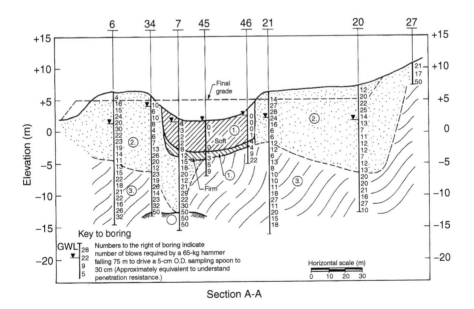

Section A-A

FIGURE 1.89

Typical geologic section across site shown in stratum descriptions in Figure 1.88: (1) recent alluvium marine deposits consisting of interbedded organic silts, sands, and clays; (2) residual soil: silt, clay, and sand mixtures; (3) micaceous saprolite: highly decomposed gneiss retaining relict structure; (4) weathered and partially decomposed gneiss. (Courtesy of Joseph S. Ward and Assoc.)

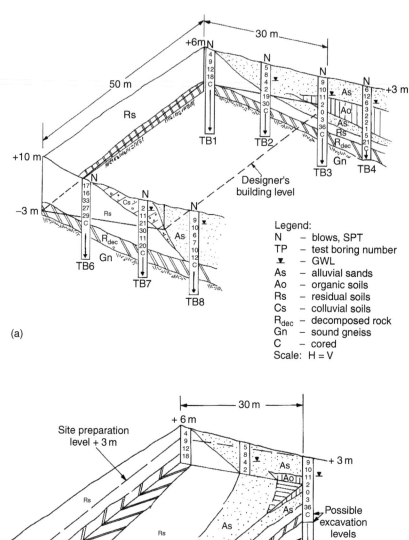

FIGURE 1.90
Geologic diagrams illustrate conditions at and below the building level proposed by the designer for a Class I structure in a nuclear power plant complex where support on material of only one type is required. The diagrams show clearly the conditions for foundations and for excavation support. To reach decomposed rock a maximum excavation of 13 m is required and to reach sound rock, 15 m. Backfill would be controlled fill or rollcrete depending on requirements of the licensing agency. (a) Geologic conditions; (b) geologic conditions at designer's building level.

Boring Logs

General

Logs are prepared to provide complete documentation on the drilling, sampling, and coring operations and on the materials and other aspects of the subsurface encountered, including groundwater conditions. They provide the basis for analysis and design, and therefore complete documentation and clear and precise presentation of all data are necessary. Normally, two sets of logs are required: field logs and report logs, each serving a different purpose.

Field Log

A field log is intended to record all of the basic data and significant information regarding the boring operation. Typical contents are indicated in the example given in Figure 1.91, which is quite detailed, including the sample description and remarks on the drilling operations. The field log is designed to describe each sample in detail as well as other conditions encountered. All of the information is necessary for the engineer to evaluate the validity of the data obtained, but it is not necessary for design analysis.

Field boring log No. 1 Type UD-C Coordinates: N10256302 E78765.50 Elevation: +5.20 Datum: MSL Sheet 1 of 1
Project: New Plant Building Location: Uptown, N.J. Client: Widget Inc Project No. 78-102
Contractor: Drill Good Inc. Driller: W. Dowell Inspector: J. Eyesharp Start: 7.21.78
Drill rig type: Strong 1200 Rod diameter: N Casing (Dia. / type / wgt. / drop): 4-Flush Jt 300lb 24in Sampler (Dia. / wgt. / drop): 2ft 140lb 30in Finish: 7.21.78

Casing blows per foot	Samples				Blows per 6 in.				Pen, in	Rec, in	GWL: At start 4.2 ft, end of Boring 3.0 ft	24 hrs 4.5 ft
	Depth, ft	No.	Type	Diameter	6	12	18	24			Sample description	Remarks on Operations
3/6	0-1.5	1	SS	2 in	2	4	4		18	9	Topsoil, DfS, l·$ to 5 in, then	7/21 — 8 a.m.-11 a.m. set rig over
Mud drilled											BC-fS, t·$, dry, stratified	hole and prepare for boring
below 2 ft	1.5-3.0	2	SS	2 in	3	5	7		18	12	BC-fs, t·$, moist, stratified,	S-1 taken from surface, then
to 21.0 ft	3.0-4.5	3	SS	2 in	1	2	4		18	0	Lost sample	2 ft section 4 in casing set
then 4 in	4.5-6.0	4	SS	2 in	1	2	2		18	10	Top of sample: 3 in Bm-fs, l-$ soft	and mixed mud
casing											Bot. " " : DO$ w/shell fgmts.	S-3 Sampling spoon lacked
stalled	6.0-8.0	5	P	3 in	Push-100 psi				24	24	Top of sple: " " , soft	core catcher
to 20 ft											Bot. of sple: " , more plastic	Stratum change at 5.5 ft
for rock	8.0-9.5	6	SS	2 in	0	1	1		18	7	DO$, clayey, w/shell fgmts.	" " at 10.5 ft
coring	9.5-11.0	7	SS	2 in	2	4	9		18	10	Top of sple: 4 in DO$ w/shell fgmts.	Below 17.0 ft hard drilling
											Bot. of sple: 6 in residual soil:	Refusal to s.s at 20.1 ft
											multicolored c-fS l·$ plastic	Cleaned hole to 21.0 ft, drove
											micaceous, compact	casing to 21.0 ft; starting
	11.0-13.0	8	D	2.5 in	Drilled				24	20	Same as S-7, compact	coring at 3 p.m. w/DS barrel
	13.0-14.5	9	SS	2 in	7	8	12		18	10	" " "	series "M" @ NX diamond bit w/
	15.0-16.5	10	SS	2 in	7	10	17		18	11	Grading to saprolite, rock	bottom discharge.
											structure apparent, mica and	
											quartz in sand sizes, silty	Boring completed: 6 p.m. 7/21
	20.0-20.1	11	SS	2 in	50/2 in				2	2	Gneiss Rock: Soft, L.G, mod. dec.	GWL at 6 p.m. - Depth = 3.0 ft
	21.0-26.0	12	DS	NX in	125 rpm @ 12 lb/in²				60	51	GN, L.G., soft-hard, mod. dec.	Hole Flushed; 5 ft casing left in hole
					(rate = 1 ft/6 in)						closely jointed, RQD = 60%	GWL at 6 p.m. 7.22 at d = 4.5 ft
	26.0-31.0	13	DS	NX in	125 rpm @ 250 lb/in²				60	57	GN, L.G., V. Hd slightly dec., mod. jointed	Field log boring No. 1
					(rate = 1 ft/12 in)						RQD = 75%	Sheet 1 of 1

FIGURE 1.91
Example of a field test boring log for soil and rock drilling.

Report Log

A preliminary report log is begun by the field inspector as the field boring log is prepared. The report log is intended to record the boring data needed for design analysis as well as some laboratory identification test data and a notation of the various tests performed. The report log also allows changes to be made in the material description column so that the descriptions agree with gradation and plasticity test results from the laboratory. The examples given in Figure 1.92 (test boring report log) for a soil and rock borehole and Figure 1.93 (core boring report log) for rock core borings illustrate the basic information required for report logs.

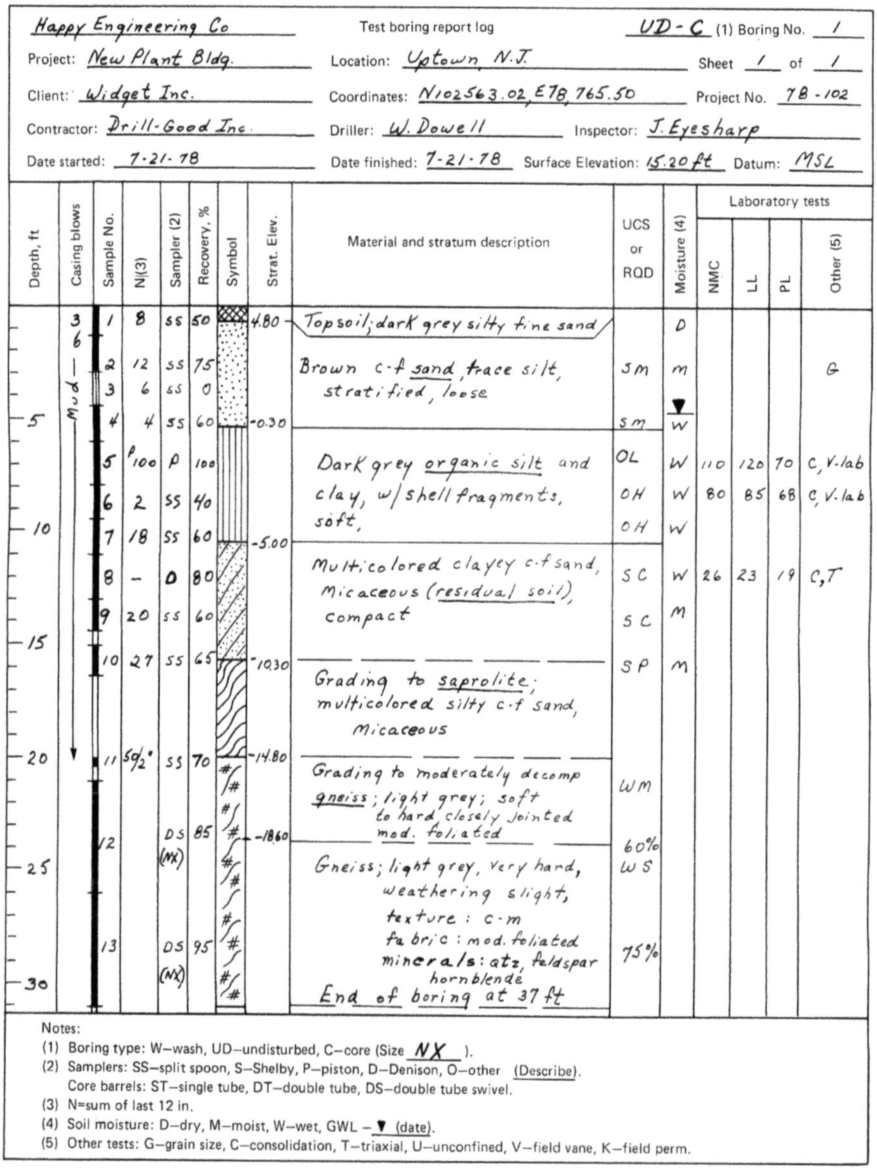

FIGURE 1.92
Example of test boring log for soil and rock boring.

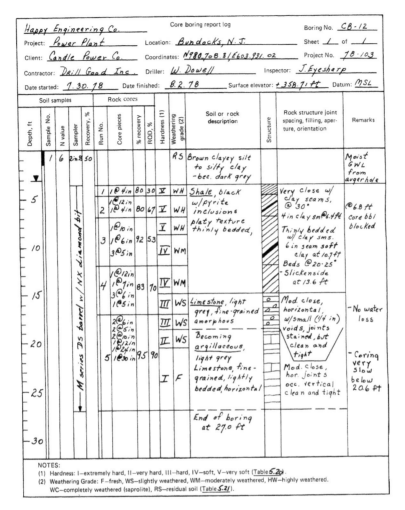

FIGURE 1.93
Example of a report log for rock-core boring.

References

ASCE, Geophysical Exploration for Engineering and Environmental Investigations, Technical Engineering and Design Guides as adapted from the US Army Corps of Engineers, No. 23, 1998.

ASP, *Manual of Photo Interpretation*, American Society of Photogrammetry, Washington, DC, 1960.

ASTM, *Symposium on Surface and Subsurface Reconnaissance*, Spec. Pub. No. 122, American Society for Testing and Materials, Philadelphia, June 1951.

Auld, B., Cross-hole and down-hole vs by mechanical impulse, *Proc. ASCE J. Geotech. Eng. Div.*, 103, 1381–1398, 1977.

Avery, T.E. and Berlin, G.L., *Fundamentals of Remote Sensing and Airphoto Interpretation*, 5th ed., Macmillan, New York, 1992.

Ballard, R. F., Jr., Method for crosshole seismic testing, *Proc. ASCE J. Geotech. Eng. Div.*, 102, 1261–1273, 1976.

Belcher, D. J., The Engineering Significance of Landforms, Pub. No. 13, Highway Research Board, Washington, DC, 1948.

Broms, B. B., Soil sampling in Europe: state-of-the art, *Proc. ASCE J. Geotech. Eng. Div.*, 106, 65–98, 1980.

Civil Engineering, Directional Drill Scouts Water Tunnel's Path, Civil Engr, ASCE, Apr. 1998.

Cotecchia, V., Systematic Reconnaissance Mapping and Registration of Slope Movements, Bull. No. 17, Intl. Assoc. Eng. Geol., June 1978, pp. 537.

Dahlin, T., Bjelm, L., and Svennsson, C., Resistivity Pre-Investigations for the Railway Tunnel through Hallandsas Sweden, *Proceedings of the 2nd European EEDS Meeting*, Nantes, France, 2–4 Sept., 1996, pp. 109–112.

Dobecki, T. L., Measurements of Insitu Dynamic Properties in Relation to Geologic Conditions, *Geology in the Siting of Nuclear Power Plants, Reviews in Engineering Geology* IV, The Geological Society of America, Boulder, CO, 1979, pp. 201–225.

Eide, O., Marine Soil Mechanics-Applications to North Sea Offshore Structures, Pub. No. 103, Norwegian Geotechnical Institute, Oslo, 1974, p. 19.

ENR, Soil Sampling Techniques, *Engineering News-Record*, Apr. 24, 1952.

ENR, Impact Drill Drives through Hard Rock Fast, *Engineering News-Record*, Nov. 3, 1977, p. 14.

Deere, D. U., Technical description of rock cores for engineering purposes, *Rock Mech. Eng. Geol.*, 1, 18–22, 1963.

DeLoach, S.R. and Leonard, J., Making Photographic History, Professional Surveyor, Apr. 2000.

Ghatge, S.L., Microgravity method for detection of abandoned mines in New Jersey, *Bull. Assoc. of Eng. Geologists*, 30, 79–85, 1993.

Gillbeaut, J.C., Lidar: Mapping a Shoreline by Laser Light, *Geotimes*, November 2003.

Greenfield, R. J., Review of geophysical approaches to the detection of Karst, *Bull. Assoc. Eng. Geol.*, 16, 393–408, 1979.

Griffiths, D. H. and King, R. F., *Applied Geophysics for Engineers and Geologists*, Pergamon Press, London, 1969.

Henderson III, F.B., Remote Sensing for Acid Mine Sites, *Geotimes*, November 2000.

Hryciw, R.D., Raschke, S.A., Ghalib, A.M., and Shin, S., A Cone With a View: The VisCPT, Geo-Strata, Geo-Institute, ASCE, July 2002.

Ladd, J. W., Buffler, R. T., Watkins, J. S., Worzel, J. L., and Carranza, A., Deep Seismic Reflection Results from the Gulf of Mexico, *Geology*, Geological Society of America, Vol. 4, No. 6, 1976, pp. 365–368.

Lueder, D. R., *Aerial Photographic Interpretation: Principles* and *Applications*, McGraw-Hill, New York, 1959.

Mooney, H. M., *Handbook of Engineering Seismology*, Bison Instruments Inc., Minneapolis, MN, 1973.

Myung, J. T. and Baltosser, R. W., Fracture Evaluation by the Borehole Logging Method, *Stability of Rock Slopes*, ASCE, New York, 1972, pp. 31–56.

Professional Surveyor, Vol. 19, October 1999.

Professional Surveyor, Component-built Aerial Sensor Means Imagery for Everyone, October 2002.

Robertson, P.K., Campanella, R.G., Gillespe, D., and Grieg, J., Use of piezometer cone data, *Proceedings In-Situ '86, ASCE Special Conference*, Blacksburg, VA, 1986.

Robertson, P.K., Seismic Cone Penetration for Evaluating Liquefaction Potential, *Conference on Recent Advances in Earthquake Design Using Laboratory and In-situ Tests*, Seminar Sponsored by ConeTec Investigations, Ltd., Feb. 5, 1990.

Robertson, P.K., Soil classification using the cone penetration test, *Can. Geotech. J.*, 27, 151–158, 1990.

Robertson, P.K., Lunne, T., and Powell, J.J.M., Geo-environmental applications of penetration testing, in *Geotechnical Site Characterization*, Robertson, P. K. and Mayne, R., Eds., Balkema, Rotterdam, 1998, pp. 35–48.

Rogers, F. C., Engineering Soil Survey of New Jersey, Report No. 1, Engineering Research Bulletin No. 15, College of Engineering, Rutgers Univ., Edwards Bros. Inc., Ann Arbor, MI, 1950.

Sanglerat, G., *The Penetrometer and Soil Exploration*, Elsevier, Amsterdam, 1972, p. 464.

Schmertmann, J. H., Guidelines for CPT Performance and Design, U.S. Dept. of Transportation, Federal Highway Admin., Offices of Research and Development, Washington, DC, 1977.

USACE, Engineering and Design — Soil Sampling, U.S. Army Corps of Engrs Pub. No. EM 1110-1-1906, 1996, 10–19, 10–25.

USBR, *Earth Manual*, U.S. Bureau of Reclamation, Denver, CO, 1974.

Way, D. S., *Terrain Analysis*, 2nd ed., Dowden, Hutchinson & Ross, Stroudsburg, PA, 1978.

Further Reading

Cook, J. C., Status of Ground Probing Radar and Some Recent Experience, *Subsurface Exploration for Underground Excavation and Heavy Construction, Proc. ASCE*, New York, 1974, pp. 175–194.

Godfrey, K. A., Jr., What Future for Remote Sensing in Space, Civil Engineering, ASCE, July, 1979, pp. 61–65.

Hvorslev, J. J., *Subsurface Exploration and Sampling of Soils for Civil Engineering Purposes*, Waterways Experimental Station, U.S. Army Engineers, Vicksburg, MS, November 1949.

Lowe III, J. and Zaccheo, P. F., Subsurface explorations and sampling, *Foundation Engineering Handbook*, Winterkorn, H. F. and Fang, H.-Y., Eds., Van Nostrand Reinhold Co., New York, 1975, chap. 1, pp. 1–66.

Lundstrom, R. and Stanberg, R., Soil-Rock Drilling and Rock Locating by Rock Indicator, *Proceedings of the 6th International Conference on Soil Mechanics and Foundation Engineering*, Montreal, 1965.

McEldowney, R. C. and Pascucci, R. F., Applications of Remote-sensing Data to Nuclear Power Plant Site Investigations, *Geology in the Siting of Nuclear Power Plants, Reviews in Engineering Geology IV*, The Geological Society of America, Boulder, CO, 1979, pp. 121–139.

Moffatt, B. T., Subsurface Video Pulse Radars, *Subsurface Exploration for Underground Excavation and Construction, Proc. ASCE*, New York, 1974, pp. 195–212.

Morey, R. M., Continuous Subsurface Profiling by Impulse Radar, *Subsurface Exploration in Underground Excavation and Heavy Construction, Proceedings of the ASCE*, New York, 1974, pp. 213–232.

Underwood, L. B., Exploration and Geologic Prediction for Underground Works, *Subsurface Exploration for Underground Excavation and Heavy* Construction, ASCE, New York, 1974, pp. 65–83.

USDA, Soil Survey of Autauga County, Alabama, U.S. Dept. of Agriculture, Soil Conservation Service, 1977, 64 pp. and maps.

2

Measurement of Properties

2.1 Introduction

2.1.1 Objectives

The properties of geologic materials are measured to provide the basis for:

1. Identification and classification.
2. Correlations between properties including measurements made during other investigations in similar materials.
3. Engineering analysis and evaluations.

2.1.2 Geotechnical Properties

Basic Properties

Basic properties include the fundamental characteristics of the materials and provide a basis for identification and correlations. Some are used in engineering calculations.

Index Properties

Index properties define certain physical characteristics used basically for classifications, and also for correlations with engineering properties.

Hydraulic Properties

Hydraulic properties, expressed in terms of permeability, are engineering properties. They concern the flow of fluids through geologic media.

Mechanical Properties

Rupture strength and deformation characteristics are mechanical properties. They are also engineering properties, and are grouped as static or dynamic.

Correlations

Measurements of hydraulic and mechanical properties, which provide the basis for all engineering analyses, are often costly or difficult to obtain with reliable accuracy. Correlations based on basic or index properties, with data obtained from other investigations in which extensive testing was employed or engineering properties were evaluated by back-analysis of failures, provide data for preliminary engineering studies as well as a check on the reasonableness of data obtained during investigation.

Data on typical basic, index, and engineering properties are given throughout the book for general reference. A summation of the tables and figures providing these data is given in Appendix E.

2.1.3   Testing Methods Summarized

General

A general summary of the significant basic, index, and engineering properties of soil and rock, and an indication of whether they are measured in the laboratory, *in situ*, or both, is given in Table 2.1.

TABLE 2.1

Measurement of Geotechnical Properties of Rock and Soil

	Laboratory Test		*In Situ*	
Property	**Rock**	**Soil**	**Rock**	**Soil**
(a) Basic Properties				
Specific gravity	X	X		
Porosity	X	X		
Void ratio		X		
Moisture content	X	X	X	X
Density	X	X	X	X
Natural		X		X
Maximum		X		
Minimum		X		
Relative		X		X
Optimum moisture density		X		
Hardness	X			
Durability	X			
Reactivity	X	X		
Sonic-wave characteristics	X	X	X	X
(b) Index Properties				
Grain-size distribution		X		
Liquid limit		X		
Plastic limit		X		
Plasticity index		X		
Shrinkage limit		X		
Organic content		X		
Uniaxial compression	X			
Point-load index	X			
(c) Engineering Properties				
Permeability	X	X	X	X
Deformation moduli: static or dynamic	X	X	X	X
Consolidation		X		X
Expansion	X	X	X	X
Extension strain	X		X	
Strength				
Unconfined	X	X		
Confined				
Static	X	X	X	X
Dynamic		X		
California bearing ratio (CBR)		X		X

Laboratory Testing

Soil samples and rock cores are, for the most part tested in the laboratory. Rock cores are occasionally field tested.

Rock cores are tested in the laboratory primarily for basic and index properties, since engineering properties of significance are not usually represented by an intact specimen. Laboratory tests of intact specimens, the property measured, and the application of the test in terms of the data obtained are summarized in Table 2.2.

Soil samples are tested for basic and index properties and for engineering properties when high-quality undisturbed samples are obtained (generally limited to soft to hard intact specimens of cohesive soils lacking gravel size or larger particles). Laboratory soil tests, properties measured, and the application of the tests in terms of the data obtained are summarized in Table 2.3.

In Situ Testing

Geologic formations are tested *in situ* within boreholes, on the surface of the ground, or within an excavation.

TABLE 2.2

Intact Rock Specimens: Laboratory Testing

Property or Test	Applications	Section
Basic Properties	*Correlations, analysis*	
Specific gravity	Mineral Identification	2.2.1
Porosity	Property correlations	2.2.1
Density	Material and property correlations	2.2.1
	Engineering analysis	
Hardness	Material correlations	2.2.1
	Tunneling machine excavation evaluation	
Durability		
LA abrasion	Evaluation of construction aggregate quality	2.2.1
British crushing		
Reactivity	Reaction between cement and aggregate	2.2.1
Sonic velocities	Computations of dynamic properties	2.5.3
Index Properties	*Classification and correlations*	
Uniaxial compression	See rupture strength	2.4.3
Point-load test	See rupture strength	2.4.3
Permeability	*Not normally performed in the lab*	
Rupture Strength	*Measurements of*	
Triaxial shear	Peak drained or undrained strength	2.4.3
Unconfined compression	Unconfined (uniaxial) compressive strength used for correlations	2.4.3
Point-load test	Tensile strength for correlation with uniaxial compression	2.4.3
Uniaxial tensile strength	Strength in tension	2.4.3
Flexural or beam strength	Strength in bending	2.4.3
Deformation (static)	*Measurements of*	
Triaxial test	Deformation moduli E_i, E_s, E_t	2.4.3
Unconfined compression	Deformation moduli E_i, E_s, E_t	2.4.3
Dynamic Properties	*Measurements of*	
Resonant column	Compression and shear wave velocities V_p, V_s	2.5.3
	Dynamic moduli E, G, D	

TABLE 2.3

Soils: Laboratory Testing

Property or Test	Applications	Section
Basic Properties	Correlations, classification	
Specific gravity	Material identification	2.2.3
	Void ratio computation	
Moisture or water content	Material correlations in the natural state	2.2.3
	Computations of dry density	
	Computations of Atterberg limits	
Density: natural (unit weight)	Material correlations	2.2.3
	Engineering analysis	
Density: maximum	Relative density computations	2.2.3
	Moisture–density relationships	
Density: minimum	Relative density computations	2.2.3
Optimum-moisture density	Moisture–density relationships for field compaction control	2.2.3
Sonic velocities	Computations of dynamic properties	2.5.3
Index Properties	Correlations, classification	
Gradation	Material classification	2.2.3
	Property correlations	
Liquid limit	Computation of plasticity index	2.2.3
	Material classification	
	Property correlations	
Plastic limit	Computation of plasticity index	2.2.3
Shrinkage limit	Material correlations	2.2.3
Organic content	Material classification	2.2.3
Permeability	Measurements of k in	
Constant head	Free-draining soil	2.3.3
Falling head	Slow-draining soil	2.3.3
Consolidometer	Very slow draining soil (clays)	2.5.4
Rupture Strength	Measurements of	
Triaxial shear (compression or extension)	Peak undrained strengh s_a, cohesive soils (UU test)	2.4.4
	Peak drained strength, $\phi, c, \bar{\phi}, \bar{c}$, all soils	
Direct shear	Peak drained strength parameters	2.4.4
	Ultimate drained strength $\bar{\phi}_r$ cohesive soils	
Simple shear	Undrained and drained parameters	2.4.4
Unconfined compression	Unconfined compressive strength for cohesive soils	2.4.4
	Approximately equals $2s_a$	
Vane shear	Undrained strength s_a for clays	2.4.4
	Ultimate undrained strengths s_r	
Torvane	Undrained strengths s_a	2.4.4
	Ultimate undrained strength s_r (estimate)	
Pocket penetrometer	Unconfined compressive strength (estimate)	2.4.4
California bearing ratio	CBR value for pavement design	2.4.4
Deformation (static)	Measurements of	
Consolidation test	Compression vs. load and time in clay soil	2.5.4
Triaxial shear test	Static deformation moduli	2.5.4
Expansion test	Swell pressures and volume change in the consolidometer	2.5.4
Dynamic Properties		
Cyclic triaxial	Low-frequency measurements of dynamic moduli (E, G, D), stress vs. strain and	2.5.5
	strength	2.4.4

(Continued)

TABLE 2.3

(*Continued*)

Property or Test	Applications	Section
Cyclic torsion	Low-frequency measurements of dynamic moduli, stress vs. strain	2.5.5
Cyclic simple shear	Low-frequency measurements of dynamic moduli, stress vs. strain and strength	2.5.5 2.4.4
Ultrasonic device	High-frequency measurements of compression- and shear-wave velocities V_p, V_s	2.5.5
Resonant column device	High-frequency measurements of compression and shear-wave velocities and the dynamic moduli	2.5.5

Rock masses are usually tested *in situ* to measure their engineering properties, as well as their basic properties. *In situ* tests in rock masses, their applications, and their limitations are summarized in Table 2.4.

Soils are tested *in situ* to obtain measures of engineering properties to supplement laboratory data, and in conditions where undisturbed sampling is difficult or not practical such as with highly organic materials, cohesionless granular soils, fissured clays, and cohesive soils with large granular particles (such as glacial till and residual soils). *In situ* soil tests, properties measured, applications, and limitations are summarized in Table 2.5.

TABLE 2.4

Rock Masses — *In Situ* Testing

Category — Tool or Method	Applications	Limitations	Section
Basic Properties			1.3.6
Gamma–gamma borehole probe	Continuous measure of density	Density measurements	
Neutron borehole probe	Continuous measure of moisture	Moisture measurements	1.3.6
Index Properties			
Rock coring	Measures the RQD (rock quality designation) used for various empirical correlations	Values very dependent on drilling equipment and techniques	1.4.5
Seismic refraction	Estimates rippability on the basis of P-wave velocities	Empirical correlations. Rippability depends on equipment used	1.3.2
Permeability			
Constant-bead test	In boreholes to measure k in heavily jointed rock masses.	Free-draining materials Requires ground saturation	2.3.4
Falling-head test	In boreholes to measure k in jointed rock masses. Can be performed to measure k_{mean}, k_v, or k_h	Slower draining materials or below water table	2.3.4
Rising-head test	Same as for falling-head test	Same as for falling-head test	2.3.4
Pumping tests	In wells to determine k_{mean} in saturated uniform formations.	Not representative for stratified formations. Measures average k for entire mass	2.3.4

(*Continued*)

TABLE 2.4

Rock Masses — *In Situ* Testing (*Continued*)

Category — Tool or Method	Applications	Limitations	Section
Pressure testing	Measures k_h in vertical borehole	Requires clean borehole walls Pressures can cause joints to open or to clog from migration of fines	2.3.4
Shear Strength			
Direct shear box	Measures strength parameters along weakness planes of rock block	Sawing block specimen and test setup costly. A surface test. Several tests required for Mohr's envelope	2.4.3
Triaxial or uniaxial compression	Measures triaxial or uniaxial compressive strength of rock block	Same as for direct shear box	2.4.3
Borehole shear device	Measures ϕ and c_c in borehole		2.4.3
Dilatometer or Goodman jack	Measures limiting pressure P_L in borehole	See Dilatometer under Deformation. Limited by rock-mass strength	2.5.3
Deformation Moduli (Static)			
Dilatometer or Goodman jack	Measures E in lateral direction	Modulus values valid for linear portion of load–deformation curve. Results affected by borehole roughness and layering	2.5.3
Large-scale foundation-load test	Measures E under footings or bored piles. Measures shaft friction of bored piles	Costly and time-consuming	2.5.4
Plate-jack test	Measures E: primarily used for tunnels and heavy structures	Requires excavation and heavy reaction or adit. Stressed zone limited by plate diameter and disturbed by test preparation	2.5.3
Flat-jack test	Measures E or residual stresses in a slot cut into the rock	Stressed zone limited by plate diameter. Test area disturbed in preparation. Requires orientation in same direction as applied construction stresses	2.5.3
Radial jacking tests (pressure tunnels)	Measures E for tunnels. Data most representative of *in situ* rock tests and usually yields the highest values for E	Very costly and time-consuming and data difficult to interpret. Preparation disturbs rock mass	2.5.3
Triaxial compression test	Measures E of rock block	Costly and difficult to set up. Disturbs rock mass during preparation	2.5.3
Dynamic Properties			
Seismic direct methods	Obtain dynamic elastic moduli E, G, K, and v in boreholes	Very low strain levels yield values higher than static moduli	2.5.3

(*Continued*)

TABLE 2.4

(*Continued*)

Category — Tool or Method	Applications	Limitations	Section
3-D velocity logger	Measures velocity of shear and compression waves (V_s, V_p) from which moduli are computed. Borehole test	Penetrates to shallow depth in borehole	2.5.3
Vibration monitor	Measure peak particle velocity, or frequency, acceleration, and displacement for monitoring vibrations from blasting, traffic, etc.	Surface measurements, low-energy level	3.2.5

TABLE 2.5

SOILS — *In Situ* Testing

Category Test or Method	Applications	Limitations	Section
Basic Properties			
Gamma-gamma borehole probe	Continuous measure of density	Density measurements	1.3.6
Neutron borehole probe	Continuous measure of moisture content	Moisture-content measures	1.3.6
Sand-cone density apparatus	Measure surface density	Density at surface	2.2.3
Balloon apparatus	Measure density at surface	Density at surface	2.2.3
Nuclear density moisture meter	Surface measurements of density and moisture	Moisture and density at surface	2.2.3
Permeability			
Constant-head test	In boreholes or pits to measure k in free-draining soils	Free-draining soils requires ground saturation	2.3.4
Falling-head test	In boreholes in slow-draining materials, or materials below GWL. Can be performed to measure k_{mean}, k_v, or k_h	Slow-draining materials or below water table	2.3.4
Rising-head test	Similar to falling-head test	Similar to falling-head test	2.3.4
Pumping tests	In wells то measure k_{mean} in saturated uniform soils	Results not representative in stratified formations	2.3.4
Shear Strength (Direct Methods)			
Vane shear apparatus	Measure undrained strength s_u and remolded strength s_r in soft to firm cohesive soils in a test boring	Not performed in sands or strong cohesive soils affected by soil anisotropy and construction time-rate differences	2.4.4
Pocket penetrometer	Measures approximate U_c in tube samples, test pits in cohesive soils	Not suitable in granular soils	2.4.4
Torvane	Measures s_a in tube samples and pits	Not suitable in sands and strong cohesive soils	2.4.4
Shear Strength (Indirect Methods)			
Static cone penetrometer (CPT)	Cone penetration resistance is correlated with s_a in clays and ϕ in sands	Not suitable in very strong soils	2.4.5

(*Continued*)

TABLE 2.5

SOILS — *In Situ* Testing (*Continued*)

Category Test or Method	Applications	Limitations	Section
Flat Dilatometer (DMT)	Correlations with pressures provide estimares of ϕ and s_u	Not suitable in very strong soils	2.4.5
Pressuremeter	Undrained strength is found from limiting pressure correlations	Strongly affected by soil anisotropy	2.5.4
Camkometer (self-boring pressuremeter)	Provides data for determination of shear modulus; shear strength, pore pressure, and lateral stress K_o	Affected by soil anisotropy and smear occurring during installation	2.5.4
Penetration Resistance			
Standard penetration test (SPT)	Correlations provide measures of granular soil D_R, ϕ, E, allowable bearing value, and clay soil consistency. Samples recovered	Correlations empirical. Not usually reliable in clay soils. Sensitive to sampling procedures	2.4.5
Static cone penetrometer test (CPT)	Continuous penetration resistance can provide measures of end bearing and shaft friction. Correlations provide data similar to SPT	Samples not recovered, material identification requires borings or previous area experience	2.4.5
California bearing ratio	CBR value for pavement design	Correlations are empirical	2.4.5
Deformation Moduli (Static)			
Pressuremeter	Measures E in materials difficult to sample undisturbed such as sands, residual soils, glacial till, and soft rock in a test boring	Modulus values only valid for linear portion of soil behavior, invalid in layered formations; not used in weak soils	2.5.4
Camkometer	See shear strength.		
Plate-load test	Measures modulus of subgrade reaction used in beam-on-elastic-subgrade problems. Surface test	Stressed zone limited to about 2 plate diameters. Performed in sands and overconsolidated clays	2.5.4
Lateral pile-load test	Used to determine horizontal modulus of subgrade reaction	Stressed zone limited to about 2 pile diameters. Time deformation in clays not considered	Not described
Full-scale foundation load tests	Obtain E in sands and design parameters for piles	Costly and time-consuming	2.5.4
Dynamic Properties			
Seismic direct methods	Borehole measurements of S-wave velocity to compute E_d, G_d, and K	Very low strain levels yield values higher than static moduli	1.3.2
Steady-state vibration method	Surface measurement of shear wave velocities to obtain E_d, G_d, and K	Small oscillators provide data only to about 3 m. Rotating mass oscillator provides greater penetration	2.5.5
Vibration monitors	Measure peak particle velocity or frequency, acceleration, and displacement for monitoring vibrations from blasting, traffic, etc.	Surface measurements at low-energy level for vibration studies	3.2.5

2.2   Basic and Index Properties

2.2.1   Intact Rock

General

Testing is normally performed in the laboratory on a specimen of fresh to slightly weathered rock free of defects.

Basic properties include volume–weight relationships, hardness (for excavation resistance), and durability and reactivity (for aggregate quality).

Index tests include the uniaxial compression test (see Section 2.4.3), the point load index test (see Section 2.4.3), and sonic velocities, that are correlated with field sonic velocities to provide a measure of rock quality (see Section 2.5.3).

Volume–Weight Relationships

Include specific gravity, density, and porosity as defined and described in Table 2.6.

Hardness

General

Hardness is the ability of a material to resist scratching or abrasion. Correlations can be made between rock hardness, density, uniaxial compressive strength, and sonic velocities, and between hardness and the rate of advance for tunneling machines and other

TABLE 2.6

Volume–Weight Relationships for Intact Rock Specimens

Property	Symbol	Definition	Expression	Units
Specific gravity (absolute)	G_s	The ratio of the unit weight of a pure mineral substance to the unit weight of water at 4°C. $\gamma_w = 1\,g/cm^3$ or 62.4 pcf	$G_s = \gamma_m/\gamma_w$	
Specific gravity (apparent)	G_s	The specific gravity obtained from a mixture of minerals composing a rock specimen	$G_s = \gamma_m/\gamma_w$	
Density	ρ or γ	Weight W per unit volume V of material	$\rho = W/V$	t/m^3
Bulk density	ρ	Density of rock specimen from field (also g/cm^3, pcf)	$\rho = W/V$	t/m^3
Porosity	n	Ratio of pore or void volume V_v to total volume V_t.	$n = V_v/V_s$	%
		In terms of density and the apparent specific gravity	$n = 1 - (\rho/G_s)$	% (metric)

Notes: *Specific Gravities:* Most rock-forming minerals range from 2.65 to 2.8, although heavier minerals such as hornblende, augite, or hematite vary from 3 to 5 and higher.

Porosity: Depends largely on rock origin. Slowly cooling igneous magma results in relatively nonporous rock, whereas rapid cooling associated with escaping gases yields a porous mass. Sedimentary rocks depend on amount of cementing materials present and on size, grading, and packing of particles.

Density: Densities of fresh, intact rock do not vary greatly unless they contain significant amounts of the heavier minerals.

Porosity and density: Typical value ranges are given in Table 2.12.

Significance: Permeability of intact rock often related to porosity, although normally the characteristics of the *in situ* rock govern rock-mass permeability. There are strong correlations between density, porosity, and strength.

excavation methods. The predominant mineral in the rock specimen and the degree of weathering decomposition are controlling factors.

Measurement Criteria

The following criteria are used to establish hardness values:

1. Moh's system of relative hardness for various minerals.
2. Field tests for engineering classification.
3. "Total" hardness concept of Deere (1970) based on laboratory tests and developed as an aid in the design of tunnel boring machines (TBMs). Ranges in total hardness of common rock types are given in Figure 2.1.
4. *Testing methods for total hardness* (Tarkoy, 1975):

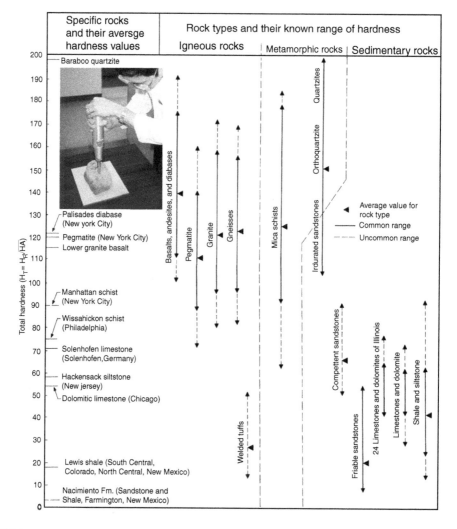

FIGURE 2.1

Range of "total" hardness for common rock types. Data are not all inclusive, but represent the range for rocks tested in the Rock Mechanics Laboratory, University of Illinois, over recent years. H_R = Schmidt hardness; H_A = abrasion test hardness. (From Tarkoy, P. J., *Proceedings of the 15th Symposium on Rock Mechanics*, Custer State Park, South Dakota, ASCE, New York, 1975, pp. 415–447. With permission.) (a) Inset: Schmidt hammer.

Total hardness H_T is defined as

$$H_T = H_R \sqrt{H_A}\, g^{-1/2} \tag{2.1}$$

where H_R is the Schmidt hardness and H_A the abrasion test hardness.

- *Schmidt rebound hardness test*: An L-type concrete test hammer (Figure 2.1a), with a spring in tension, impels a known mass onto a plunger held against the specimen (energy = 0.54 ft lb. or 0.075 m kg). The amount of energy reflected from the rock–hammer interface is measured by the *amount that* the hammer mass is caused to rebound (ASTM C805).
- *Shore (C-2) sclerescope* is also used to measure rebound hardness. The rebound height of a small diamond-tipped weight falling vertically down a glass tube is measured and compared with the manufacturer's calibration.
- *Abrasion hardness test* is performed on a thin disk specimen which is rotated a specific number of times against an abrading wheel, and the weight loss recorded.

Durability

General

Durability is the ability of a material to resist degradation by mechanical or chemical agents. It is the factor controlling the suitability of rock material used as aggregate for roadway base course, or in asphalt or concrete. The predominant mineral in the specimen, the microfabric (fractures or fissures), and the decomposition degree are controlling factors.

Test Methods

Los Angeles abrasion test (ASTM C535-03 and C131-03): specimen particles of a specified size are placed in a rotating steel drum with 12 steel balls (1 7/8 in. in diameter). After rotation for a specific period, the aggregate particles are weighed and the weight loss compared with the original weight to arrive at the LA abrasion value. The maximum acceptable weight loss is usually about 40% for bituminous pavements and 50% for concrete.

British crushing test: specimen particles of a specified size are placed in a 4-in.-diameter steel mold and subjected to crushing under a specified static force applied hydraulically. The weight loss during testing is compared with the original weight to arrive at the British crushing value. Examples of acceptable value ranges, which may vary with rock type and specifying agency, are as follows: particle size (maximum weight loss), 3/4–1 in. (32%), 1/2–3/4 in. (30%), 3/8–1/2 in. (28%); and 1/8–3/16 in. (26%).

Slake durability test (ASTM D4644): determines the weight loss after alternate cycles of wetting and drying shale specimens. High values for weight loss indicate that the shale is susceptible to degradation in the field when exposed to weathering processes.

Reactivity: Cement–Aggregate

Description

Crushed rock is used as aggregate to manufacture concrete. A reaction between soluble silica in the aggregate and the alkali hydroxides derived from portland cement can produce abnormal expansion and cracking of mortar and concrete, often with severely detrimental effects to pavements, foundations, and concrete dams. There is often a time delay of about 2 to 3 years after construction, depending upon the aggregate type used.

The Reaction

Alkali–aggregate reaction can occur between hardened paste of cements containing more than 0.6% soda equivalent and any aggregate containing reactive silica. The soda equivalent is calculated as the sum of the actual Na_2O content and 0.658 times the K_2O content of the clinker (NCE, 1980). The alkaline hydroxides in the hardened cement paste attack the silica to form an unlimited-swelling gel that draws in any free water by osmosis and expands, disrupting the concrete matrix. Expanding solid products of the alkali–silica reaction help to burst the concrete, resulting in characteristic map cracking on the surface. In severe cases, the cracks reach significant widths.

Susceptible Rock Silicates

Reactive silica occurs as opal or chalcedony in certain cherts and siliceous limestones and as acid and intermediate volcanic glass, cristobolite, and tridymite in volcanic rocks such as rhyolite, dacites, and andesites, including the tuffs. Synthetic glasses and silica gel are also reactive. All of these substances are highly siliceous materials that are thermodynamically metastable at ordinary temperatures and can also exist in sand and gravel deposits. Additional descriptions are given in Krynine (1957).

Reaction Control

Reaction can be controlled (Mather, 1956) by:

1. Limiting the alkali content of the cement to less than 0.6% soda equivalent. Even if the aggregate is reactive, expansion and cracking should not result.
2. Avoiding reactive aggregate.
3. Replacing part of the cement with a very finely ground reactive material (a pozzolan) so that the first reaction will be between the alkalis and the pozzolan, which will use up the alkalis, spreading the reaction and reaction products throughout the concrete.

Tests to Determine Reactivity

Tests include:

- The mortar-bar expansion test (ASTM C227-03) made from the proposed aggregate and cement materials.
- Quick chemical test on the aggregates (ASTM C289-01).
- Petrographic examination of aggregates to identify the substances (ASTM C295).

2.2.2 Rock Masses

General

The rock mass, often referred to as *in situ* rock, may be described as consisting of rock blocks, ranging from fresh to decomposed, and separated by discontinuities. Mass density is the basic property. Sonic-wave velocities and the rock quality designation (RQD) are used as index properties.

Mass Density

Mass density is best measured *in situ* with the gamma–gamma probe (see Section 1.3.6), which generally allows for weathered zones and the openings of fractures and small voids, all serving to reduce the density from fresh rock values.

Rock Quality Indices

Sonic wave velocities from seismic direct surveys (see Section 1.3.2) are used in evaluating rock mass quality and dynamic properties.

Rock quality designation may be considered as an index property (see Section 1.4.5).

Rippability

Rippability refers to the ease of excavation by construction equipment. Since it is related to rock quality in terms of hardness and fracture density, which may be measured by seismic refraction surveys (see Section 1.3.2), correlations have been made between rippability and seismic P wave velocities as given in Table 2.7. If the material is not rippable by a particular piece of equipment, then jack-hammering and blasting are required.

2.2.3 Soils

General

The basic and index properties of soils are generally considered to include volume–weight and moisture–density relationships, relative density, gradation, plasticity, and organic content.

TABLE 2.7
Rock Rippability as Related to Seismic p-Wave Velocities (Courtesy of Caterpillar Tractor Co.)

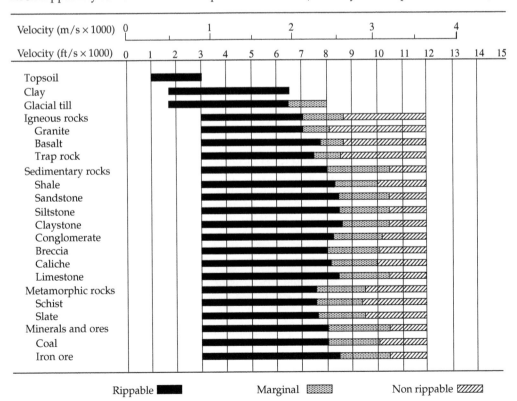

Rippable ■ Marginal ▨ Non rippable ▨

Rippability based on caterpillar D9 with mounted hydraulic No.9 ripper

Volume–Weight Relationships

Definitions of the various volume–weight relationships for soils are given in Table 2.8.

Commonly used relationships are void ratio *e*, soil unit weight (also termed density or mass density and reported as total or wet density γ_t, dry density γ_d, and buoyant density γ_b), moisture (or water) content *w*, saturation degree *S*, and specific gravity G_s of solids.

Determinations of basic soil properties are summarized in Table 2.9. A nomograph for the determination of basic soil properties is given in Figure 2.2.

Sand Cone Density Device (Figure 2.3a)

A hole 6 in. deep and 6 in. in diameter is dug and the removed material is stored in a sealed container. The hole volume is measured with calibrated sand and the density is calculated from the weight of the material removed from the hole (ASTM D1556-00).

Rubber Balloon Device

A hole is dug and the material is stored as described above. The hole volume is measured by a rubber balloon inflated by water contained in a metered tube (ASTM D2167-94).

Nuclear Moisture-Density Meter (ASTM D2922)

A surface device, the nuclear moisture-density meter, measures wet density from either the direct transmission or backscatter of gamma rays; and, moisture content from the transmission or backscatter of neutron rays (Figure 2.3b). The manner of measurement is similar to that of the borehole nuclear probes (see Section 1.3.6). In the direct transmission mode a rod containing a Celsium source is lowered into the ground to a desired depth. In the backscatter mode, the rod is withdrawn and gamma protons are scattered from the surface contact. A rapid but at times approximate method, measurement with the meter yields satisfactory results with modern equipment and is most useful in large projects where soil types used as fills do not vary greatly. Frequent calibration is important to maintain accuracy.

Borehole Tests

Borehole tests measure natural density and moisture content. Tests using nuclear devices are described in Section 1.3.6.

Moisture content (w)

The moisture meter is used in the field (ASTM D4444-92). Calcium carbide mixed with a soil portion in a closed container generates gas, causing pressure that is read on a gage to indicate moisture content. Results are approximate for some clay soils.

For cohesive soils, moisture content is most reliably determined by drying in the laboratory oven for at least 24 h at 104°C.

Moisture–Density Relationships (Soil Compaction)

Optimum moisture content and *maximum dry density relationships* are commonly used to specify a standard degree of compacting to be achieved during the construction of a load-bearing fill, embankment, earth dam, or pavement. Specification is in terms of a percent of maximum dry density, and a range in permissible moisture content is often specified as well (Figure 2.4).

Description

The density of a soil can be increased by compacting with mechanical equipment. If the moisture content is increased in increments, the density will also increase in increments

TABLE 2.8

Volume–Weight Relationships for Soils[a]

Property	Saturated Sample (W_s, W_w, G_s, are Known)	Unsaturated Sample (W_s, W_w, G_s, V are Known)	Illustration of Sample
Volume Components			
Volume of solids V_s	$\dfrac{W_s}{G_s \gamma_w^{\text{b}}}$		
Volume of water V_w	$\dfrac{W_w}{\gamma_w^{\text{b}}}$		
Volume of air or gas V_a	Zero	$V - (V_s + V_w)$	
Volume of voids V_v	$\dfrac{W_w}{\gamma_w^{\text{c}}}$	$V - \dfrac{W_s}{G_s \gamma_w}$	
Total volume of sample V	$V_s + V_w$	Measured	
Porosity n	$\dfrac{V_v}{V}$ or $\dfrac{e}{1+e}$		
Void ratio e	$\dfrac{V_v}{V_s}$ $(G_{ra}s)\text{-}1$		
Weights for Specific Sample			
Weight of solids W_s	Measured		
Weight of water W_w	Measured		
Total weight of sample W_t	$W_s + W_w$		
Weights for Sample of Unit Volume			
Dry-unit weight γ_d	$\dfrac{W_s + W_w}{V_s + V_w}$	$\dfrac{W_s}{V}$	
Wet-unit weight γ_t	$\dfrac{W_s + W_w}{V_s + V_w}$	$\dfrac{W_s + W_w}{V}$	
Saturated-unit weight γ_s	$\dfrac{W_s + W_w}{V_s + V_w}$	$\dfrac{W_s + W_w \gamma_w}{V}$	
Submerged (buoyant) unit weight γ_b	$\gamma_s - \gamma_w^{\text{c}}$		
Combined relations			
Moisture content w	$\dfrac{W_w}{W_s}$		
Degree of saturation S	1.00	$\dfrac{V_w}{V_v}$	$\gamma_d = \dfrac{\gamma_t}{1 + W}$ $\gamma_s = \gamma_d + \gamma_w \left(\dfrac{e}{1+e}\right)$
Specific gravity G_s	$\dfrac{W_s}{V_s \gamma_w}$		

a   After NAVFAC, Design manual DM-7.1, *Soil Mechanics, Foundations and Earth Structures*, Naval facilities Engineering Command, Alexandria, VA, 1982.

b γ_w is unit weight of water, which equals 62.4 pcf for fresh water and 64 pcf for sea water (1.00 and 1.025 g/cm³).

c   The actual unit weight of water surrounding the soil is used. In other cases use 62.4 pcf. Values of w and s are used as decimal numbers.

TABLE 2.9

Determination of Basic Soil Properties

	Determination	
Basic Soil Property	**Laboratory Test**	**Field Test**
Unit weight or density, γ_d, γ_t, γ_s, γ_b	Weigh specimens	Cone density device Figure 3.3a, ASTM U1556
		Rubber balloon device, ASTM D2167
		Nuclear moisture-density meter, ASTM D2922
Specific gravity G_s	ASTM D854	None
Moisture content w	ASTM D4444	Moisture meter
	ASTM D2922	Nuclear moisture-density meter
Void ratio e	Computed from unit dry weight and specific gravity	

under a given compactive effort, until eventually a peak or maximum density is achieved for some particular moisture content. The density thereafter will decrease as the moisture content is increased. Plotting the values of $w\%$ vs. γ_t, or $w\%$ vs. γ_d will result in curves similar to those given in Figure 2.5; 100% saturation is never reached because air remains trapped in the specimen.

Factors Influencing Results

The shape of the moisture–density curve varies for different materials. Uniformly graded cohesionless soils may undergo a decrease in dry density at lower moisture as capillary forces cause a resistance to compacting or arrangement of soil grains (bulking). As moisture is added, a relatively gentle curve with a poorly defined peak is obtained (Figure 2.5). Some clays, silts, and clay–sand mixtures usually have well-defined peaks, whereas low-plasticity clays and well-graded sands usually have gently rounded peaks (Figure 2.6). Optimum moisture and maximum density values will also vary with the compacted energy (Figure 2.7).

Test Methods

Standard compaction test (Proctor Test) (ASTM D698): An energy of 12,400 ft lb is used to compact 1 ft^3 of soil, which is accomplished by compacting three sequential layers with a 5 1/2-lb hammer dropped 25 times from a 12-in. height, in a 4-in.-diameter mold with a volume of 1/30 ft^3.

Modified compaction test (ASTM D1557): An energy of 56,250 ft lb is used to compact 1 ft^3 of soil, which is accomplished by compacting five sequential layers with a 10 lb hammer dropped 25 times from an 18 in. height in a standard mold. Materials containing significant amounts of gravel are compacted in a 6-in.-diameter mold (0.075 ft^3) by 56 blows on each of the five layers. Methods are available for correcting densities for large gravel particles removed from the specimen before testing.

Relative Density D_R

Relative density D_R refers to an *in situ* degree of compacting, relating the natural density of a cohesionless granular soil to its maximum density (the densest state to which a soil can be compacted, $D_R = 100\%$) and the minimum density (the loosest state that dry soil grains can attain, $D_R = 0\%$). The relationship is illustrated in Figure 2.8, which can be used to find D_R when γ_N (natural density), γ_D (maximum density), and γ_L (loose density) are known. D_R may be expressed as

$$D_R = (1/\gamma_L - 1/\gamma_N)/(1/\gamma_L - 1/\gamma_D) \tag{2.2}$$

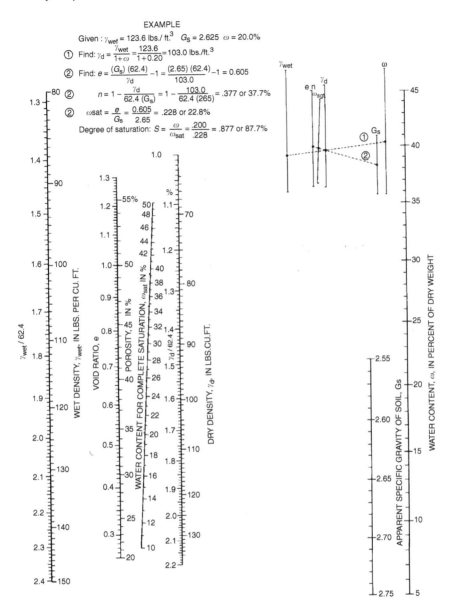

FIGURE 2.2
Nomograph to determine basic soil properties. (From USBR, *Earth Manual*, U. S. Burean of Reclamation, Denver, CO, 1974. With permission.)

Significance

D_R is used for classification of the degree of *in situ* compactness as given in Figure 2.8 or, more commonly, to classify *in situ* density as follows: very loose (0–15%), loose (15–35%), medium dense (35–65%), dense (65–85%), and very dense (85–100%) (see Table 2.23 for correlations with N values of the Standard Penetration Test (SPT)). Void ratio and unit weight are directly related to D_R and gradation characteristics. Permeability, strength, and compressibility are also related directly to D_R and gradation characteristics.

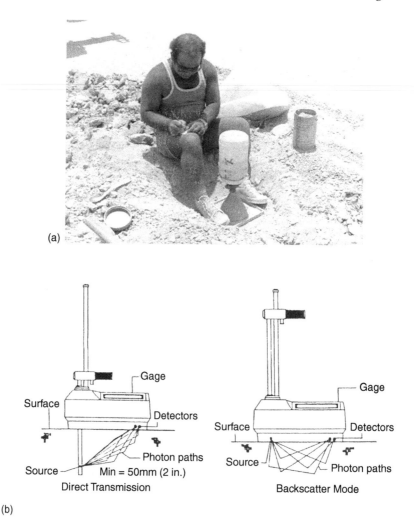

FIGURE 2.3

(a) Sand cone density device being used to measure *in situ* density of a compacted subgrade test section for an airfield pavement. (b) Nuclear moisture density meter used to measure *in situ* density.

Measurements of D_R

Laboratory testing: See ASTM D4254-00 and Burmister (1948). Maximum density is determined by compacting tests as described in the above section, or by vibrator methods wherein the dry material is placed in a small mold in layers and densified with a hand-held vibrating tool. Minimum density is found by pouring dry sand very lightly with a funnel into a mold. D_R measurements are limited to material with less than about 35% nonplastic soil passing the No. 200 sieve because fine-grained soils falsely affect the loose density. A major problem is that the determination of the natural density of sands cannot be sampled undisturbed. The shear-pin piston (see Section 1.4.2) has been used to obtain values for γ_N, or borehole logging with the gamma probe is used to obtain values (see Section 1.3.6).

Field testing: The SPT and Cone Penetrometer Test (CPT) methods are used to obtain estimates of D_R.

Correlations: Relations such as those given in Figure 2.10 for various gradations may be used for estimating values for γ_D and γ_L.

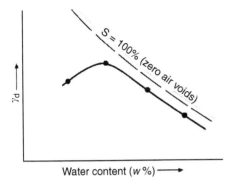

FIGURE 2.4
The moisture–density relationship. The soil does not become fully saturated during the compaction test.

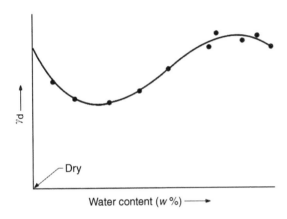

FIGURE 2.5
Typical compaction curve for cohesionless sands and sandy gravels. (From Foster, C. R., *Foundation Engineering*, G. A. Leonards, Ed., McGraw-Hill Book Co., New York, 1962, pp. 1000–1024. With permission. Reprinted with permission of the McGraw-Hill Companies.)

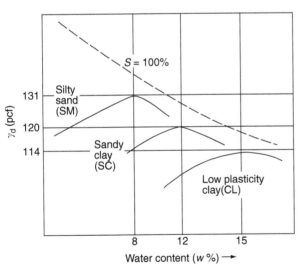

FIGURE 2.6
Typical standard Proctor curves for various materials.

Gradation (Grain Size Distribution)

Gradation refers to the distribution of the various grain sizes in a soil specimen plotted as a function of the percent by weight passing a given sieve size (Figure 2.9):

- *Well-graded* — a specimen with a wide range of grain sizes.
- *Poorly graded* — a specimen with a narrow range of grain sizes.
- *Skip-graded* — a specimen lacking a middle range of grain sizes.

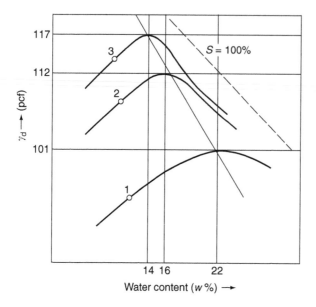

3. Mod. AASHO – 56 blows,10-lb hammer,
          18-in drop, five layers

2. Mod. AASHO – 25 blows,10-lb hammer,
          18-in drop, five layers

1. Std. AASHO – 25 blows,5 1/2-lb hammer,
          12-in drop, three layers

(All in 6-in molds)

FIGURE 2.7
Effect of different compactive energies on a silty clay. (After paper presented at Annual ASCE Meeting, January 1950.)

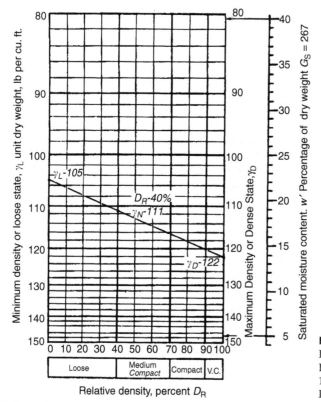

FIGURE 2.8
Relative density diagram. (From Burmister, D. M., ASTM, Vol. 48, Philadelphia, PA, 1948. Copyright ASTM International. Reprinted with permission.)

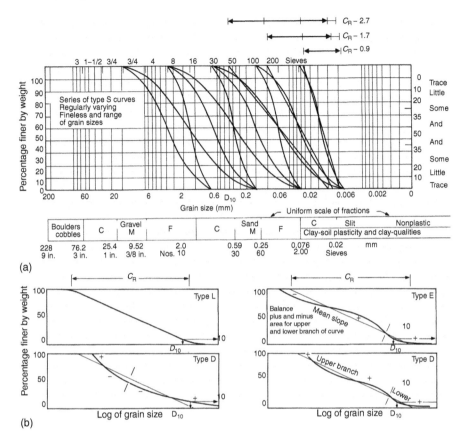

FIGURE 2.9
Distinguishing characteristics of grain size curves: fineness, range of grain sizes, and shape: (a) type S grain size curves and (b) type of grain size curve. (From Burmister, D. M., ASTM, Vol. 48, Philadelphia, PA, 1948. Copyright ASTM International. Reprinted with permission.)

- *Coefficient of uniformity* C_u — the ratio between the grain diameter at 60% finer to the grain diameter corresponding to the 10% finer line, or

$$C_u = D_{60}/D_{10} \tag{2.3}$$

Significance

Gradation relationships are used as the basis for soil classification systems. Gradation curves from cohesionless granular soils may be used to estimate γ_D and γ_L, and, if γ_N or D_R is known, estimates can be made of the void ratio, porosity, internal friction angle, and coefficient of permeability.

Gradation Curve Characteristics (Burmister, 1948, 1949, 1951a)

The gradation curves and characteristic shapes, when considering range in sizes, can be used for estimating engineering properties. The range of sizes C_R represents fractions of a uniform division of the grain size wherein each of the divisions 0.02 to 0.06, 0.06 to 0.02, etc., in Figure 2.9 represents a $C_R = 1$. Curve shapes are defined as L, C, E, D, or S as given in Figure 2.9 and are characteristic of various types of soil formations as follows:

- S shapes are the most common, characteristic of well-sorted (poorly graded) sands deposited by flowing water, wind, or wave action.

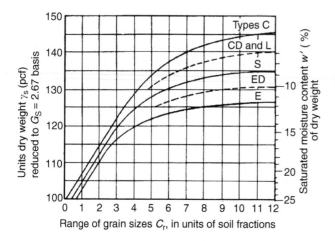

(b) Approximate Minimum Densities, 0% D_r		
	Decrease in Density (pcf)	
Range in grain sizes (C_r)	Coarser soils	Finer soils
1 – 3	10	to 20
3 – 5	20	to 25
5 and greater	25	to 30+

(c) Approximate Influence of Grain Shape on Density	
Grain shape	Change in density (pcf)
	–
Very angular	10 to –15
Subangular	0 to normal
Rounded or waterworn	+2 to +5
0.5% mica	– 2 to – 5

FIGURE 2.10

Maximum compacted densities, approximate minimum densities, and influence of grain shape on density for various gradations. (From Burmister, D. M., ASTM, Vol. 48, Philadelphia, PA, 1948. Reprinted with permission of the American Society for Testing and Materials.)

- C shapes have a high percentage of coarse and fine particles compared with sand particles and are characteristics of some alluvial valley deposits in an arid climate where the native rocks are quartz-poor.
- E and D shapes include a wide range of particle sizes characteristics of glacial tills and residual soils.

Relationships

General relationships among gradation characteristics and maximum compacted densities, minimum densities, and grain angularity are given in Figure 2.10 (note the significance of grain angularity). Gradation characteristics for soils of various geologic origins, as deposited, are given in Figure 2.11.

Test Methods

Gradations are determined by sieve analysis (ASTM D422) and hydrometer analysis (ASTM D422), the latter test being performed on material finer than a no. 200 sieve. For sieve analysis, a specimen of known weight is passed dry through a sequence of sieves of decreasing size of openings and the portion retained is weighed, or a specimen of known weight is washed through a series of sieves and the retained material dried and weighed. The latter procedure is preferred for materials with cohesive portions because dry sieving is not practical and will yield erroneous results as fines clog the sieves.

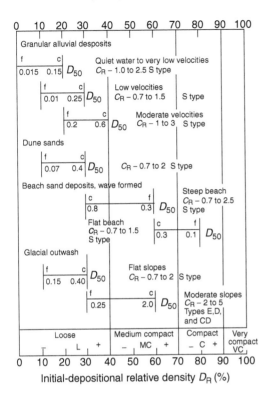

FIGURE 2.11
Probable initial depositional relative densities produced by geologic process of granular soil formation as a tentative guide showing dependence on grain-size parameters, grading-density relations, and geological processes. (From Burmister, D. M., ASTM Special Technical Publication No. 322, 1962a, pp. 67–97. Reprinted with permission of the American Society for Testing and Materials.)

Plasticity

Definitions and Relationships

Atterberg limits, which include the liquid limit, plastic limit, and the shrinkage limit, are used to define plasticity characteristics of clays and other cohesive materials.

Liquid limit (LL) is the moisture content at which a soil passes from the liquid to the plastic state as moisture is removed. At the LL, the undrained shear strength $s_u \approx 0.03$ tsf.

Plastic limit (PL) is the moisture content at which a soil passes from the plastic to the semisolid state as moisture is removed.

Plasticity index (PI) is defined as PI = LL − PL.

Shrinkage limit (SL) is the moisture content at which no more volume change occurs upon drying.

Activity is the ratio of the PI to the percent by weight finer than 2 μm (Skempton, 1953)

Liquidity index (LI) is used for correlations and is defined as

$$LI = (w - PL)/(LL - PL) = (w - PL)/PI \qquad (2.4)$$

Significance

A plot of PI vs. LL provides the basis for cohesive soil classification as shown on the plasticity chart (Figure 2.12). Correlations can be made between test samples and characteristic values of natural deposits. For example, predominantly silty soils plot below the A line, and predominantly clayey soils plot above. In general, the higher the value for the PI and LL, the greater is the tendency of a soil to shrink upon drying and swell upon wetting. The relationship between the natural moisture content and LL and PI is an indication of the soil's consistency, which is related to strength and compressibility (see Table 2.37). The

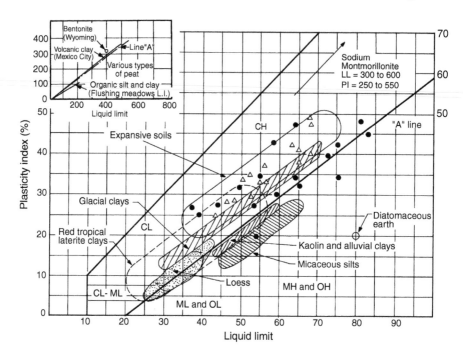

FIGURE 2.12
Plasticity chart for Unified Classification System.

liquidity index expresses this relationship quantitatively. The controlling factors in the values of PI, PL and LL for a given soil type are the presence of clay mineral, and the percentages of silt, fine sand, and organic materials.

Test Methods

Liquid limit (ASTM D4318-00) is performed in a special device containing a cup that is dropped from a controlled height. A pat of soil (only material passing a no. 40 sieve) is mixed thoroughly with water and placed in the cup, and the surface is smoothed and then grooved with a special tool. The LL is the moisture content at which 25 blows of the cup are required to close the groove for a length of 1 cm. There are several test variations (Lambe, 1951).

Plastic limit (ASTM D4318-00) is the moisture content at which the soil can just be rolled into a thread 1/8 in. in diameter without breaking.

Shrinkage limit (ASTM D427) is performed infrequently. See Lambe (1951) for discussion.

Organic Content

General

Organic materials are found as pure organic matter or as mixtures with sand, silt, or clay.

Basic and Index Properties

Organic content is determined by the *loss by ignition test* that involves specimen combustion at 440°F until constant weight is attained (Arman, 1970). Gradation is determined after loss by ignition testing. Plasticity testing (PI and LL) provides an indication of organic matter as shown in Figure 2.12 (see also ASTM D2914-00).

2.3 Hydraulic Properties (Permeability)

2.3.1 Introduction

Flow-Through Geologic Materials

Definitions and Relationships

Permeability, the capacity of a material to transmit water, is only summarized in this chapter. Flow through a geologic medium is quantified by a material characteristic termed the *coefficient of permeability k* (also known as coefficient of hydraulic conductivity), expressed in terms of Darcy's law, valid for laminar flow in a saturated, homogeneous material, as

$$k = q/iA \text{ (cm/sec)} \tag{2.5}$$

where q is the quantity of flow per unit of time (cm^3/sec), i the hydraulic gradient, i.e., the head loss per length of flow h/L (a dimensional number) and, A the area (cm^2).

Values for k are often given in units other than cm/sec. For example, 1 ft/day $\times$ 0.000283 = cm/sec; cm/sec $\times$ 3528 = ft/day.

Secondary permeability refers to the rate of flow through rock masses, as contrasted with that through intact rock specimens, and is often given in Lugeon units (see Section 2.3.4).

Factors Affecting Flow Characteristics

Soils: In general, gradation, density, porosity, void ratio, saturation degree, and stratification affect k values in all soils. Additional significant factors are relative density in granular soils and mineralogy and secondary structure in clays.

Rocks: k values of intact-rock relate to porosity and saturation degree. k values of *in situ* rock relate to fracture characteristics (concentration, opening width, nature of filling), degree of saturation, and level and nature of imposed stress form (compressive or tensile). Tensile stresses, for example, beneath a concrete dam can cause the opening of joints and foliations, significantly increasing permeability.

Permeability Considerations

Determinations of k values

k values are often estimated from charts and tables (see Section 2.3.2) or can be measured in laboratory tests (see Section 2.3.3) or *in situ* tests (see Section 2.3.4).

Applications

k values as estimated or measured in the laboratory, are used for:

- Flow net construction and other analytical methods to calculate flow quantities and seepage forces.
- Selection of groundwater control methods for surface and underground excavations.
- Design of dewatering systems for excavations.
- Evaluation of capillary rise and frost susceptibility.
- Evaluation of yield of water-supply wells.

In situ measurements of *k* values are made for evaluations of:

- Percolation rates for liquid-waste disposal systems.
- Necessity for canal linings (as well as for designing linings).
- Seepage losses beneath and around dam foundations and abutments.
- Seepage losses in underground-cavern storage facilities.
- Groundwater control during excavation.

Associated Phenomena: Capillary, Piping, and Liquefaction

Capillary is the tendency of water to rise in "soil tubes," or connected voids, to elevations above the groundwater table. It provides the moisture that results in heaving of foundations and pavements from freezing (frost heave) and swelling of expansive soils. Rating criteria for drainage, capillary, and frost heave in terms of soil type are given in Table 2.10.

Piping refers to two phenomena: (1) water seeping through fine-grained soil, eroding the soil grain by grain and forming tunnels or pipes; and (2) water under pressure flowing upward through a granular soil with a head of sufficient magnitude to cause soil grains to lose contact and capability for support. Also termed boiling or liquefaction, piping is the cause of a "quick" condition (as in quicksand) during which the sand essentially liquefies.

"Cyclic" liquefaction refers to the complete loss of supporting capacity occurring when dynamic earthquake forces cause a sufficiently large temporary increase in pore pressures in the mass.

2.3.2   Estimating the Permeability Coefficient *k*

General

Basis

Since *k* values are a function of basic and index properties, various soil types and formations have characteristic range of values. Many tables and charts have been published by various investigators relating k values to geologic conditions, which are based on numerous laboratory and field investigations and which may be used for obtaining estimates of *k* of sufficient accuracy in many applications.

Partial Saturation Effects

In using tables and charts, one must realize that the values given are usually for saturated conditions. If partial saturation exists, as often obtained above the groundwater level, the voids will be clogged with air and permeability may be only 40 to 50% of that for saturated conditions.

Stratification Effects

In stratified soils, lenses and layers of fine materials will impede vertical drainage, and horizontal drainage will be much greater than that in the vertical direction.

Relationships

Permeability characteristics of soils and their methods of measurement are given in Table 2.11. Typical permeability coefficients for various conditions are given in the following

Measurement of Properties

TABLE 2.10

Tentative Criteria for Rating Soils with Regard to Drainage, Capillarity, and Frost Heaving Characteristics[a]

Fineness identification[b]	"Trace fine sand"	"Trace silt"	"Little silt" (coarse and fine)	"Some fine silt" "Little clayey silt" (fissured clay soils)	"Some clayey silt" (clay soils dominating)
Approx. effective size, D_{10} (mm)[c]	0.4 0.2	0.2 0.074	0.074 0.02	0.02 0.01	0.01
Drainage	Free drainage under gravity excellent	Drainage by gravity good	Drainage good to fair	Drains slowly, fair to poor	Poor to Impervious
Approx. range of k (cm/s)	0.5 0.10 0.2	0.04 0.020	0.006 0.0010	0.0004 0.0002	0.0001
Capillarity	Deep wells — Negligible	Slight	Well points successful — Moderate	Moderate to high	High
Approx. rise in feet, H_c	0.5	1.0 1.5	3.0 7.0	10.0 15.0	25.0
Frost heaving susceptibility	Nonfrost-heaving	Slight	Moderate to objectionable	Objectionable	Objectionable to moderate

Groundwater within 6 ft or $H_c/2$

[a] Criteria for soils in a loose to medium-compact state. From Burmister D.M., ASTM Special Publication, 113, American Society for Testing and Materials, Philadelphia, PA, U.S.A.

[b] Fineness classification is in accordance with the ASSE Classification System.

[c] Hazen's D_{10}: The grain size for which 10% of the material is finer.

TABLE 2.11

Permeability Characteristics of Soils and Their Methods of Measurement[a]

Coefficient of Permeability k (cm/s) (log scale)

Coefficient of permeability k (cm/s) (log scale)

	10^2	10^1	10	10^{-1}	10^{-2}	10^{-3}	10^{-4}	10^{-5}	10^{-6}	10^{-7}	10^{-8}	10^{-9}
Drainage			Good drainage				Poor drainage			Practically impervious		
Types of soil		Clean gravel		Clean sand / Clean sand and gravel mixtures			Very fine sands; organic and inorganic silts; mixtures of sand, silt, and clay; glacial till stratified clay deposits; etc. / "Impervious soils" which are modified by the effects of vegetation and weathering				"Impervious soils, e.g.,homogeneous clays below zone of weathering	
Direct determination of coefficient of permeability		Direct testing of soil in its original position (e.g. field-pumping tests)										
			Constant-head permeameter									
					Falling-head permeameter							
Indirect determination of coefficient of permeability		Computations from grain size distribution, porosity, etc.										
					Horizontal capillarity test				Computations from time rate of consolidation and rate of pressure drop at constant volume			

[a] After Casagrande, A. and Fadum, R.E., Soil Mechanics Series, Cambridge, MA, 1940 (from Leonards, G.A., Foundation Engineering, McGraw-Hill Book Co., New York, 1962, ch. 2).

tables: rock and soil formations, Table 2.12; some natural soil formations, Table 2.13; and various materials for turbulent and laminar flow, Table 2.14. Values of k for granular soils in terms of gradation characteristics (D_{10}, C_R, curve type) are given in Figures 2.13 and 2.14, with the latter figure giving values in terms of D_R.

Rock masses: Permeability values for various rock conditions are given in Table 2.12. A useful chart for estimating the effect of joint spacing and aperture on the hydraulic conductivity is given in Figure 2.15.

2.3.3 Laboratory Tests

Types and Applications

Constant-head tests are used for coarse-grained soils with high permeability. *Falling-head tests* are used for fine-grained soils with low permeability. *Consolidometer tests* may be used for essentially impervious soils as described in Section 2.5.4.

Constant- and Falling-Head Tests

The two types of laboratory permeameters are illustrated in Figure 2.16. In both cases, remolded or undisturbed specimens, completely saturated with gas-free distilled water, are used. Falling head tests on clay specimens are often run in the triaxial compression

TABLE 2.12

Typical Permeability Coefficients for Rock and Soil Formations[a]

	k (cm/s)	Intact Rock	Porosity n (%)	Fractured Rock	Soil
Practically impermeable	10^{-10} 10^{-9} 10^{-8} 10^{-7}	Massive low-porosity rocks	0.1–0.5 0.5–5.0		Homogeneous clay below zone of weathering
Low discharge, poordrainage	10^{-6} 10^{-5} 10^{-4} 10^{-3} (Sandstone)	Weathered granite Schist	5.0–30.0	Clay-filled joints	Very fine sands, organic and inorganic silts, mixtures of sand and clay, glacial till stratified clay deposits
High discharge, free draining	10^{-2} 10^{-1} 1.0 10^{1} 10^{2}			Jointed rock Open-jointed rock Heavily fractured rock	Clean sand, clean sand and gravel mixtures Clean gravel

[a] After Hoek, E. and Bray, J.W., *Rock Slope Engineering*, Institute of Mining and Metallurgy, London, 1977.

TABLE 2.13

Permeability Coefficients for Some Natural Soil Formations[a]

Formation	Value of k (cm/s)
River Deposits	
Rhone at Genissiat	Up–0.40
Small streams, eastern Alps	0.02–0.16
Missouri	0.02–0.20
Mississippi	0.02–0.12
Glacial Deposits	
Outwash plains	0.05–2.00
Esker, Westfield, Mass.	0.01–0.13
Delta, Chicopee. Mass.	0.0001–0.015
Till	Less than 0.0001
Wind Deposits	
Dune sand	0.1–0.3
Loess	0.001 ±
Loess loam	0.0001 ±
Lacustrine and Marine Offshore Deposits	
Very fine uniform sand, C_u = 5 to 2[b]	0.0001–0.0064
Bull's liver, Sixth Ave, N.Y., C_u = 5 to 2	0.0001–0.0050
Bull's Liver, Brooklyn, C_u = 5	0.00001–0.0001
Clay	Less than 0.0000001

[a] From Terzaghi, K. and Peck, R.B., *Soil Mechanics in Engineering Practice*, 2nd ed., Wiley, New York, 1967. Reprinted with permission of John Wiley & Sons, Inc.

[b] C_u = uniformity coefficient.

TABLE 2.14

Typical Permeability Coefficients for Various Materials[a]

| | Particle-Size Range | | | | "Effective" Size | | Permeability Coefficient k | | |
| | Inches | | Millimeters | | D_{14}, in | D_{15}, (mm) | ft/year | ft/month | cm/sec |
	D_{max}	D_{min}	D_{max}	D_{min}					
				Turbulent Flow					
Derrick atone	120	36			48		100×10^6	100×10^5	100
One-man stone	12	4			6		30×10^6	30×15^5	30
Clean, fine to coarse gravel	3	¼	80	10	½		10×10^6	10×10^5	10
Fine, uniform gravel	3/8	1/16	8	1.8	1/2		5×10^6	5×10^5	5
Very coarse, clean, uniform tend	1/8	1/32	3	0.8	1/16		3×10^6	3×10^5	3
				Laminar Flow					
Uniform, coarse sand	1/8	1/64	2	0.5		0.6	0.4×10^6	0.4×10^5	0.4
Uniform, medium sand			0.5	0.25		0.3	0.1×10^5	0.1×10^5	0.1
Clean, well-graded sand and gravel			10	0.05		0.1	0.01×10^5	0.01×10^5	0.01
Uniform, fine sand			0.25	0.05		0.06	4000	400	40×10^{-4}
Well-graded, silty sand and gravel			5	0.01		0.02	400	40	4×10^{-4}
Silty sand			2	0.005		0.01	100	10	10^{-4}
Uniform silt			0.05	0.005		0.006	50	5	0.5×10^{-4}
Sandy clay			1.0	0.001		0.002	5	0.5	0.05×10^{-4}
Silty clay			0.05	0.001		0.0015	1	0.1	0.01×10^{-4}
Clay (30–50% clay sizes)			0.05	0.0005		0.0008	0.1	0.01	0.001×10^{-4}
Colloidal clay (−2μm≤50%)			0.01	10Å		40Å	0.001	10^{-4}	10^{-9}

[a] From Hough, K.B., *Basic Soils Engineering*, The Ronald Press, New York, 1957.

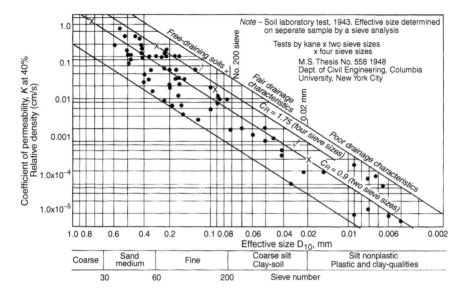

FIGURE 2.13
Relationships between permeability and Hazen's effective size D_n, Coefficient of permeability reduced to basis of 40% D_R by Figure 2.15. (From Burmister, D. M., ASTM, Vol. 48, Philadelphia, PA, 1948. Reprinted with permission of the American Society for Testing and Materials.)

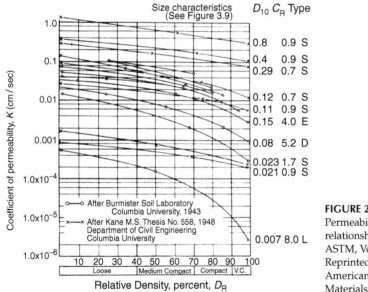

FIGURE 2.14
Permeability–relative density relationships. (From Burmister, D. M., ASTM, Vol. 48, Philadelphia, PA, 1948. Reprinted with permission of the American Society for Testing and Materials.)

device in which the time required for specimen saturation is substantially shortened by applying backpressure (Section 2.4.4).

Constant-Head Test (ASTM 2434)

A quantity of water is supplied to the sample, while a constant head is maintained and the discharge quantity q is measured. From Darcy's law,

$$k = qL/Ah \qquad (2.6)$$

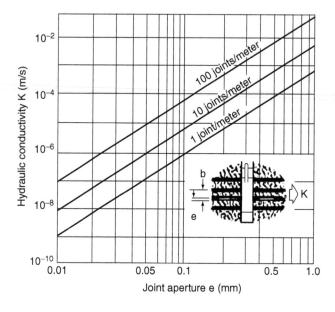

FIGURE 2.15
Effect of joint spacing and aperture on hydraulic conductivity. (From Hock, E. and Bray, J. W., *Rock Slope Engineering*, Institute of Mining and Metallurgy, London, 1977. With permission.)

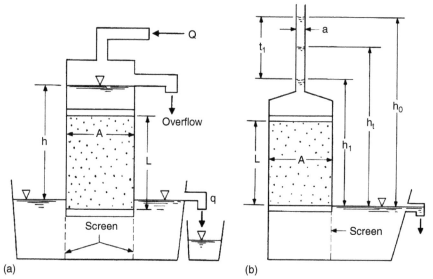

FIGURE 2.16
Two types of laboratory permeameters: (a) constant-head test; (b) falling-head test.

Falling-Head Test

Flow observations are made on the rate of fall in the standpipe (Figure 2.16b). At time t, the water level drops from h_0 to h_1, and

$$k = (aL/At_1)(\ln h_0/h_1) \tag{2.7}$$

2.3.4 *In Situ* Testing

Seepage Tests in Soils

Tests include constant head, falling or variable head, and rising head. They are summarized in terms of applicable field conditions, method, and procedure in Table 2.15.

TABLE 2.15

Seepage Tests in Soils

Test	Field Conditions	Method	Procedure
Constant head	Unsaturated granular soils	(a) Shallow-depth small pit, 12 in deep and square (percolation test)	1. Uncased holes, backfill with fine gravel or coarse
		(b) Moderate depth, hand-auger hole	2. Saturate ground around hole
		(c) Greater depth, install casing (open-end pipe test)[a]	3. Add metered quantities of water to hole until quantity decreases to constant value (saturation)
			4. Continue adding water to maintain constant level, recording quantity at 5-min intervals
			5. Compute k as for laboratory test
Falling or variable head	Below GWL, or in slow-draining soils	Performed in cased hole[a]	1. Fill casing with water and measure rate of fall
			2. Computations[b]
Rising-head test	Below GWL in soil of moderate k	Performed in cased hole[a]	1. Bail water from hole
			2. Record rate of rise in water level until rise becomes negligible
			3. After testing, sound hole bottom to check for quick condition as evidenced by rise of soil in casing computations[b]

[a] Tests performed in casing tan have a number of bottom-flow conditions. These are designed according to geologic conditions, to provide measurements of k_{mean}, k_v, or k_h:
- k_{mean}: determined with the casing flush with the end of the borehole in uniform material, or with casing flush on the interface between an impermeable layer over a permeable layer.
- k_v: determined with a soil column within the casing, similar to the laboratory test method, in thick, uniform material.
- k_h: determined by extending an uncased hole some distance below the casing and installing a well-point filter in the extension.

[b] References for computations of k with various boundary conditions:
- NAVFAC Design Manual DM-7.1 (1982).
- Hoek and Bray (1977).
- Lowe and Zaccheo (1975).
- Cedergren (1967).

Soil Penetration Tests (CPTU, DMT)

The piezocone CPT (Sections 1.3.4 and 2.4.5) and the flat dilatometer test (Section 2.4.5) provide estimates of the horizontal coefficient of permeability, k_h.

Pumping Tests

Tests are made from gravity wells or artesian wells in soils or rock masses as described in Table 2.16.

Pressure Testing in Rock Masses

General Procedures

The general arrangement of equipment is illustrated in Figure 2.17, which shows two packers in a hole. One of the two general procedures is used, depending on rock quality.

The *common procedure*, used in poor to moderately poor rock with hole collapse problems, involves drilling the hole to some depth and performing the test with a single packer. Casing is installed if necessary, and the hole is advanced to the next test depth.

The *alternate procedure*, used in good-quality rock where the hole remains open, involves drilling the hole to the final depth, filling it with water, surging it to clean the walls of fines, and then bailing it. Testing proceeds in sections from the bottom–up with two packers.

Packer spacing depends on rock conditions and is normally 1, 2, or 3 m, or at times 5 m. The wider spacings are used in good-quality rock and the closer spacings in poor-quality rock.

Testing Procedures

1. Expand the packers with air pressure.
2. Introduce water under pressure into the hole, first between the packers and then below the lower packer.
3. Record elapsed time and volume of water pumped.
4. Test at several pressures, usually 15, 30, and 45 psi (1, 2, and 3 tsf) above the natural piezoelectric level (Wu, 1966). To avoid rock-mass deformation, the excess

TABLE 2.16

Pumping Tests

Test	Field Conditions	Method	Procedure	Disadvantages
Gravity well	Saturated, uniform soil (unconfined aquifer)	Pump installed in screened and filtered well and surrounded by a pattern of observation wells	Well is pumped at constant rate until cone of drawdown measured in observation wells has stabilized (recharge equals pumping rate)	Provides values for k_{mean}
Gravity well	Rock masses	Similar to above	Similar to above	Flow from entire hole measured. Provides an average value
Artesian well	Confined aquifer (pervious under thick impervious layer)	Similar to above	Similar to above	Provides values for $k_{mean}k_{mean}$ in aquifer

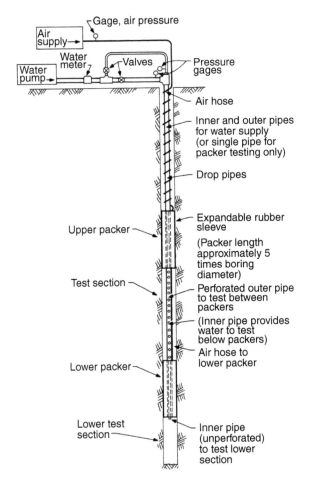

FIGURE 2.17
Apparatus for determining rock permeability *in situ* using pressure testing between packers.

pressure above the natural piezoelectric level should not exceed 1 psi for each foot (23 kPa/m) of soil and rock above the upper packer.

Data Evaluation

Curves of flow vs. pressure are plotted to permit evaluation of changes in the rock mass during testing:

- Concave-upward curves indicate that fractures are opening under pressure.
- Convex curves indicate that fractures are being clogged (permeability decreasing with increased pressure).
- Linear variation indicates that no change is occurring in the fractures.

Approximate values for k are computed from the expressions (USBR, 1974)

$$k = (Q/2\pi LH)(\ln L/r) \text{ for } L \geqslant 10r \qquad (2.8)$$

$$k = (Q/2\pi LH) \sinh^{-1}(L/2r) \text{ for } 10r > L \geqslant r \qquad (2.9)$$

where k is the coefficient of permeability, Q the constant flow rate in the hole, L the length of the test section, H the differential head on the test section (see explanation below), r the hole radius and $\sinh^{-1}$ the inverse hyperbolic sine.

Head losses in the system should be accounted for in determining the value of *H*. Since most head losses occur in the drop pipe, they can be minimized by using as large a diameter pipe as practical. Head loss can be estimated from the relationship (Davis et al., 1970)

$$h_f = f(L/D)(v^2/2g) \tag{2.10}$$

where *L* is the pipe length, *D* the pipe diameter, *v* the flow velocity, *g* the gravitational acceleration and *f* the frictional component (obtained from charts for various pipe diameters, materials, and discharges).

Lugeon Test

In the Lugeon test, used commonly in Europe, the hole is drilled to test depth and a packer installed about 15 ft (5 m) from the bottom. Flow is measured after 5 or 20 min of test under pressure, and the test is performed under several pressures. The standard measurement of pressure is 10 kg/cm² and the results are given in Lugeon units.
Lugeon unit is defined as a flow of 1 *l* of water/min of borehole length at a pressure of 10 kg/cm² (1 Lugeon unit is about 10^{-5} cm/sec).

Disadvantages of Pressure Testing

Values can be misleading because high pressures cause erosion of fines from fractures as well as deformation of the rock mass and closure of fractures (Serafin, 1969).

Additional References

Dick (1975) and Hoek and Bray (1977) provide additional information.

2.4   Rupture Strength

2.4.1   Introduction

Basic Definitions

- *Stress (σ)* is force *P* per unit of area, expressed as

$$\sigma = P/A \tag{2.11}$$

System	Equivalent units for stress
English	13.9 psi = 2000 psf = 2 ksf = 1 tsf = 1 bar
Metric	1 kg/cm² = 10 T/m² (≈1 tsf)
SI	100 kN/m² = 100 kPa = 0.1 MPa (≈1 tsf)

- *Strain (ε)* is change in length per unit of length caused by stress. It can occur as compressive or tensile strain. Compressive and tensile strain are expressed as

$$\varepsilon = \Delta L/L \tag{2.12}$$

- *Shear* is the displacement of adjacent elements along a plane or curved surface.
- *Shear strain (ξ)* is the angle of displacement between elements during displacement.

- *Shear stress (τ)* is the stress causing shear.
- *Shear strength (S or s)* is a characteristic value at which a material fails in rupture or shear under an applied force.
- *Dilatancy* is the tendency of the volume to increase under increasing shear or stress difference.

Strength of Geologic Materials

Components: Friction and Cohesion

Friction is a resisting force between two surfaces as illustrated in Figure 2.18. It is often the only source of strength in geologic materials and is a direct function of the normal force. *Cohesion* results from a bonding between the surfaces of particles. It is caused by electrochemical forces and is independent of normal forces.

Influencing Factors

Strength is not a constant value for a given material, but rather is dependent upon many factors, including material properties, magnitude and direction of the applied force and the rate of application, drainage conditions in the mass, and the magnitude of the confining pressure.

Stress Conditions In Situ

Importance

A major factor in strength problems is the existence of stress conditions in the ground, primarily because normal stresses on potential failure surfaces result from overburden pressures.

Geostatic Stresses

Overburden pressures, consisting of both vertical and lateral stresses, exist on an element in the ground as a result of the weight of the overlying materials. Stress conditions for

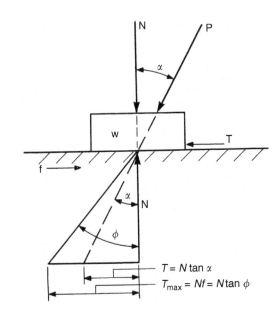

FIGURE 2.18
Frictional force f resisting shearing force T [P = force applied in increments until slip occurs; N = normal force component including block weight W; T = shearing stress component; f = frictional resistance; α = angle of obliquity [resultant of N and T]; ϕ = friction angle, or α_{max} at slip; S_{max} = maximum shearing resistance = T_{max}].

level ground are illustrated in Figure 2.19; sloping ground results in more complex conditions. (Changes in geostatic stresses are invoked by surface foundation loads, surface and subsurface excavations, lowering of the groundwater level, and natural phenomena such as erosion and deposition). Vertical earth pressures from overburden weight alone are found by summing the weights from the various strata as follows:

$$\sigma_v = \sum_0^z \gamma_n Z_n \qquad (2.13)$$

Coefficient of lateral earth pressure "at-rest" K_0 is the ratio of the lateral to vertical stress in a natural deposit that has not been subject to lateral strain, the values for which vary substantially with material types and properties (see Section 2.4.2). It is expressed as

$$K_0 = \sigma_h / \sigma_v \qquad (2.14)$$

or

$$\sigma_h = K_0 \, \sigma_v \qquad (2.15)$$

For an elastic solid

$$K_0 = v/1 - v \qquad (2.16)$$

In the above expressions, γ is the material unit weight (γ_n above groundwater level, γ_b below groundwater level) and v is the Poisson's ratio (see Section 2.5.1).

Total and Effective Stresses

The total stress on the soil element in Figure 2.19 at depth z is

$$\sigma_v = \gamma Z \qquad (2.17)$$

If the static water table is at the surface, however, and the soil to depth z is saturated, there is pressure on the water in the pores because of a piezoelectric head h_w and the unit weight of water γ_w. This is termed the neutral stress (acting equally in all directions), or the pore-water pressure u_w or u and is given as

$$u = \sigma_w h_w \qquad (2.18)$$

The effective stress σ'_v, or actual intergranular stresses between soil particles, results from a reduction caused by the neutral stress and is equal to the total stress minus the pore-water pressure, or

$$\sigma'_v = \sigma_v - u \qquad (2.19)$$

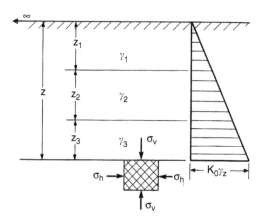

FIGURE 2.19
The geostatic stress condition and "at-rest" earth pressures.

or

$$\sigma_v' = \gamma_b z \qquad (2.20)$$

where $\gamma_b = \gamma_t - \gamma_w$ is the effective or submerged soil weight.

In calculations, therefore, above the groundwater level, the effective soil weight is the total weight γ_t, and below the groundwater level (or any other water surface), the effective soil weight is the submerged soil weight γ_b.

Prestress in Soil Formations

General: Soils compress naturally under the weight of overlying materials or some other applied load, resulting in strength increase over values inherent as deposited, or shortly thereafter. Three categories of prestress are defined according to the degree of compression (termed consolidation) that has occurred.

Normally consolidated (NC): The soil element has never been subjected to pressures greater than the existing overburden pressures.

Overconsolidated (OC): The soil element has at some time in its history been subjected to pressures in excess of existing overburden, such as resulting from glacial ice loads, removal of material by erosion, desiccation, or lowering of the groundwater level.

Underconsolidated (UC): The soil element exists at a degree of pressure less than existing overburden pressures. This case can result from hydrostatic pressures reducing overburden load as illustrated in Figure 2.20. Such soils are normally relatively weak. Weakening of strata can also occur due to removal of a cementing agent or other mineral constituents by solution.

Principal Stresses and the Mohr Diagram

Importance

Fundamental to the strength aspects of geologic materials are the concepts of principal stresses and the Mohr diagram on which their relationships may be illustrated.

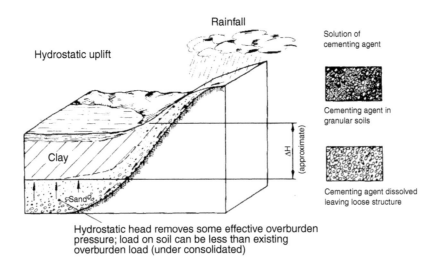

FIGURE 2.20
Soil profile weakening processes.

Principal Stresses

Stresses acting on any plane passed through a point consist of a normal stress σ (compression or tension) and a shearing stress τ. (Soil mechanics problems are normally concerned with compressive stresses.) On one particular plane, the normal stress will be the maximum possible value and the shearing stress will be equal to zero. On one plane perpendicular to this plane, the normal stress will be the minimum possible value, with shear stress also equal to zero. On a second plane perpendicular to this plane, the normal stress will have an intermediate value and the shearing stress will also be zero. These planes are termed the principal planes.

The principal stresses are the stresses acting perpendicular to the principal planes including the maximum (major) principal stress σ_1, the minimum (minor) principal stress σ_3, and the intermediate principal stress σ_2. The relationship between principal stresses and the normal stress and the shear stress acting on a random plane through a point is shown in Figure 2.21. The intermediate principal stress is the plane of the paper and, in soil mechanics problems, is normally considered to be equal to σ_3.

The Mohr Diagram

To attain equilibrium, the sum of the forces given in Figure 2.21 should be zero. Therefore, σ_n and τ can be expressed in terms of the principal stresses and the angle θ as

$$\sigma_n = [\,(\sigma_1 + \sigma_2)/2 + (\sigma_1 - \sigma_3)/2\,]\cos 2\theta \qquad (2.21)$$

$$\tau = [(\,\sigma_1 - \sigma_3)/2]\sin 2\theta \qquad (2.22)$$

If points are plotted to represent coordinates of normal and shearing stresses acting on a particular plane for all values of θ given in equations 2.21 and 2.22, their loci form a circle which intersects the abscissa at coordinates equal to the major (σ_1) and minor (σ_3) principal stresses. The circle is referred to as the Mohr diagram, or Mohr's circle, given in Figure 2.22.

Applications of Strength Values

Stability Analysis

The values for strength are used in stability analyses; the discussion is beyond the scope of this book, except for evaluations of slopes. In general terms, stability is based on plastic equilibrium or a condition of maximum shear strength with failure by rupture imminent. When the imposed stresses cause the shear strength to be exceeded, rupture occurs in the mass along one or more failure surfaces. Analyses are normally based on the *limit equilibrium*

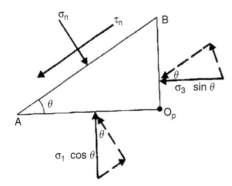

FIGURE 2.21
Stresses on a random plane through a point (σ_2 is the plane of the paper).

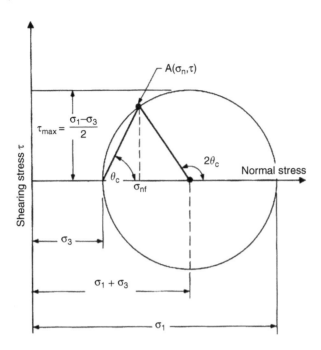

FIGURE. 2.22
The Mohr diagram relating τ, σ_{nf}, σ_3, and σ_1 ($\theta_{cr} = \theta$ at failure; σ_{nf} = normal stress at failure).

approach, i.e., a limiting value that can be reached when the forces acting to cause failure are in balance with the forces acting to resist failure. Resistance to failure is provided by the shear strength mobilized along the failure surface.

Typical Problems

Some field conditions involving failure by rupture are illustrated in Figure 2.23, showing the relationships between the force acting to cause failure, the strength acting along the failure surface, and the principal and normal stresses.

2.4.2   Shear Strength Relationships

Basic Concepts

Shear strength may be given in several forms, depending on various factors, including the drained strength, the undrained strength, the peak strength, the residual or ultimate strength, and strength under dynamic loadings. In addition, strength is the major factor in determining active and passive Earth pressures.

Under an applied force, a specimen will strain until rupture occurs at some peak stress; in some materials, as strain continues, the resistance reduces until a constant minimum value is reached, termed the ultimate or residual strength.

Factors Affecting Strength

Material type: Some materials exhibit only a frictional component of resistance ϕ; others exhibit ϕ as well as cohesion c. In soft clays, at the end of construction, it is normally the undrained strength s that governs.

Confining pressure: In materials with ϕ acting, the strength increases as the confining pressure increases.

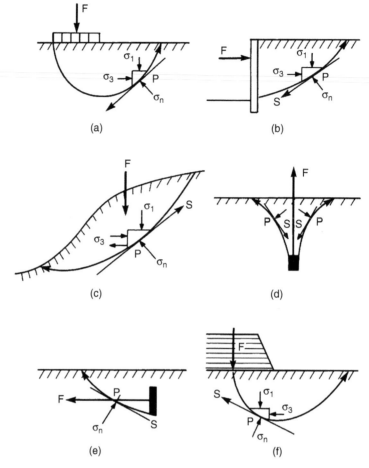

FIGURE 2.23
Some field conditions involving failure by rupture: (a) foundations; (b) retaining structures; (c) slopes; (d) ground anchors; (e) wall anchors; (f) embankments.

Undrained or drained conditions: It relates to the ability of a material to drain under applied stress and determines whether total or effective stresses act.

Loading direction: Forces applied parallel to weakness planes, as represented by stratification in soils, foliation planes, or joints in rock masses, will result in lower strengths than if the force is applied perpendicularly. Compressive strengths are much higher than tensile strengths.

Displacement and normal stress: In some cohesive materials, such as overconsolidated fissured clays and clay shales, the strain at which the peak stress occurs depends on the normal stress level and the magnitude of the peak strength varies with the magnitude of normal stress (Peck, 1969) as shown in Figure 2.24. As strain continues the ultimate strength prevails. These concepts are particularly important in problems with slope stability.

Angle of Internal Friction ϕ

The stresses acting on a confined specimen of dry cohesionless soil, either in the ground or in a triaxial testing device (see Section 2.4.4), are illustrated in Figure 2.25. p is the applied stress, σ_3 is the confining pressure, and $\sigma_1 = p + \sigma_3$ (in testing p is termed the deviator stress σ_d). As stress p is gradually increased, it is resisted by the frictional forces acting between

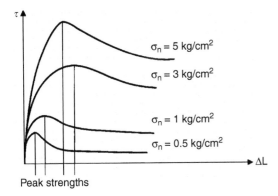

Peak strengths

FIGURE 2.24
Peak strength vs. displacement and normal stress. (From Peck, R. B., *Proceedings of ASCE, Stability and Performance of Slopes* Embankments, Berkeley, CA, 1969, pp. 437–451. With permission.)

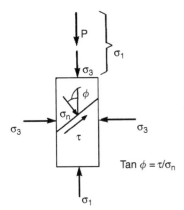

FIGURE 2.25
Stresses on a specimen in a confined state.

grains until some characteristic stress level is reached at which resistance is exceeded and rupture occurs. This stress is termed the *peak strength*.

If σ_3 and σ_1 (for the peak strength) are plotted to define a Mohr's circle at failure, a line drawn tangentially to the circle passes through the origin at an angle ϕ as shown in Figure 2.26 and Figure 2.27. Other specimens of the same material loaded to failure, but at different confining pressures, will have Mohr's circles tangent to the line defined by ϕ. The circle tangent line represents the limits of stability and is termed *Mohr's envelope*. It defines shear strength in terms of the friction angle ϕ, and the normal or total stresses, and it expressed as

$$s = \tau_{max} = \sigma_n \tan \phi \qquad (2.23)$$

(In actuality, the envelope line will not be straight, but will curve downward slightly at higher confining pressures.)

Total vs. Effective Stresses (ϕ vs. ϕ')

In a fully saturated cohesionless soil, the ϕ value will vary with the drainage conditions prevailing during failure. In undrained conditions, total stresses prevail and in drained conditions, only effective stresses act. The friction angle is expressed as ϕ'.

Total Stresses

Saturated Soils

If no drainage is permitted from a fully saturated soil as load is applied, the stress at failure is carried partially by the pore water and partially by the soil particles which thus develop some intergranular stresses. The peak "undrained" strength parameters mobilized depend on the soil type:

- Cohesionless soils — the envelope passes through the origin (Figure 2.26 and Figure 2.27) and strength is expressed in terms of ϕ as in equation 2.23.
- Soils with cohesion — the envelope intercepts the shear stress ordinate at a value taken as cohesion c, similar to the case shown in Figure 2.28, and the strength is expressed by the *Mohr–Coulomb* failure law:

$$s = c + \sigma_n \tan \phi \tag{2.24}$$

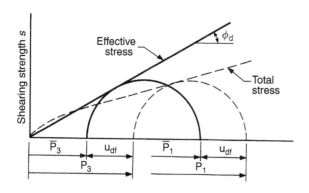

FIGURE 2.26
Mohr's envelopes for total and effective stresses from CU triaxial tests on loose saturated sands.

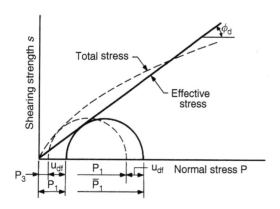

FIGURE 2.27
Mohr's envelopes for total and effective stresses from CU triaxial tests on dense saturated sands.

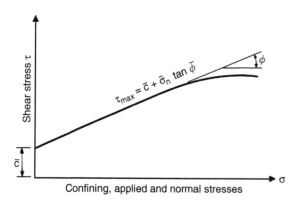

FIGURE 2.28
Drained triaxial test on soil with cohesion.

- Normally to slightly overconsolidated clays have no frictional component ($\phi = 0$) and the strength is expressed as the undrained strength s_u, as will be discussed later.

Applicable Field Conditions

- Initial phases of slope excavations or retaining wall construction in slowly draining soils, where time is too short to permit drainage. (Results are conservative if drainage occurs.)
- Sudden drawdown conditions in the upstream slope of a dam embankment, or when flood waters recede rapidly from stream banks and the soils remain saturated.
- Rapid placement of embankments or loading of storage tanks over soft clays, or high live loads applied to slow-draining soils.

Effective Stresses

Saturated Soils

If the soil is permitted to drain as load is applied, the stress is initially carried partly by the pore water, but as drainage occurs the load is transferred to the soil grains, and the effective stresses (intergranular stresses) are mobilized. The effective stress is equal to the total normal stress σ_n minus the pore-water pressure u, and the peak "drained" shear strength is expressed as

$$s = (\sigma_n - u)\tan \phi' = \sigma_n' \tan \phi' \tag{2.25}$$

As shown in Figures 2.26 and 2.27, ϕ values based on effective stresses are often higher than those based on total stresses as long as u is positive. In dense sands, u is negative and the undrained strength is higher, i.e., ϕ is higher. If pore pressures are measured during *undrained* loadings, effective stresses can be computed.

For soils with a cohesion intercept (Figure 2.28), the drained strength is expressed by the Coulomb–Terzaghi equation as

$$s = c' + (\sigma_n - u)\tan \phi' = c' + \sigma_n' \tan \phi' \tag{2.26}$$

Applicable Field Conditions

Drained strength prevails in the field under relatively slow loading conditions during which pore pressures can dissipate. Such conditions generally exist in:

- Most foundations, except for cases involving rapid load application.
- Natural slopes, except for the sudden drawdown case.
- Cut slopes, embankments, and retaining structures some time after construction completion.

Partially Saturated Soils

In partially saturated soils, strength is controlled by effective stresses, but the effective stress concept cannot be applied directly because of pressures in the air or gas in the partially saturated voids. Strength should be estimated from tests performed to duplicate *in situ* conditions as closely as possible in terms of percent saturation, total stress, and pressure on the liquid phase (Lambe and Whitman, 1969).

Apparent cohesion results from capillary forces in partially saturated fine-grained granular soils, such as fine sands and silts, and provides a temporary strength which is lost upon saturation or drying. The apparent cohesion has been expressed in terms of the depth D to the water table (Lambe and Whitman, 1969) as

$$c_a = D\gamma_w \tan \phi' \tag{2.27}$$

Pore-Pressure Parameters

Definition

Pore-pressure parameters express the portion of a stress increment carried by the pore fluid in terms of the ratio of the pore-pressure increment (Δu) to the total stress increment ($\Delta\sigma$). As indicated in Lambe and Whitman (1969) and Bishop and Henkel (1962), the parameters are

$$C = \Delta u / \Delta\sigma_1 \tag{2.28}$$

for loading in the odeometer (one-dimensional compression),

$$B = \Delta u / \Delta\sigma \tag{2.29}$$

for isotopic loading (three-dimensional compression),

$$A = (\Delta u - \Delta\sigma_3)/(\Delta\sigma_1 - \Delta\sigma_3) \tag{2.30}$$

for triaxial loading and

$$A = \Delta u / \Delta\sigma_1 \tag{2.31}$$

for the normal undrained test where $\sigma_3 = 0$.

Pore-pressure parameter A is the most significant in practice. Values depend on soil type, state of stress, strain magnitude, and time. Typical values are given in Table 2.17 for conditions at failure but important projects always require measurement by testing.

High values occur in soft or loose soils. Negative values indicate negative pore pressures, which occur in dense sands and heavily preconsolidated clays as the result of volume increase during shear (dilatancy). Pore pressures are most responsive to applied

TABLE 2.17

Typical Values of Pore-Pressure Parameter A at Failure[a]

Soil Type	Parameter A
Sensitive clay	1.5–2.5
Normally consolidated clay	0.7–1.3
Overconsolidated clay	0.3–0.7
Heavily overconsolidated clay	−0.5–0.0
Very loose fine sand	2.0–3.0
Medium fine sand	0.0
Dense fine sand	−0.3
Loess	−0.2

[a] From Lambe, T.W., *Proc. ASCE, J. Soil Mech. Found. Eng. Div.*, 88, 19–47, 1962.

stress in sensitive clays and loose fine sands that are subject to liquefaction. In soft or loose soils, higher shear strain magnitudes result in higher values for parameter A.

Applications

Pore-pressure parameter A is used to estimate the magnitude of initial excess pore pressure produced at a given point in the subsoil by a change in the total stress system. With piezometers used to measure *in situ* pore pressures, the validity of the estimates can be verified and the loading rate during surcharging or embankment construction can be controlled to avoid failure caused by exceeding the undrained shear strength.

Undrained Shear Strength s_u

Concepts

During undrained loading in soft to firm saturated clays, the applied stress is carried partly by the soil skeleton and partly by the pore water. Increasing the confining pressure does not increase the diameter of Mohr's circle, since the pore pressure increases as much as the confining pressure. The undrained strength, therefore, is independent of an increase in normal stress ($\phi = 0$), and as shown in Figure 2.29, is given by the expression

$$s_u = \tfrac{1}{2}\,\sigma_d \tag{2.32}$$

where $\sigma_d = (\sigma_1 - \sigma_3)$ is the applied or deviator stress. In the figure, U_c represents the unconfined compressive strength where $\sigma_3 = 0$. As shown, $s_u = 1/2U_c$.

For NC clays, s_u falls within a limited fraction of the effective overburden pressure (σ_v' or p'), usually in the range $s_u/p' = 0.16$ to 0.4 (based on field vane and K_0 triaxial tests). Therefore, if $s_u > 0.5p'$ the clay may be considered as overconsolidated (see Section 2.5.2).

Factors Affecting s_u Values

Time rate of loading: For soft clays, s_u, measured by vane shear tests, has been found to be greater than the field strength mobilized under an embankment loading, and the difference increases with the plasticity of the clay (Casagrande, 1959; Bjerrum, 1969, 1972). The difference is attributed to the variation in loading time rates; laboratory tests are performed at much higher strain rates than those that occur during the placement of an embankment in the field, which may take several months. A correction factor, therefore, is often applied to test values for s_u (see Section 2.4.4).

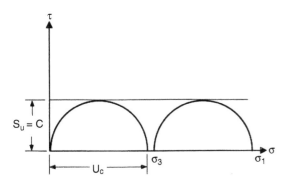

FIGURE 2.29
Undrained tests on saturated soil with cohesion.

Direction of loading: Soil anisotropy may cause shear strength measured in a horizontal direction to be significantly higher than when measured in a vertical direction (see *In Situ Vane Shear Test,* in Section 2.4.4).

Residual or Ultimate Shear Strength s_r or ϕ_r'

Concepts

The lowest strength that a cohesive material can attain in the confined state is termed the ultimate strength. In many cohesive materials, if strain continues past the peak strength under continued stress, the strength will decrease until some minimum value is reached. Thereafter, strength remains constant with increasing strain as shown in Figure 2.30.

Residual strength is a strength lower than peak strength remaining after failure has occurred. It has become generally accepted that the residual strength is the lowest strength that can be obtained during shear. In a normally consolidated clay the *remolded* undrained shear strength is considered to be equal to the ultimate or residual strength s_r.

The envelope for the ultimate drained strength passes through the origin as a straight line on the Mohr diagram and has no cohesion intercept (even in cohesive materials) as shown in Figure 2.31 (Lambe and Whitman, 1969). The drained ultimate shear strength is expressed as

$$s = \sigma_n' \tan \phi_r' \qquad (2.33)$$

Other Factors

Natural slopes: The residual strength, rather than the peak strength, often applies. An approximate relationship between ϕ_r and plasticity index for rock gouge materials is given in Figure 2.32.

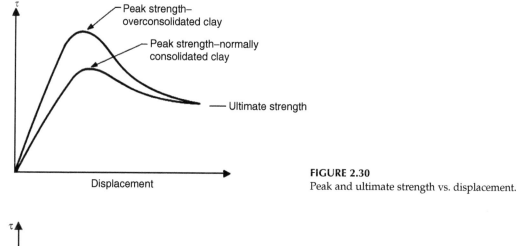

FIGURE 2.30
Peak and ultimate strength vs. displacement.

FIGURE 2.31
Mohr's envelope for ultimate drained strength in clay.

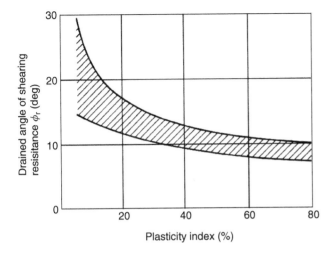

FIGURE 2.32
Approximate relationship between the drained angle of residual shearing resistance and plasticity index for rock gouge material. (From Patton, F. D. and Hendron Jr., A. J., *Proceedings of the 2nd International Congress*, International Association Engineering Geology, Sao Paulo, 1974, p. V-GR. 1. Kanji, M. A., M.S. Thesis, Department of Geology, University of Illinois, Urbana, 1970. With permission.)

Sensitivity S_t: The ratio of the natural peak strength to the ultimate undrained strength when a sample is completely remolded at unaltered water content is referred to as the soil sensitivity, expressed as

$$S_t = s_u / s_r \qquad (2.34)$$

A clay classification based on S_t is given in Table 2.18. Soils termed as "quick" have high sensitivities and are extremely sensitive to vibrations and other forms of disturbance. They can quickly lose their strength, change to a fluid, and flow, even on very flat slopes.

Thixotropy refers to the regain in strength occurring in remolded soils because of a "rehabilitation of the molecular structure of the adsorbed layers" (Lambe and Whitman, 1969).

Dynamic Shear

Concepts

Under dynamic or cyclic loading a soil specimen subject to shear initially undergoes deformations that are partially irreversible, irrespective of the strain amplitude; hence, stress–strain curves in loading and unloading do not coincide. If strain amplitude is small, the difference between successive reloading curves tends to disappear after a few loading

TABLE 2.18
Clay Classification by Sensitivity S_t^a

Sensitivity s_u/s_r^b	Classification
2	Insensitive
2–4	Moderately sensitive
4–8	Sensitive
8–16	Very sensitive
16–32	Slightly quick
32–64	Medium quick
64	Quick

[a] From Skempton, A.W., *Proceedings of the 3rd International Conference on Soil Mechanics and Foundation Engineering*, Switzerland, Vol. I, 1953, pp. 57–61.

[b] s_u = peak undrained strength, s_t = remolded strength.

cycles of a similar amplitude and the stress–strain curve becomes a closed loop that can be defined by two parameters as shown in Figure 2.33:

- *Shear modulus* (Section 2.5.2) is defined by the average slope and has been found to decrease markedly with increasing strain amplitude.
- *Internal damping* refers to energy dissipation and is defined by the enclosed areas as shown in the figure (damping ratio λ).

In *clay soils*, if strain amplitude is large there is a significant reduction in the undrained strength.

In *cohesionless soils*, pore pressure increases almost linearly with the number of cycles until failure occurs. The simple shear device used for dynamic testing in the laboratory is shown on Figure 2.34. *Liquefaction* occurs when pore pressures totally relieve effective stresses.

Field Occurrence

Wave forces against offshore structures cause low-frequency cyclic loads. Seismic waves from earthquakes cause low- to high-frequency cyclic loads.

At-Rest, Passive, and Active Stress States

At-Rest Conditions (K_0)

The coefficient of lateral at-rest earth pressure has been defined as the ratio of lateral to vertical stress, $K_0 = \sigma_h / \sigma_v$ (Equation 2.15).

For *sands and NC clays*, K_0 is normally in the range of 0.4 to 0.5 and is a function of ϕ in accordance with

$$K_0 = 1.0 - \sin \phi \tag{2.35}$$

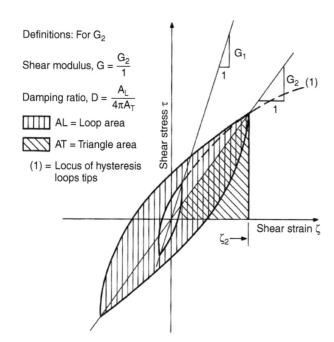

Definitions: For G_2

Shear modulus, $G = \dfrac{G_2}{1}$

Damping ratio, $D = \dfrac{A_L}{4\pi A_T}$

▦ AL = Loop area

▨ AT = Triangle area

(1) = Locus of hysteresis loops tips

FIGURE 2.33
Hysteretic stress-strain relationship from cyclic shear test at different strain amplitudes. (*After USAEC*, National Technical Information Service Publication TID-25953, U. S. Department of Commerce, Oak Ridge National Laboratory, Oak Ridge, TN, 1972.)

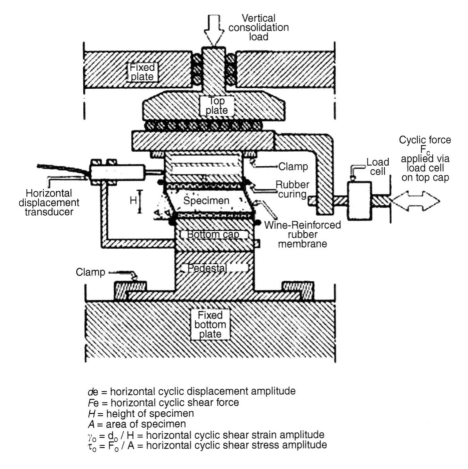

de = horizontal cyclic displacement amplitude
Fe = horizontal cyclic shear force
H = height of specimen
A = area of specimen
$\gamma_0 = d_0 / H$ = horizontal cyclic shear strain amplitude
$\tau_0 = F_0 / A$ = horizontal cyclic shear stress amplitude

FIGURE 2.34
NGI-type simple shear device at the UCLA Soil Dynamics Laboratory. (From Vucetic, 2004.)

In clay soils, K_0 has been found to be directly related to the amount of prestress or pre-consolidation in the formation, ranging from 0.5 to almost 3. The degree of prestress is given in terms of the overconsolidation ratio (OCR), i.e., the ratio of the maximum past pressure to the existing overburden pressure (see Section 2.5.4).

Passive State of Stress (K_p)
The passive state exists when a force pushes against a soil mass and the mass exerts its maximum resistance to the force. The principal stresses for the passive state are shown in Figure 2.35a; as the soil element is pushed, the vertical stress remains unchanged but the horizontal stress increases ($\sigma_h > \sigma_v$). As movement continues the shear stress increases from the at-rest condition (1) until $\sigma_h = \sigma_v$ (2), then continues to increase as slip lines (rupture planes) form and finally failure occurs at (4). At this point of plastic equilibrium, the passive state has been reached.

The relationship between ϕ and the principal stresses at failure, as shown on the Mohr diagram of Figure 2.35a, may be expressed for a cohesionless granular soil with a horizontal ground surface as

$$\sigma_1/\sigma_3 = \sigma_v/\sigma_h = (1 + \sin \phi)/(1 - \sin \phi) \tag{2.36}$$

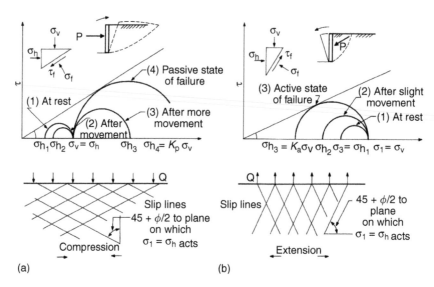

FIGURE 2.35
Mohr diagrams for the (a) passive and (b) active states of stress.

or

$$\sigma_v/\sigma_h = \tan^2 (45^\circ + \phi/2) = K_p \tag{2.37}$$

where K_p is the coefficient of passive stress.

Active State of Stress

The active state exists when a soil mass is allowed to stretch, for example, when a retaining wall tilts and $\sigma_v > \sigma_h$. In Figure 2.35b, as the mass stretches, σ_v remains unchanged and σ_h decreases (2) until the induced shear stress is sufficient to cause failure (3). At this point of plastic equilibrium, the active state has been reached.

The coefficient of active stress K_a, (the ratio σ_h/σ_v) represents a minimum force, expressed for a cohesionless soil with a horizontal ground surface as

$$K_a = \tan^2 (45^\circ - \phi/2) \tag{2.38}$$

and

$$K_a = 1/K_p \tag{2.39}$$

Applications

The *at-rest coefficient* K_0 has a number of practical applications. It is used to compute lateral thrusts against earth-retaining structures, where lateral movement is anticipated to be too small to mobilize K_a. It is fundamental to the reconsolidation of triaxial test specimens according to an anisotropic stress path resembling that which occurred *in situ* (CK$_0$U tests). It is basic to the computation of settlements in certain situations (Lambe, 1964). It has been used for the analysis of progressive failure in clay slopes (Lo and Lee, 1973), the prediction of pore-water pressure in earth dams (Pells, 1973), and the computation of lateral swelling pressures against friction piles in expansive soils (Kassif and Baker, 1969).

Coefficients of lateral earth pressure is the term often used to refer to K_p and K_a. When rupture occurs along some failure surface, earth pressures are mobilized. The magnitudes of the pressures are a function of the weight of the mass in the failure zone and the strength acting along the failure surface. These pressures (P_p or P_a) are often expressed in terms of the product of the weight of the mass times the active or passive coefficients, as follows:

$$P_p = K_p \gamma z \tag{2.40}$$

$$P_a = K_a \gamma z \tag{2.41}$$

Examples of the occurrence of P_p and P_a in practice are given in Figure 2.36.

2.4.3 Rock Strength Measurements

General

Intact Rock

The confined strength of fresh, intact rock is seldom of concern in practice because of the relatively low stress levels imposed.

Brittle shear failure occurs under very high applied loads and moderate to high confining pressures, except for the softer rocks such as halite, foliated and schistose rocks, and lightly cemented sandstones. In softer rocks, rupture occurs in a manner similar to that in soils, and the parameters described in Section 2.4.2 hold.

Under very high confining pressures (approximately 45,000 psi or 3000 bar), some competent rocks behave ductilely and failure may be attributed to plastic shear (Murphy,

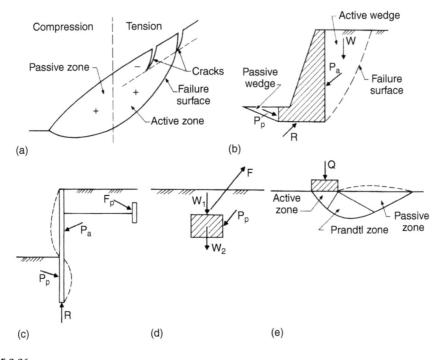

FIGURE 2.36
Examples of the occurrence of active and passive pressures encountered in practice: (a) slope; (b) retaining wall; (c) anchored bulkhead; (d) anchor block; (e) foundation.

1970). Intact specimens are tested in uniaxial compression or tension to provide data for classification and correlations. Other tests include those for flexural strength and triaxial compressive strength. Applied forces for the various tests are shown in Figure 2.37.

Rock Masses

Rock-mass strength is normally controlled either by the joints and other discontinuities or by the degree of decomposition, and the strength parameters described in Section 2.4.2 here. Strength is measured *in situ* by direct shear equipment or special triaxial shear equipment.

Uniaxial Compressive Strength U_c (ASTM D2938)

Procedure

An axial compressive force is applied to an unconfined specimen (Figure 2.38) until failure occurs.

Data Obtained

A stress–strain curve and the unconfined or uniaxial compressive strength (in tsf, kg/cm^2, kPa) result from the test. Stress–strain curves for various rock types are given in Table 2.24.

Data Applications

Primarily used for correlations as follows:

- Material "consistency" vs. U_c — Figure 2.39.
- Schmidt hardness vs. U_c — Figure 2.40. The Schmidt hardness instrument (Figure 3.1a) is useful for field measurements of outcrops to correlate variations in U_c. Corrections are available for inclinations from the vertical.
- Hardness classification.

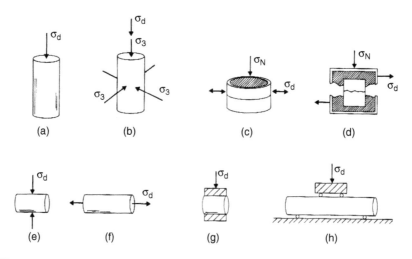

FIGURE 2.37
Common laboratory tests to measure strength of rock cores: (a) uniaxial compression; (b) triaxial compression; (c) direct shear for soft specimens; (d) direct shear for joints; (e) point load; (f) direct tension; (g) splitting tension (Brazilian); and (h) four-point flexural.

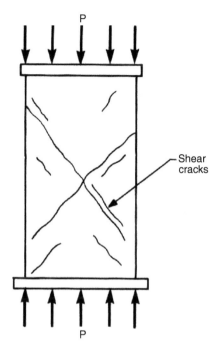

FIGURE 2.38
Uniaxial compression test.

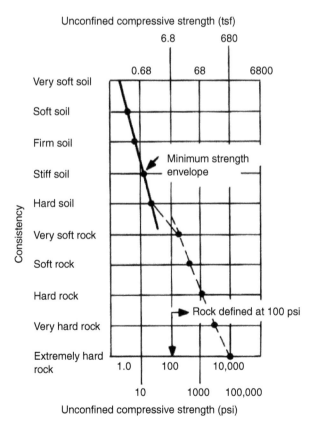

FIGURE 2.39
Relationship between "consistency" and U_c (100 psi = 6.8 kg/cm² = 689.5 kN/m²). (After Jennings, J. E., *Proceedings of ASCE, 13th Symposium on Rock Mechanics,* University of Illinois, Urbana, 1972, pp. 269–302.)

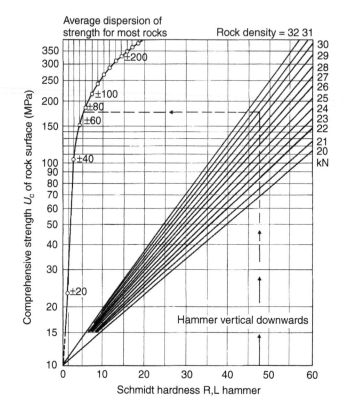

FIGURE 2.40

Correlation chart for Schmidt L. hammer, relating rock density, uniaxial compressive strength, and rebound number R (Schmidt hardness). Hammer vertical downward; dispersion limits defined for 75% confidence. (*Note*: 100 MPa = 14.5 × 10³ psi = 1021 tsf; 1 kN/m² = 6.3 pcf.) (From ISRM, *Rock Characterization and Monitoring*, E. T. Brown, Ed., Pergomon Press, Oxford, 1981. With permission. After Deere, D. U. and Miller, R. P., Technical Report AFWL-TR-65-116, AF Special Weapons Center, Kirtland Air Force Base, New Mexico, 1966.)

Uniaxial Tensile Strength

Cable-Pull Test

Caps are attached to the ends of a cylindrical specimen with resins. The specimen is then pulled apart by cables exerting tension axially (Figure 2.37f). The method yields the lowest values for tensile strength, which generally ranges from 5 to 10% of the uniaxial compression strength.

Point-Load Test (Broch and Franklin, 1972) (ASTM D5731-95)

Compressive loads P are applied through hardened conical points to diametrically opposite sides of a core specimen of length of at least $1.4D$ until failure occurs. The equipment is light and portable (Figure 2.41) and is used in the field and the laboratory.

   Point-load index is the strength factor obtained from the test, and is given by the empirical expression (Hoek and Bray, 1977)

$$I_s = P/D^2 \tag{2.42}$$

where D is the diameter.

FIGURE 2.41
Point load strength test apparatus.

Values for I_s are used to estimate U_c through various correlations as shown in Figure 2.42, and, for a core diameter of 50 mm,

$$U_c = 24\ I_s \tag{2.43}$$

Flexural Strength or Modulus of Rupture

Procedure

A rock beam is supported at both ends and loaded at midpoint until failure (Figure 2.37h).

Data Obtained

The flexural strength is proportional to the tensile strength but is about three times as great (Leet, 1960).

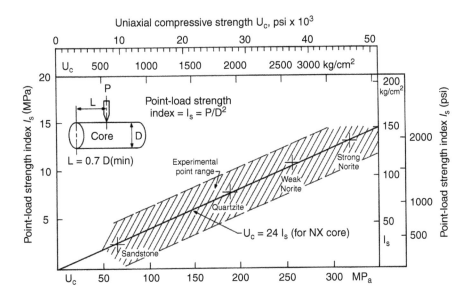

FIGURE 2.42
Relationship between point load strength index I_s and uniaxial compressive strength U_c. (After Bieniawski, Z. T., *Proceedings 3rd International Congress for Rock Mechanics*, International Society for Rock Mechanics, Vol. IIA, Denver, 1974, pp. 27–32. Reprinted with permission of National Academy Press.)

Triaxial Shear Strength

Apparatus and Procedures

General description is given in Section 2.4.4 and, as applicable to rock testing, in Section 2.5.3 and in Table 2.28.

Strength Values

Studies have been made relating analysis of petrographic thin sections of sandstone to estimates of the triaxial compressive strength (Fahy and Guccione, 1979). Relationships have been developed for approximating peak strengths for rock masses (Hoek and Brown, 1980).

Direct Shear Strength

Purpose

The purpose is to obtain measurements of the parameters ϕ and c *in situ*. It is particularly useful to measure strength along joints or other weakness planes in rock masses.

In Situ Test Procedure

A diamond saw is used to trim a rock block from the mass with dimensions 0.7 to 1.0 m^2 and 0.3 m in height, and a steel box is placed over the block and filled with grout (Haverland and Slebir, 1972). Vertical load is imposed by a hydraulic jack, while a shear force is imposed by another jack (Figure 2.43) until failure. All jack forces and block movements are measured and recorded. Deere (1976) suggests at least five tests for each geologic feature to be tested, each test being run at a different level of normal stress to allow the construction of Mohr's envelope.

Laboratory Direct Shear Tests (See Section 2.4.4)

If the specimen is decomposed to the extent that it may be trimmed into the direct shear ring (Figures 2.37c and 2.51) the test is performed similar to a soil test (Section 2.4.4). If possible, the shearing plane should coincide with the weakness planes of the specimen. Tests to measure the characteristics of joints in fresh to moderately weathered rock are performed by encapsulating the specimen in some strong material within the shear box as shown in Figure 2.37d (ISRM, 1981). The specimen is permitted to consolidate under a normal force and then sheared to obtain measures of peak and residual strength as described in Section 2.4.4. The normal stress is increased, consolidation permitted, and the specimen sheared again. The process is repeated until five values of shear stress vs. normal stress is obtained, from which a graph for peak and residual strength is constructed as shown in Figure 2.44.

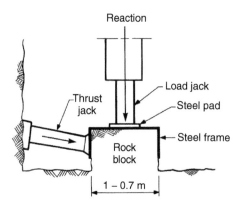

FIGURE 2.43
In situ direct shear test.

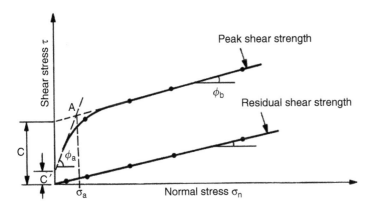

FIGURE 2.44
Shear strength–normalized stress graph for direct shear test on rock specimen.

Shear strength parameters ϕ_a, ϕ_b, ϕ_r, c' and c are abstracted from the graph, which often is a composite of several tests, as shown in Figure 2.44 where

- ϕ_r = residual friction angle.
- ϕ_a = apparent friction angle below stress σ_a; point A is a break in the peak shear strength curve resulting from the shearing off of the major irregularities (asperities) on the shear surface. Between points O and A, ϕ_a will vary slightly and is measured at the stress level of interest ($\phi_a = \phi_u + j$ where ϕ_u is the friction angle obtained for smooth surface of rock and angle j is the inclination of surface asperities) (Figure 2.45).
- ϕ_b = the apparent friction angle above stress level σ_a; it is usually equal to or slightly greater than ϕ_r and varies slightly with the stress level. It is measured at the level of interest.
- c' = cohesion intercept of peak shear strength which may be zero.
- c = apparent cohesion at a stress level corresponding to ϕ_b.

Borehole Shear Test (BST) (ASTM D4917-02)

Purpose

The borehole shear test measures peak and residual values of ϕ and c *in situ*. Initially developed at Iowa State University by R.L. Handy and N.S. Fox for the U.S. Bureau of

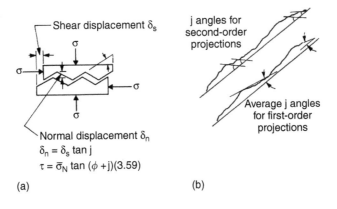

(a)   (b)

FIGURE 2.45
The joint roughness angle *j*: (a) experiments on shearing regular projection and (b) measurements of *j* angles for first- and second-order projections on rough rock surface. (From Patton, F. D., *Proceedings of the 1st International Congress of Rock Mechanics*, Lisbon, Vol. 1, 1966, pp. 509–513. With permission.)

Mines, it was designed for near-surface or in-mine testing of coal and other fractured rocks that are difficult to core. It has been used in all soft to medium-hard rocks.

Procedure

The apparatus is shown in Figure 2.46. A shear head, consisting of opposing plates with two carbide teeth, is lowered into a 3 in. (75 mm) borehole. Normal stress is applied by pushing the shear plates into the sides of the hole using a hand-operated pump. The pressure is then valved off so it remains constant, while the same pump is used to pull the expanded shear head a short distance upward along the hole by means of a hollow-ram jack. Both the expansion pressure and the pulling resistance are recorded, and the test is repeated with different preselected normal stresses. Up to four tests can be conducted at the same depth by rotating the shear head 45° between tests. A plot of each test is made to obtain a Mohr's envelope of shear stress vs. normal stress providing measurements of the angle of internal friction (ϕ) and cohesion (c) (Handy et al., 1976). Comparison with data from *in situ* direct shear tests indicate that cohesion values from the BST are lower although the friction angles are close (R. L. Handy, personal communication, 2004).

Applications

The apparatus is used in both rock (RBST) and soil (BST) in vertical, inclined, or horizontal boreholes. The entire apparatus, including the shear head and pulling device, is easily portable. A very significant advantage over other methods to measure shear strength is that many tests can be run in a short interval of time, as many as ten per day, and yield results on-site that can indicate if more tests are needed.

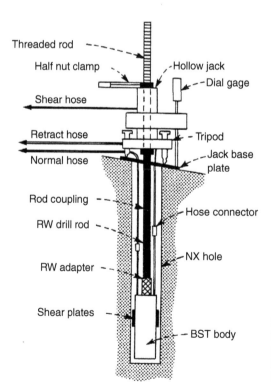

FIGURE 2.46

Schematic of borehole shear strength tester (BST) in the borehole. (From USBM, Bureau of Mines, U.S. Department of Interior, New Technology No. 122, 1981. With permission.)

2.4.4 Soil Strength Measurements

General

Selection of Test Method

A number of factors required for the selection of the method for testing soils, including the following:

- Loading conditions: static or dynamic.
- Loading duration in the field: long-term (drained conditions) or short-term (undrained conditions).
- Parameter desired: peak or ultimate (residual) strength.
- Material suitability for undisturbed sampling and the necessity or desirability for *in situ* testing.
- Orientation of the field failure surface with that in the test; some cases are shown in Figure 2.47 and Figure 2.48. Stability analysis is often improperly based on compression tests only, whereas direct shear and extension tests are often required. Their strength values may differ significantly from the compressive parameters.

Testing Methods Summarized
- Soil laboratory static strength tests — Table 2.19.
- *In situ* static strength tests — Table 2.20.
- Laboratory dynamic strength tests — Table 2.33.

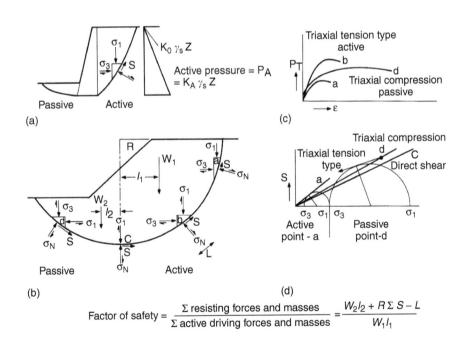

$$\text{Factor of safety} = \frac{\Sigma \text{ resisting forces and masses}}{\Sigma \text{ active driving forces and masses}} = \frac{W_2 l_2 + R \Sigma S - L}{W_1 l_1}$$

FIGURE 2.47
Probable natural stress and strain restraint conditions: (a) Retaining wall influence of lateral yielding on stresses. (b) Mass slide of excavated slope. Influence of lateral yielding. (c) Stress–strain relations corresponding to lateral yield conditions in (b). (d) Angle of friction relations corresponding to lateral yield conditions in (b). (From Burmister, D. M., ASTM Special Technical Publication No. 131, 1953. Reprinted with permission of the American Society for Testing and Materials.)

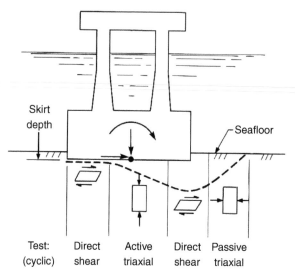

| Test:
(cyclic) | Direct
shear | Active
triaxial | Direct
shear | Passive
triaxial |

FIGURE 2.48
Relevance of laboratory tests to shear strength along potential slip surface beneath offshore gravity structure. (From Kierstad and Lunne, *International Conference on Behavior of Offshore Structure*, London, 1979. With permission.)

TABLE 2.19

Summary of Soil Laboratory Static Strength Tests

Test (F = also field test)	Parameter Measured	Reference	Comments
Triaxial compression			
CD	$\bar{\phi}, \bar{c}$	Figure 2.28	Most reliable method for effective stresses
CU	$\phi, c, \bar{\phi}, \bar{c}$	Figure 2.49	Strength values higher than reality because
	s_u	Figure 2.50	disturbance causes lower $w\%$ upon reconsol-
		Table 2.21	dation (see footnote c in Table 2.21)
UU	s_u	Figure 2.29	Most representative laboratory value for undrained shear strength in compression
Triaxial extension	$\bar{\phi}, \bar{c}, s_u$	Table 2.21	Normally consolidated clays yield values approximately one-third those of compression tests because of soil anisotropy (Bjerrum et al., 1972)
Plain strain compression or extension	ϕ	Table 2.21	Values are a few degrees higher than those of normal triaxial test except for loose sands; more closely approach reality for retaining structure (Lambe and Whitman, 1969)
Direct shear box	$\bar{\phi}, \bar{c}, \bar{\phi}_r$	Figure 2.51	Values most applicable where test failure surface has same orientation with field failure surface. Values generally lower than triaxial compression values for a given soil, but higher than triaxial extension. Most suitable test for determination of residual strength $\bar{\phi}_r$ from UD samples
Simple shear	$s_u, \bar{\phi}, \bar{c}$	Figure 2.34	Horizontal plane becomes plane of maximum shear strain at failure
Unconfined compression	$S_u = 1/2\, U_c$	Table 2.22	Strength values generally lower than reality
Vane shear (F)	s_u, s_r	Table 2.22	Applies shear stress on vertical planes
Torvane (F)	s_u, s_r	Table 2.22	Shear occurs in a plane perpendicular to the axis of rotation
Pocket penetrometer (F)	$S_u = 1/2\, U_c$	Table 2.22	Yields approximate values in clays. Used primarily for soil classification by consistency
California bearing ratio (F)	CBR value	Figure 2.64	Used for pavement design. Empirical strength correlates roughly with U_c

TABLE 2.20

Summary of Soil *In Situ* Static Strength Tests

Test	Parameters Measured	Reference	Comments
Vane shear	s_u, s_r (direct test)	Figure 2.56	Measures undrained strength by shearing two circular horizontal surfaces and a cylindrical vertical surface: therefore, affected by soil anisotropy
SPT	$\bar{\phi}$, s_u (indirect test)	Section 2.4.5	D_r is estimated from N and correlated with soil gradation to obtain estimates of ϕ (Figure 2.93, Table 2.36) Consistency is determined from N and correlated with plasticity to obtain estimates of U_c (Figure 2.94, Table 2.37)
CPT	$\bar{\phi}$, s_u (indirect test)	Section 2.4.5	Various theoretical and empirical relationships have been developed relating q_c to $\bar{\phi}$, (Figure 2.61) s_u is expressed as in Equation (2.50) where N_{kt} is the cone-bearing capacity factor (deep foundation depth correction factor) Pore-water pressure u is measured by some cones (piezocones)
Pressuremeters	s_u (indirect test)	Section 2.5.4	Affected by material anisotropy, s_u is expressed as in Equation (2.77)
California bearing ratio	CBR value	Section 2.4.5	Field values generally less than lab values because of rigid confinement in the lab mold

Triaxial Shear Test

Purpose

Total or effective stress parameters, either in compression or extension, are measured in the triaxial shear apparatus. The test method is generally unsuited for measuring ultimate strength because displacement is limited and testing parallel to critical surfaces is not convenient.

Apparatus includes a compression chamber to contain the specimen (Figure 2.49) and a system to apply load under controlled stress or strain rates, and to measure load, deflection, and pore pressures (Figure 2.50).

Specimens are usually 2.78 in. in diameter as extruded from a Shelly tube, or 1.4 in. in diameter as trimmed from an undisturbed sample. Specimen height should be between 2 and 2.6 times the diameter.

General Procedures

Rate of loading or strain is set to approximate field-loading conditions and the test is run to failure (Bishop and Henkel, 1962).

Confining pressures: Tests are usually made on three different specimens with the same index properties to permit defining a Mohr diagram. Test method variations are numerous. Specimens can be preconsolidated or tested at their stress conditions as extruded. Tests can be performed as drained or undrained, or in compression or extension. The various methods, parameters measured, and procedures are summarized in Table 2.21. CU tests are covered in ASTM D4767-02.

Direct Shear Test (ASTM D5607-02 for rock, ASTM D3080-03 for soils)

Purpose

The purpose normally is to measure the drained strength parameters ϕ', c', and ϕ_r'.

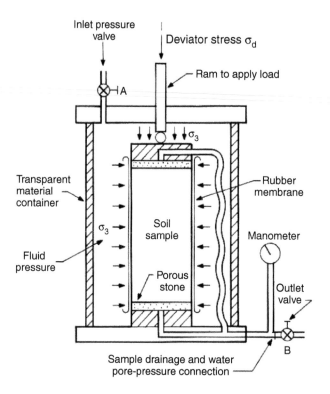

FIGURE 2.49
Triaxial compression chamber arrangement.

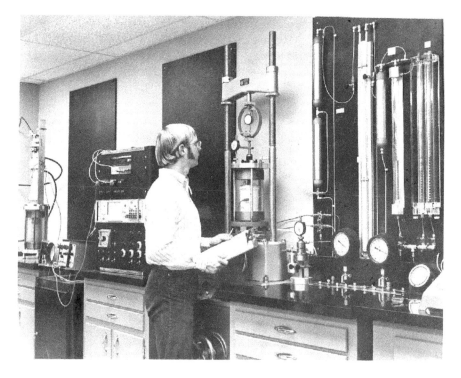

FIGURE 2.50
The triaxial compression chamber, load application, and measurement system.

TABLE 2.21

Triaxial Test Methods[a]

Test	To Measure	Procedure
Consolidated-drained (CD) or (S) compression test	Effective stress parameters $\bar{\phi}, \bar{c}$	Specimen permitted to drain and consolidate under confining pressure until $u = 0$. Deviator stress applied slowly to failure while specimen drains during deformation[b]
Consolidated-undrained (CU) or (R) compression test	Total stress parameters ϕ, c (Figure 2.28)	Specimen permitted to drain and consolidate under confining pressure until $u = 0$. Deviator stress applied slowly to failure, but specimen drainage not permitted
	Effective stress parameters $\bar{\phi}, \bar{c}$	Pore pressures are measured during test (see Pore-Pressure Parameters)
$CK_0 U$ test	s_u	See Notes [c] and [d] below
Unconsolidated-undrained (UU) or (Q) compression test	Undrained strength s_u (Figure 2.29)	Confining pressure applied but no drainage or consolidation permitted to reduce test time. Deviator stress applied slowly to failure with no drainage permitted
Extension tests as CD, CU, UU	Lateral shear strength	Maintain confining pressure constant and reduce axial stress, or maintain axial stress constant and increase confining pressure until failure
Plane strain compression or extension test	Parameter ϕ in cohesionless granular soils	Modified triaxial apparatus in which specimen can strain only in axial direction and one lateral direction while its dimension remains fixed in the other lateral direction

[a] See Table 2.19 for comments on test results and test comparisons.

[b] Backpressure is applied to the pore water to simulate *in situ* pore-water pressures, or to saturate partially saturated specimens.

[c] CK_0U test: specimen anisotropically consolidated with lateral pressure at K_0p_0 and vertical pressure at p_0.

[d] SHANSEP procedure (Ladd and Foott, 1974) attempts to minimize the effects of sample disturbance and assumes normalized behavior of clay. Specimens are consolidated to p_c stresses higher than p_0, rebounded to selected values of OCR, and then tested in undrained shear to establish the relationship between s_u/p_0 vs. OCR for different modes of failure. Normalized behavior requires this relationship to be independent of p_c. The s_u profile is then calculated front the values of p_0 and OCR vs. depth (Baligh et al., 1980).

Apparatus

The test apparatus is illustrated in Figure 2.51.

Procedure

The specimen is trimmed to fit into the shear box between two plates, which can be pervious or impervious, depending upon the drainage conditions desired, and a normal load applied which remains constant throughout the test. The test is normally run as a consolidated drained (CD) test (sample permitted to consolidate under the normal load), but it

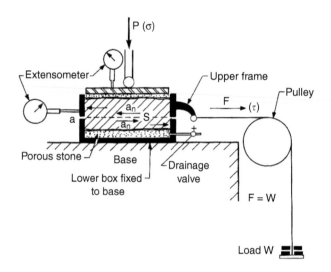

FIGURE 2.51
Laboratory direct shear box.

has been performed as a consolidated undrained (CU) test when load application rates are high, although some drainage always occurs during shear. The shear force is applied either by adding deadweights (stress-controlled) or by operating a motor acting through gears (controlled strain) and gradually increasing the load until failure occurs.

Mohr's Envelope

Plotting the results for a sequence of tests under different normal stresses as points (τ, σ) results in a Mohr's envelope. Because of the form of the test, only the normal and shear stresses on a single plane are known. If it is assumed that the horizontal plane through the shear box is identical with the theoretical failure plane (usually an inexact assumption), it can be assumed that the stresses at failure are in the ratio of $\tau/\sigma = \tan \phi$ to provide a point on the envelope.

Residual or Ultimate Shear Strength

The direct shear test is often used for measuring the residual strength. The test is performed by a back-and-forth sequential movement of the shear box after the failure of an undisturbed specimen has occurred, or by testing a remolded specimen.

An alternate method, the *rotational shear test*, is considered to provide the lowest possible values for the residual strength. The specimen is ground down to produce all particles smaller than 0.74 mm (#200 sieve) and placed between two circular plates which are rotated and the shearing resistance measured.

The relationship between peak and ultimate strength for normally consolidated and overconsolidated clays is given in Figure 2.30, and Mohr's envelope for the ultimate drained strength ϕ_r, in overconsolidated clay, is given in Figure 2.31.

Simple Shear Test (ASTM D6528-00 for CU direct simple shear)

Purpose

The purpose is to measure drained or undrained strength parameters.

Apparatus

The soil specimen is contained either in a cylindrical rubber membrane reinforced by wire, which permits shear deformation to be distributed fairly uniformly throughout the specimen

(NGI apparatus-Kjellman, 1951), or in a box which provides a rigid confinement as lateral stress is applied (Roscoe, 1953). A vertical stress is applied to cause consolidation and a lateral stress is applied to induce shear failure (Figure 2.34).

Procedure

The specimen is consolidated anisotropically under a vertical stress and then sheared by the application of lateral stress. In the undrained test, zero volume change during shear can be maintained by adjusting the vertical stress continuously during the test (Bjerrum and Landva, 1966).

Principal Axis Orientation

Initially, the principal axes are in the vertical and horizontal directions. At failure, the horizontal plane becomes the plane of maximum shear strain (the lower point on the failure circle in Figure 2.47b).

Undrained Shear Tests: Laboratory

Laboratory tests used to measure the undrained shear strength, and in some cases the residual or remolded shear strength, are described in Table 2.22. Included are unconfined compression (Figure 2.52), miniature vane (Figure 2.53) (ASTM D4648-00), torvane

TABLE 2.22

Laboratory Tests for Undrained Shear Strength

Test	Parameters	Procedure	Comments
Triaxial compression Unconfined compression (Figure 2.52)	s_u $U_c = 2s_u$ (Figure 3.29)	See Table 2.21 Unconfined specimen ($\sigma_3 = 0$) strained to failure by axially applied load	UU or CK_0U test Test yields stress–strain curve and U_c. Slight drainage may occur during shear. Values usually conservative because of nonrecovered decrease in effective stress occurring from disturbance during extraction
Vane shear (miniature vane) (Figure 2.53)	s_u, s_r	Small two-bladed vane is pushed about 2 in. into a UD-tube specimen and then rotated, while the torque indicated on a gage is recorded to measure peak and ultimate strength	Test limited to clay soils of very soft to firm consistency (Table 2.37)
Torvane[a] (Figure 2.54)	s_u, s_r	One-inch-diameter disk containing eight small vanes is pushed about ¼ in. penetration in UD sample and rotated while torque is recorded	Test limited to clay soils of very soft to firm consistency. Tested zone may be disturbed by sample trimming and cutting
Pocket penetrometer[a] (Figure 2.55)	$U_c = 2s_u$	Calibrated spring-loaded rod (¼ in. diameter) is pushed into soil to a penetration of 6 mm and the gage read for U_c	Penetration is limited to soils with $U_c < 4.5$ tsf. Data are representative for soils with PI$>$12. Below this value the ϕ of granular particles increases strength to above the measured value of s_u
Simple shear	s_u	See Section 2.4.4	

[a] These tests are considered generally as classification tests to provide an index of soil consistency (Table 2.37).

FIGURE 2.52
Unconfined compression apparatus.

(Figure 2.54), pocket penetrometer (Figure 2.55), and the simple shear device. The unconfined compression test (ASTM D2166-00) is described in Section 2.4.3. It provides very conservative values used primarily for correlations.

In Situ Vane Shear Test (ASTM D2573-01)

Purpose

The purpose is to measure s_u and s_r *in situ* in normally consolidated to slightly overconsolidated clays in the field.

Apparatus

The field test apparatus is illustrated in Figure 2.56.

Procedure

A standard 4-in.-diameter boring is made and cased to the desired test depth. A 3-in.-wide steel vane on a 1-in.-diameter rod is lowered to the test depth (other vane sizes are available for smaller casing). Spacers at 30 ft intervals center the rod in the casing. The vane is pushed to a depth 2 1/2 ft below the casing and torsion applied at a constant rate through a friction-free mechanism, while the torque is measured on a proving ring. Failure is evidenced by a sudden loss of torsion; the test is stopped for a short interval and then rerun to measure s_r. Charts give values for s_u and s_r for the measured torque and vane size used.

FIGURE 2.53
Miniature vane-shear apparatus.

FIGURE 2.54
Torvane applied to the end of a Shelby
tube specimen.

Comments

The test is generally considered to give the most reliable values for s_u and s_r, but since the test applies shear directionally, the values are affected by soil anisotropy. Failure actually takes place by the shearing of two circular horizontal surfaces and a cylindrical vertical surface. Vanes of different height-to-diameter ratios have been used by the Norwegian

FIGURE 2.55
Pocket penetrometer applied to a Shelby tube specimen.

Geotechnical Institute (Aas, 1965) to evaluate anisotropy approximately. Relating s_u as measured by the field vane to s_u as measured in triaxial tests, Bjerrum et al. (1972) found that:

- s_u (compression) $\approx 1.5\, s_u$ (vane).
- s_u (extension) $\approx 0.5\, s_u$ (vane) in low- to medium-PI clays.
- s_u (extension) $\approx 1.0\, s_u$ (vane) in highly plastic clays.

Field vane loading time rate is very rapid compared with actual field loadings, and during construction values are generally lower (Section 2.4.2). Rate correction factor μ is proposed as a function of plasticity (Figure 2.57) by Bjerrum et al. (1972) as found by the analyses of a number of embankment failures constructed over soft clays.

Borehole Shear Test (BST)

The borehole shear test (See Section 2.4.3) in soils is similar to the rock test, except that it is a staged test and the soil reconsolidates after it is sheared, moving the shear surface outward into fresh soil. In addition, the shear plate teeth are continuous rather than two carbide teeth, and regulated pressure from refillable CO_2 cylinders is used to expand the plates tightly to the hole sides (Wineland, 1975). For greater portability, a hand pump may also be used. Drainage is rapid, and the test is designed to be a drained test that yields effective stress shear strength parameters. BST applications for bridge foundation design is described by Handy et al. (1985) and for slope stability by Handy (1986). Because the strength is measured vertically along a borehole it is particularly useful in obtaining values applicable to shaft friction along a pile. Very hard soils are tested using a two-teeth adaption of the rock tester shear plates.

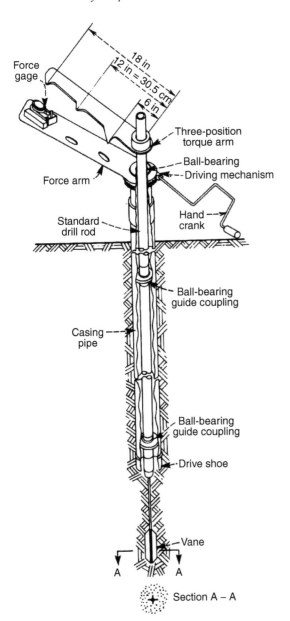

FIGURE 2.56
The *in situ* vane shear test arrangement. (Courtesy of Acker Drill Co.)

Pressuremeter Testing

See Section 2.5.4.

Dynamic Strength Tests

Cyclic Triaxial Test (ASTM D5311-96, ASTM D3999-96)

Purpose: The purpose is to measure strength parameters under low-frequency loadings (intermediate to large strain amplitudes). Cyclic compression modulus E_d of cohesive soil (dynamic shear modulus G_d is computed from E_d, Table 2.27) and the liquefaction resistance of undisturbed and reconstituted specimens of cohesionless soils under low-frequency loadings.

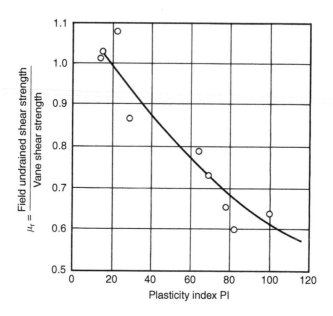

FIGURE 2.57
Loading time-rate correction factor vs. plasticity index for undrained shear strength as measured by the field vane. (From Bjerrum, L. et al., *Proceedings of the 5th European Conference on Soil Mechanics and Foundation Engineering*, Madrid, Vol. II, 1972, pp. 169–196. With permission.)

Apparatus: The apparatus is similar to standard triaxial equipment except that the deviator stress is applied cyclically. Sophisticated equipment allows variation in the frequency of applied load as well as control of the waveform (sinusoidal or square, ramp or random). In random loadings, actual earthquake histories are imposed.

Procedure: The specimen is placed in the chamber and subjected to all-around confining pressure. Backpressure may be applied to the pore water to saturate partially saturated specimens and to simulate *in situ* pore pressures. Constant-amplitude cyclic axial load is applied while specimen drainage is prevented, and pore-water pressure, axial load, and axial deformations are recorded. For many soils, after a number of stress cycles, the pore pressure increases to a value approximately equal to the cell pressure at which point the soil will begin to deform excessively.

Limitations: The test only approximately duplicates field stress conditions during earthquake loadings (USAEC, 1972).

Cyclic Simple Shear Test

Purpose: The purpose is to measure strength parameters under intermediate to large strain amplitudes and to measure the dynamic shear modulus G_d and the damping ratio of cohesive and cohesionless soils.

Apparatus: The simple shear device, described previously, in which the lateral load is applied cyclically. Loading frequency and waveform are controlled.

Procedure: The specimen is placed in the chamber and subjected to confining pressure. Backpressure may be applied. A vertical stress is applied, and a cyclic shear force is applied laterally while no drainage is permitted.

Comment: The test most nearly duplicates field conditions for soils subject to strong earthquake motion (large strains), although test boundary conditions will cause values to be somewhat lower than reality (USAEC, 1972).

2.4.5 Soil Penetration Tests

Standard Penetration Test (SPT) (ASTM D1586)

Purpose

SPT values are correlated with the compactness of granular soils, from which ϕ values are estimated and correlated with the consistency of cohesive soils.

Test Method

The split-barrel sampler (see Section 1.4.2), 2-in. O.D. and 1-in. I.D., is driven into the ground by a 140-lb hammer dropping in free fall from a height of 30 in. (energy = 4200 in. lb). The number of blows required to advance the sampler each 6 in. of penetration is recorded.

Standard penetration resistance, N is taken normally as the penetration of the second plus the third 6 in. to provide "blows per foot of penetration." The first 6 in. is disregarded because of the possibility of cuttings in the hole or disturbance during washing. The penetration resistance is composed of end resistance plus shaft friction. End resistance is the larger component in granular soils, and shaft resistance is the larger component in clay soils.

Operational Factors Affecting N Values

Driving energy: Standard hammer weight and drop should always be used. The weight is lifted by various methods: rope or cable on a drum, block and tackle, or by automatic hammers. Common weights include a mass with a "stinger" rod that fits into the drill rod, or a "doughnut"-shaped weight that slides over the drill rod. A quick-release grab is shown in Figure 2.58. The weight is "grabbed," raised 30 in., and automatically released. To ensure that the fall is always completely free of any resistance requires care by the operator, especially with the rope-and-drum method.

In recent years, automatic drop hammers have been developed. They are activated by a hydraulically powered chain lift device.

Sampler: The sampler is illustrated in Figure 1.61. Diameters larger than the standard 2 in. sampler require correction. The adjusted blow count B for the measured blow count B' can be determined by the equation (Burmester, 1962a).

$$B=B'\ (4200/WH)\ [(D_o^2-D_i^2)/(2.0^2-1.375^2)]=B=B'(4200/WH)\ [(D_o^2-D_i^2)/2.11] \quad (2.44)$$

where W is the hammer weight variation, H the drop height variation and D_o and D_i are the sampler diameter variations.

The damaged or blunt cutting edge on a sampler will increase resistance. Liner-designed tubes will produce lower N values when used without liners (Schmertmann, 1979).

Drill Rods and Casing: One study indicates that the difference in weight between N and A rods has negligible effect on N values (Brown, 1977), although there are differences of opinion in the literature regarding the effects. Drill rods that are loosely connected or bent, or casing that is out of plumb, will affect penetration resistance because of energy absorption and friction along casing walls. The rod length, i.e., weight, also affects the N value.

Hole bottom conditions: (see Section 1.4.2 and Figure 1.60): Soil remaining in casing, even as little as 2 in. of sand, will cause a substantial increase in N (plugged casing). The boiling and rising of fine sands in casing when the tools are removed, which causes a decrease in N, is common below the water table and is prevented by keeping the casing filled with water unless conditions are artesian. Gravel particles remaining in the mud-cased hole will increase N values by 30 to 50, and can result in a loose sand that should only be ten blows. Overwashing or washing with a bottom-discharge bit with high pressure will loosen the soils to be sampled.

FIGURE 2.58
Quick-release drop hammer used in Israel and Europe for SPT sampling.

Natural Factors Affecting N Values

Soil type and gradation: Gravel particles in particular have a very significant effect on N as shown in Figure 2.59.

Compactness and consistency: Compactness of granular soils (D_R) and consistency of clay soils affect N values.

Saturation: In gravels and coarse sands, saturation has no significant effect. In very fine or silty sands, pore-pressure effects during driving are corrected approximately for test values $N' > 15$ by the expression (Terzaghi and Peck, 1948; Sanglerat, 1972)

$$N = 15 + 1/2(N' - 15) \tag{2.45}$$

Depth correction factor (C_N): Effective overburden pressures cause a fictitious increase in N values with increasing depth as shown by the studies of various investigators (Gibbs and Holtz, 1957; Bazaraa, 1967; Marcuson and Bieganousky, 1977a, 1977b). Various investigators have proposed relationships to correct N for depth (C_N) (Peck et al., 1973; Seed, 1976, 1979; Liao and Whitman, 1986; Kayen et al., 1992):

$$N1 = C_N N, \quad C_N = (1/\sigma'_{vo})^{0.5} \text{ (Liao and Whitman, 1986)} \tag{2.46}$$

$$C_N = 2.2/(1.2 + \sigma'_{vo}/P_a) \text{ (Kayen et al., 1992)} \tag{2.47}$$

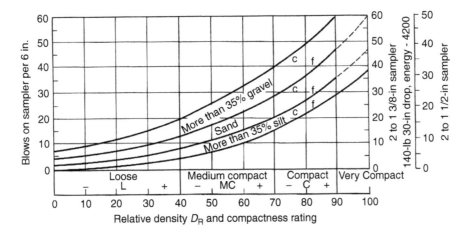

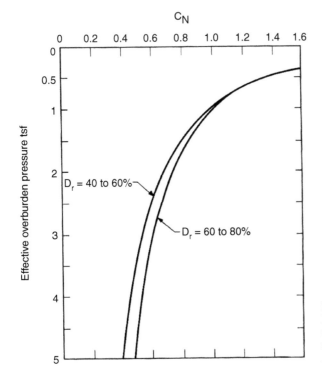

FIGURE 2.59
Relationship between sampler blows per 6 in, sampler diameter, gradation, and compactness. (From Burmister, D. M., ASTM Special Technical Publication No. 322, 1962, pp. 67–97. With permission. Reprinted with permission of the American Society for Testing and Materials.)

where C_N is a correction factor, σ'_{vo} the effective overburden pressure $(\gamma_t - \gamma_w)z$, (in tsf) and $P_a = 1$ tsf (95 kN/m²). N value at a depth corresponding to σ_1 is considered to be a standard.

Seed (1979) presented recommended curves for the determination of C_N based on averages for results obtained in various studies (Marcuson and Bieganousky, 1977a, 1977b) as given in Figure 2.60. For practical applications, the curves resulting from equations 2.46 and 3.47 approximate the curves of Figure 2.60. Investigators have proposed that the maximum value for C_N should not exceed 1.7.

FIGURE 2.60
Recommended curves for determination of C_N based on averages from WES tests. (From Seed, H. B., *Proceedings ASCE, J. Geotech. Eng. Div.*, 105, 201–255, 1979. With permission.)

SPT N-Value Corrections

In recent years, it has become a practice to adjust the N value by a hammer-energy ratio or hammer efficiency of 60%, and much attention has been given to N-values because of its use in liquefaction studies. Additional corrections have been proposed (Skempton, 1986; Robertson and Wride, 1998) for hammer type (donut and safety), borehole diameter, rod lengths, and sampler. The largest correction factor appears to be for the donut hammer ($C_E = 0.5$–1.0). Therefore

$$(N_1)_{60} = N\, C_N\, C_E \tag{2.48}$$

Applications of the N value

- *Correlations with compactness and D_R* — see Table 2.23. Marcuson and Bieganousky (1977) give an expression for computing D_R which accounts for N, σ'_o, and C_u the coefficient of uniformity. (equation. (2.3)) as follows:

$$D_R = 11.7 + 0.76\{|222N + 1600 - 53\sigma_o' - 50C_u^2|\}^{1/2} \tag{2.49}$$

From D_R and gradation curves estimates can be made for E (Young's modulus) from Table 2.25, k from Figure 2.14 and ϕ from Table 2.36.

- *Correlations with consistency and U_c* — Table 2.37 and Figure 2.94.
- *Allowable soil-bearing value for foundations* — correlations with N (see, e.g.,Terzaghi and Peck, 1967, p. 491)

Limitations

The value of N must always be considered as a rough approximation of soil compactness or consistency in view of the various influencing factors. Even under controlled laboratory testing conditions, a single value for N can represent a spread of $\pm15\%$ in D_R (Marcuson and Bieganousky, 1977b).

Cone Penetrometer Test (CPT) (ASTM D5778)

Purpose

The correlations for strength properties are made between the cone tip resistance q_c, ϕ and s_u. Determinations of deformation property are discussed in Section 2.5.4. *Test method and soil classification* are discussed in Section 1.3.4.

TABLE 2.23

Correlations for Cohesionless Soils between Compactness, D_R, and N

Compactness	Relative Density D_R[a]	N (SPT)
Very loose	<0.15	<4
Loose	0.15–0.35	4–10
Medium dense	0.35–0.65	10–30
Dense (compact)	0.65–0.85	30–50
Very dense	0.85–1.0	>50

[a] From Gibbs, H.J. and Holtz, W.G., Proceedings of the 4th International Conference on Soil Mechanics and Foundation Engineering, London, Vol. I, 1957, pp. 35–39.

Estimating ϕ

Numerous studies have been published for assessing ϕ from the CPT in clean sands. Robertson and Campanella (1983) reviewed calibration chamber test results to compare cone resistance to measured peak secant friction angle from drained triaxial compression tests. The triaxial tests were performed at the confining stress approximately equal to the horizontal stresses in the calibration chamber before the CPT.

The recommended correlation for uncemented, unaged, moderately compressible, predominantly quartz sands proposed by Robertson and Campanella (1983) is shown in Figure 2.61 (Robertson, 2000).

Estimating Undrained Shear Strength s_u

As discussed in Section 2.4.2, the measured value of s_u depends on a number of factors including loading direction, soil anisotropy, strain rate and stress history. Since anisotropy and strain rate influence the results of all *in situ* tests, their interpretation requires some empirical measure.

Robertson (2000) gives

$$s_u = (q_t - \sigma_v)/N_{kt} \tag{2.50}$$

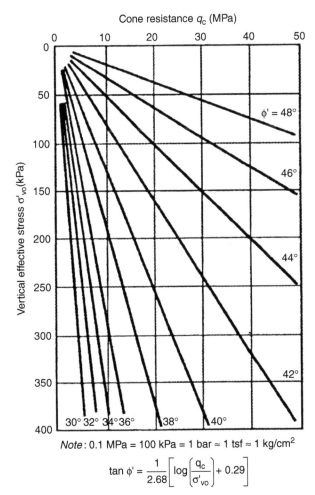

Cone resistance q_c (MPa)

Vertical effective stress σ'_{vo} (kPa)

$\phi' = 48°$

$46°$

$44°$

$42°$

$30°\ 32°\ 34°36°$ $38°$ $40°$

Note: 0.1 MPa = 100 kPa = 1 bar ≈ 1 tsf ≈ 1 kg/cm²

$$\tan\phi' = \frac{1}{2.68}\left[\log\left(\frac{q_c}{\sigma'_{vo}}\right) + 0.29\right]$$

FIGURE 2.61
Friction angle, ϕ', from CPT in uncemented silica sand. (After Robertson, P. K. and Campanella, R. G., *Can. Geotech. J.*, 20, 718–733, 1983. Permission of National Research Council of Canada.) (Courtesy of ConeTec Inc.)

where q_t is the total cone resistance (Section 1.3.4), σ_v the total overburden pressure at depth z and N_{kt} the depth correction factor varying from 10 to 20. N_{kt} tends to increase with increasing plasticity and decrease with increasing soil sensitivity. For a more conservative estimate, higher values of N_{kt} are selected; N_{kt} should be confirmed by other strength tests for each new location.

Correlations between q_c and SPT N value

Correlations provide approximate values because many factors affect the SPT N values Robertson et al. (1983) proposed the correlations given in Figure 2.62 relating the ratio q_c/N_{60} with the mean grain size D_{50}. N_{60} represents an average energy ratio of about 60%, which reflects the average energy efficiency of the SPT test.

Flat-Plate Dilatometer Test (DMT)

Purpose

Correlations for strength properties are made to provide estimates of ϕ and s_u. Deformation property determinations are discussed in Section 2.5.2.

Test Method

A flat blade with a steel membrane on one face (Figure 2.63) is pushed into the ground using a CPT or drill rig as a reaction. At test depth, the operator inflates the membrane and applies gas pressure to expand the flat-lying membrane. In about 1 min, two pressure readings are taken and electronically recorded: A, the pressure required just to begin to move the membrane; and B, the pressure required to move the center of the membrane. The blade is then advanced typically for a distance of 20 cm and the test repeated.

The pressure readings A and B are corrected by values determined by calibration to account for membrane stiffness. Corrected values for A and B are given as p_0 and p_1. These values are used to define several dilatometer indices:

$$E_D = 34.7(p - p_1) \text{ (dilatometer index)} \tag{2.51}$$

$$K_D = (p_0 - u_0)/(\sigma'_{v0}) \text{ (horizontal stress index)} \tag{2.52}$$

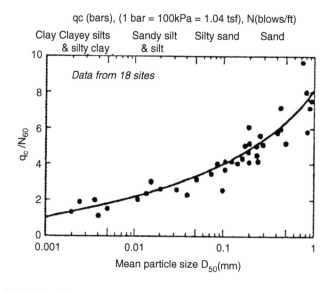

FIGURE 2.62
CPT-SPT correlations with mean grain size. Based on an SPT energy ratio of 60% (After Robertson, P. K. and Campanella, R. G., *Can. Geotech. J.*, 20, 718–733, 1983. Permission of National Research Council of Canada.)

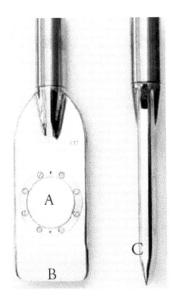

FIGURE 2.63
The flat dilatometer. (From Manchetti, S. et al., *IN SITU 2001, International Conference on In Situ Measurement of Soil Properties*, Bali, Indonesia, 2001. With permission.)

where σ'_{vo} is the overburden stress and u_0 the pore pressure, both preinsertion.

$$I_D = (p_0 - p_1)/ (p_0 - u_0) \text{ (material index)} \tag{2.53}$$

Estimating ϕ

The friction angle

$$\phi_{safeDMT} = 28° + 14.6° \log K_D - 2.1°\log^2 K_D \tag{2.54}$$

Estimating s_u (c_u)

The undrained shear strength

$$c_{uDMT} = 0.22\,\sigma'_{v0}\,(0.5\,K_D)^{1.25} \tag{2.55}$$

See Manchetti (1980), Totani et al. (2001), and Manchetti et al. (2001).

Pressuremeter

The pressuremeter is used to estimate the undrained shear strength (see Section 2.5.4).

California Bearing Ratio (CBR) Test

Purpose

A penetration test is performed to determine the California bearing ratio (CBR) value for base, subbase, and subgrade materials upon which pavement design thicknesses are based.

Laboratory Test Procedure (ASTM D1883)

A series of specimens are compacted in 6-in.-diameter molds to bracket either the standard or modified optimum moisture (see Section 2.2.3). A surcharge weight is placed on the soil surface and the specimens immersed in a water tank for 4 days to permit saturation while swelling is measured. A mold is placed in the testing apparatus (Figure 2.64) and a plunger 1.91 in. in diameter is forced to a penetration of 0.1 to 0.2 in. (penetrations to 0.5 in. are

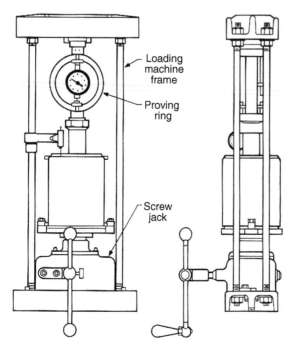

FIGURE 2.64
Laboratory CBR loading device (electric motor is
often used to drive jack at a rate of 0.05
in./min).

used) while the penetration resistance is recorded. The CBR is determined as the ratio of
the test resistance at 0.1 in. penetration to the standard resistance of crushed stone at the
same penetration (1000 psi or 70.3 tsf) (Asphalt Institute, 1969).

Field Test Procedure (ASTM D4429-93)

The plunger is jacked against the reaction of a truck weight. The test should not be performed during the dry season because resistances will be higher than those to be relied
upon for pavement support. Field values are usually lower than laboratory values, probably because of the effect of confinement in the lab.

Soil Rating System

A soil rating system for subgrade, subbase, or base for use in the design of light traffic
pavements in terms of CBR values is given in Figure 2.65. Typical CBR values for compacted materials are given in Table 2.39.

2.5   Deformation without Rupture

2.5.1   Introduction

Forms of Deformation

Ideal Materials

Elastic deformation: Stress is directly proportional to strain; the material recovers all deformation upon the removal of stress (Figure 2.66a).

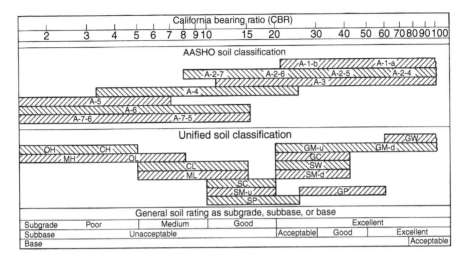

FIGURE 2.65
Approximate correlation of soil ratings based on CBR values for use in design of light-traffic pavements. (From Asphalt Institute, *Thickness Design*, The Asphalt Institute, Manual Series No. 1, College Park, MD, 1970. With permission.)

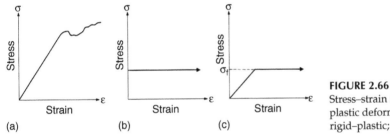

FIGURE 2.66
Stress–strain relationships of elastic and plastic deformation: (a) elastic; (b) rigid–plastic; (c) elastic–plastic.

Plastic deformation: Permanent and continuous deformation occurs when the applied stress reaches a characteristic stress level (Figure 2.66b). Geologic materials often combine deformation modes under stress (Figure 2.66c).

Viscous deformation: The rate of deformation is roughly proportional to applied stress.

Geologic Materials

Compression: Varies from essentially ideal to less than ideal, i.e., elastic compression under relatively low stress levels followed by plastic compression resulting from the closure of soil voids or rock fractures. Some materials may deform only plastically, others may exhibit plastic–elastic–plastic deformation at different stress levels.

Creep: A time-dependent deformation at constant stress level below failure level.

Expansion: An increase in volume from swelling, or elastic or plastic extension strain.

Geotechnical Parameters Used to Define Deformations

Deformation moduli are quantities expressing the measure of change in the form of the dimensions of a body occurring in response to changing stress conditions. The quantities include the elastic modulus (static and dynamic), moduli from the stress–strain curve, compression modulus from pressure-meter testing, and the modulus of subgrade reaction.

Consolidation parameters define various aspects of the slow process of compression under applied stress that occurs as water is extruded from the voids of clay soils.

Influencing Factors

Magnitude and form of deformation in geologic materials are influenced by material properties; *in situ* stress conditions; the level, direction, and rate of application of applied stress; temperature of environment; and time interval under stress.

Deformation in Practice

Foundations undergo settlement or heave; bending occurs in mats, continuous footings, and piles from differential deflection.

   Flexible retaining structures undergo bending; deflections may result in significant back-slope subsidence.

   Tunnels, particularly in rock masses, result in arching, which affects loads on support systems; in soils, deformations may result in ground surface subsidence.

   Pavements undergo settlement or heave, and bending.

Elastic Body Characteristics

Elastic Media

In an elastic medium, strain is instantly and totally recoverable. Stress is directly proportional to strain as related by Hooke's law, expressed by *Young's modulus E* (or *modulus of elasticity*) as

$$E = \sigma / \varepsilon \text{ (in tsf, kPa)} \tag{2.56}$$

Poisson's ratio v is the inverse ratio between strain in the direction of applied stress $E\sigma$ and the induced strain εL in a perpendicular direction:

$$v = \varepsilon L / \varepsilon = \varepsilon L / E\sigma \tag{2.57}$$

Young's modulus and Poisson's ratio are referred to as the elastic constants.

Dependent Factors for Elasticity

For a body to exhibit elastic characteristics defined by only a single value of E and v it must be isotropic, homogeneous, and continuous.

   Isotropic: Particles are oriented so that the ratio of stress to strain is the same regardless of the direction of applied stress, i.e., the elastic properties in every direction through any point are identical.

   Homogeneity: The body has identical properties at every point in identical directions.

   Continuity: Refers to structure; if a mass is continuous it is free of planes of weakness or breaks.

Validity for Geologic Materials

Sound, intact, massive rock approaches an elastic material under most stress levels prior to rupture.

   Generally, *most rocks* are, to some extent, anisotropic, nonhomogeneous, and discontinuous, and are termed as quasielastic, semielastic, or nonelastic. In intact rock specimens, deformation varies with the rock type as shown in Table 2.24 with regard to mineral hardness, grain bonding, and fabric. Nonintact rock or rock masses are basically discontinuous, usually undergoing plastic deformation as fractures close, then elastic deformation, often followed by plastic deformation.

TABLE 2.24

Deformation Characteristics of Intact Rock in Uniaxial Compression to Failure[a]

Deformation Stress vs. Strain	Rock Type	Failure Form
I. Elastic	Fine-grained, massive, Basalt, quartzite, diabase dolomite, some strong limestones	Sudden, explosive, brittle
II. Elastic–plastic	Fine-grained sedimentary. Softer limestones, siltstone, tuff	Plastic yielding
III. Plastic–elastic	Sandstone, granite, some diabases, schist cored parallel to foliation	Brittle fracture — stiffness increases as microfissures close
IV. Plastic–elastic–plastic	Metamorphic — marble, gneiss	Inelastic yielding — stiffness increases as microfissures close
V. Plastic–elastic-plastic	Schist cored perpendicular to foliation	Inelastic yielding — stiffness increases as microfissures close
VI. Elastic–plastic creep	Rock salt	Inelastic yielding and continuous creep

[a] After Miller, R.P., Ph.D. thesis, University of Illinois, Urbana, 1965.

Creep deformation occurs in some rocks over long time intervals at stress levels substantially less than those required to cause short-term deformation or failure. In rock masses, the problem has most practical significance in soft rocks such as halite or in overstressed rocks (high residual stresses), where relaxation occurs along joints (Hendron, 1969).

Soils are essentially nonelastic, usually plastic, and occasionally viscous. They demonstrate pseudo-elastic properties under low stress levels, as evidenced by initial stress-strain linearity. "Elastic" deformation, however, is immediate deformation, and in many soil types does not account for the total deformation occurring over long time intervals because of consolidation.

Elastic Constant Values

Typical values for the elastic constants for a variety of materials are given in Table 2.25. Typical values for various rock types are also given in Figure 2.92.

Poisson's ratio for soils is evaluated from the ratio of lateral strain to axial strain during a triaxial compression test with axial loading. Its value varies with the strain level and becomes constant only at large strains in the failure range (Lambe and Whitman, 1969). It is generally more constant under cyclic loading: in cohesionless soils it ranges from 0.25 to

TABLE 2.25

Typical Ranges for Elastic Constants of Various Materials[a]

Material	Young's Modulus Es,[b] (tsf)	Poisson's Ratio v[c]
Soils		
Clay		
Soft sensitive	25–150	
Firm to stiff	150–500	0.4–0.5
Very stiff	500–1000	(undrained)
Loess	150–600	0.1–0.3
Silt	20–200	0.3–0.35
Fine sand:		
Loose	80–120	
Medium dense	120–200	0.25
Dense	200–300	
Sand		
Loose	100–300	0.2–0.35
Medium dense	300–500	
Dense	500–800	0.3–0.4
Gravel		
Loose	300–800	
Medium dense	800–1000	
Dense	1000–2000	
Rocks		
Sound, intact igneous and metamorphics	$6\text{–}10 \times 10^5$	0.25–0.33
Sound, intact sandstone and limestone	$4\text{–}8 \times 10^5$	0.25–0.33
Sound, intact shale	$1\text{–}4 \times 10^5$	0.25–0.30
Coal	$1\text{–}2 \times 10^5$	
Other Materials		
Wood	$1.2\text{–}1.5 \times 10^5$	
Concrete	$2\text{–}3\,3 \times 10^5$	0.15–0.25
Ice	7×10^5	0.36
Steel	21×10^5	0.28–0.29

[a] Lambe and Whitman (1969).

[b] E_s (soil) usually taken as secant modulus between a deviator stress of 0 and ½ to 1/3 peak deviator stress in the triaxial test (Figure 2.70) (Lambe and Whitman, 1969). E_r (rock) usually taken as the initial tangent modulus (Farmer, 1968) (*Section 2.5.2*). E_s (clays) is the slope of the consolidation curve when plotted on a linear $\Delta h/h$ vs. p plot (CGS, 1978).

[c] Poisson's ratio for soils is evaluated from the ratio of lateral strain to axial strain during a triaxial compression test with axial loading. Its value varies with the strain level and becomes constant only at large strains in the failure range (Lambe and Whitman, 1969). It is generally more constant under cyclic loading; cohesion less soils range from 0.25 to 0.35 and cohesive soils from 0.4 to 0.5.

Note: kPa = 100 tsf.

Source: From Lambe, T.W. and Whitman, R.V., Soil Mechanics, Wiley, New York, 1969. Reprinted with permission of Wiley.

0.35 and in incohesive soils from 0.4 to 0.5. Values above 0.5 can indicate dilatant material, i.e., the lateral strain under applied vertical stress can exceed one half the vertical strain.

Induced Stresses

Evaluation of deformation under imposed loads requires the determination of the magnitude of stress increase (or decrease) at some depth below a loaded (or unloaded) area with respect to the existing *in situ* stress conditions. The determination of these stresses is based on elastic theory.

Equations were presented by Boussinesq in 1885 expressing stress components caused by a perpendicular point surface load at points within an elastic, isotopic, homogeneous mass extending infinitely in all directions below a horizontal surface, as shown in Figure 2.67. The stresses are directly proportional to the applied load and inversely proportional to the depth squared (a marked decrease with depth).

Distributions of stresses beneath a uniformly loaded circular area are given in Figure 2.68 and have been determined for other shapes, including square and rectangular areas, and uniform and triangular loadings (Lambe and Whitman, 1969). Distributions vary from the Boussinesq with thinly layered soil formations as given by the Westergaard distribution (Taylor, 1948). Other stress distributions include layered soil formations (strong over weak layer [Burmister 1962b], rigid base under weak layer [Burmister 1956], and jointed rock masses [Gaziev and Erlikhman, 1971]).

Although geologic materials are essentially nonelastic, the theory is considered applicable for relatively low stress levels, although the magnitude and distribution of stresses are significantly affected by soil layering and fracture patterns in rock masses.

2.5.2 Deformation Relationships

Deformation Moduli

General

Deformation moduli are expressed in a number of ways that relate to either the form of deformation or the stress–strain relationships obtained from a particular test method, as summarized in Table 2.26.

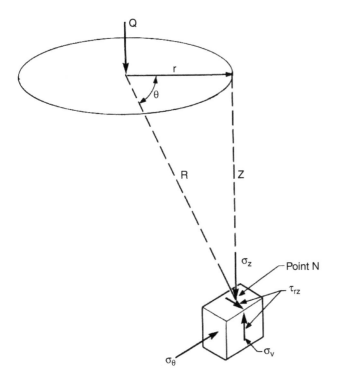

$$\sigma_z = \frac{Q}{z^2}\frac{3/2\pi}{[1+(r/z)^2]^{5/2}}; N_B = \frac{3/2\pi}{[1+(r/z)^2]^{5/2}}; \sigma_z = \frac{Q}{z^2}N_B$$

FIGURE 2.67
Stresses in an elastic medium caused by a surface, vertical point load.

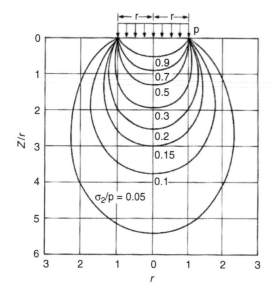

FIGURE 2.68
The "pressure bulb" of stress distribution: contours of vertical normal stress beneath a uniformly loaded circular area on a linear elastic half-space.

Elastic Moduli

Young's modulus, shear modulus, bulk modulus, and constrained modulus are defined in Figure 2.69. In general, settlement may be expressed crudely in terms of elastic moduli as

$$S = \Delta H = \Delta p \, (H/E) \text{ (in, cm)} \qquad (2.58)$$

where Δp is the pressure increase over effective overburden pressure at the midpoint of the "pressure bulb" (zone of influence or stress distribution zone beneath the loaded area as shown in Figure 2.68, H the depth of pressure bulb upto some arbitrary boundary such as $\sigma_z/p = 0.05$ (insignificant stress), or the stratum thickness, if less and $E = \sigma/\varepsilon$ as determined by *in situ* or laboratory testing with $\varepsilon = \Delta L/L$; in *in situ* testing ΔL is measured but L is not known; it is estimated, on the basis of Boussinesq stresses, as the depth to which the significant stress occurs.

  Note: If Δp is applied over a large area compared with the stratum thickness, the constrained modulus (D) is used to account for the effect of confining pressures (Figure 2.69d).

Dynamic Elastic Moduli

Compression and shear wave velocities of sonic waves (see Section 1.3.2) are functions of the elastic properties and mass density of the transmitting medium. Measurements of V_p and V_s, therefore, provide the basis for computing the dynamic properties of Poisson's ratio, Young's modulus, shear modulus, and bulk modulus as given in Table 2.27.

Moduli from the Stress–Strain Curve

Derived from the triaxial compression or uniaxial compression test, these moduli include the initial tangent modulus, tangent modulus, and secant modulus defined in Figure 2.70. The initial tangent modulus E_i is usually reported as the elastic modulus for rock (given as E_r), whereas the secant modulus E_{se} is usually reported as the elastic modulus for soils, E_s, to be used for calculating deformations. The secant modulus E_{se} is used also for defining a stress or limiting deformation.

TABLE 2.26

Parameters of Deformation

Parameter[a]	Reference	Definition		Normal Application
Elastic Moduli (ST) Expression				
Young's modulus	Figure 2.69	Relates stress to strain	$E = \sigma/\varepsilon$	Rock masses, sands, strong granular cohesive soils
Shear modulus	Figure 2.69	Relates shear strain to shear force (modulus of rigidity)	$G = \dfrac{E}{2(1+v)}$	Not commonly used statically
Bulk modulus	Figure 2.69	Ratio of all-around pressure to change in volume per unit volume (modulus describing incompressibility)	$B = \dfrac{E}{3(1-2v)}$	Not commonly used statically
Constrained modulus	Figure 2.69	Deformation occurring in confined compression	$D = \dfrac{E(1-v)}{(1+v)(1-2v)}$	Structure of large areal extent underlain by relatively thin compressible soil deposit
Dynamic Elastic Parameters — See Table 2.27				
Moduli From The Stress–Strain Curve				
Initial tangent modulus	Figure 2.70	Initial portion of curve	$E_i = \sigma/\varepsilon$	Usually taken as E for geologic materials
Secant modulus	Figure 2.70	σ and ε taken between two particular points	$E_{se} = \dfrac{\Delta\sigma}{\Delta\varepsilon}$	Define E for a particular stress limit
Tangent modulus	Figure 2.70	Modulus at specific point. Often taken at point of maximum curvature before rupture	$E_t = \dfrac{d\sigma}{d\varepsilon}$	The lowest value for E usually reported
Other Moduli				
Compression modulus	Figure 2.87	Lateral deformation caused by applied stress from pressure meter	$E_c = E_\alpha$ (tons/ft³)	*In situ* measures of E for materials difficult to sample
Modulus of subgrade reaction	Figure 2.90	A unit of pressure to produce a unit of deflection	$k_s = p/y$	Beam or plate on an elastic subgrade problems

a Unless noted, units are tsf, kg/cm², kN/m².

Other Moduli

These include the compression modulus, as determined by pressuremeter testing, which varies in its relationship to Young's modulus with soil type and the modulus of subgrade reaction, measured by plate load test.

Plastic Deformation

Compression in Clays

Compression in clay soils is essentially plastic deformation defined in terms of consolidation, i.e., the decrease in volume occurring under applied stress caused primarily by the expulsion of water from the interstices (see Section 2.5.4). Most soils exhibit primary and

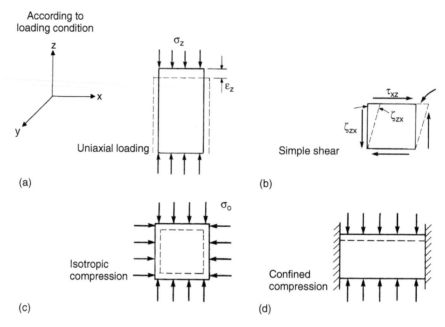

FIGURE 2.69

Various types of elastic moduli: (a) Young's modulus — $E = \sigma_z / \varepsilon_z = \sigma / \varepsilon; \varepsilon_x = \varepsilon_y = -v\varepsilon_x$; (b) shear modulus — $G = \tau_{zx}/\zeta_{zx} = E/2(1 + V)$; (c) bulk modulus — $B = \sigma_0/3\varepsilon_x = E / 3(1 - 2V)$; (d) constrained modulus $-D = \sigma_z/\varepsilon_z = E(1 - v) / (1 + v)(1 - 2v)$. (From Lambe, T. W. and Whitman, R. V., *Soil Mechanics*, Wiley, New York, 1969. Reprinted by permission of John Wiley & Sons, Inc.)

TABLE 2.27

Dynamic Elastic Parameters

Parameter	Expression
Compression-wave velocity	$V_p = \{[K + (4/3)G]/\rho\}^{1/2}$ m/s (*Section 2.3.2*)
Shear-wave velocity	$V_s = (G/\rho)^{1/2}$ m/s (*Section 2.3.2*)
Mass density of materials	$\rho = \gamma/g$ kg/m^3 (determined by gamma probe, *Section 2.3.6*)
Dynamic Poisson's ratio	$v = (V^2_p/2V^2_s - 1)/(V^2_p/V^2_s - 1)$ *Appropriate values*
	Igneous rocks — 0.25
	Sedimentary rocks — 0.33
	Soils — *see Section 11.3.2*
Dynamic Young's modulus[a]	$E_d = p(3V^2_p - 4\,V^2_s)/(V^2_p/V^2_s - 1)$ or $E_d + 2p\,V^2_s(1 + m)$
Dynamic shear modulus[a]	$G_d = pV^2_s = E^d/2(1 + v)$
Dynamic bulk modulus[a]	$K = p(V^2_p - 4V^2_s/3) = E_d/3(1 - 2v)$

Note: Units are tsf, kg/cm^2, kN/m^2.

secondary consolidation and, in some soils, the latter phenomenon, which is not well understood, can be of very significant magnitude. A substantial time delay in compression occurs in clay soils under a given applied stress that increases generally as the plasticity of the clay increases.

Compression in Sands

Sands and other cohesionless granular materials undergo a decrease in void volume under applied stress, caused primarily by the rearrangement of grains (Section 2.5.4). Small elastic

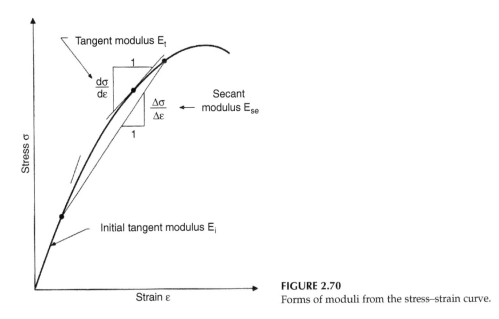

FIGURE 2.70
Forms of moduli from the stress–strain curve.

compression of quartz grains may occur. In most cases, the greater portion of compression is essentially immediate upon the application of load.

Expansion

An increase in volume occurs as a result of reduction in applied stress, increase in moisture content, or mineralogical changes in certain soil and rock materials (see Section 2.5.4).

2.5.3 Rock Deformation Measurements

Methods Summarized

Laboratory Testing

Intact specimens are statically tested in the laboratory in the triaxial and unconfined compression apparatus; dynamic properties are measured with the resonant column device or by ultrasonic testing (ASTM 2845). A summary of parameters measured, apparatus description, and test performance for intact rock specimens is given in Table 2.28. The data are normally used for correlations with *in situ* test data.

In Situ Testing

In situ testing provides the most reliable data on the deformation characteristics of rock masses because of the necessity of accounting for the effects of mass defects from discontinuities and decomposition.

The determination of moduli *in situ* requires that the deformation and the stress producing it are measurable and that an analytical method of describing the geometry of the stress deformation relationship is available. Analytical methods are governed by the testing method. Modulus is the ratio of stress to strain, and since strain is the change in length per total length, the deflection that is measured during *in situ* testing must be related to the depth of the stressed zone to determine strain. The depth of the stressed zone may be determined by instrumentation (see Chapter 3), or the Boussinesq equations may

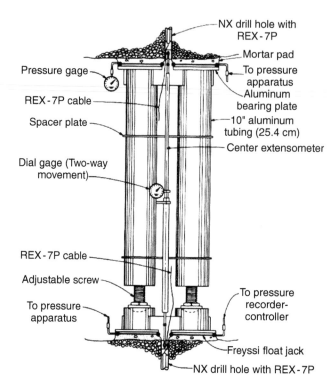

FIGURE 2.71
Arrangement for uniaxial jacking test in an adit. (From Wallace, B. et al., STP 477, American Society for Testing & Materials, Philadelphia, PA, 1970, pp. 3–26. Reprinted with permission of the American Society for Testing and Materials.)

be used to determine stress distributions. The values for the modulus E are given in terms of the test geometry, the applied pressure, the deflection, and Poisson's ratio.

Static moduli are determined from plate-jack tests, radial jacking and pressure tunnel tests, flat-jack tests, borehole tests (dynamometer and Goodman jack), and triaxial compression tests.

Dynamic moduli are determined from seismic direct velocity tests (see Section 1.3.2) and the 3-D velocity probe (sonic logger) (see Section 1.3.6). Relationships between seismic velocities and dynamic moduli are given in Table 2.27. In moduli computations, the shear wave velocity V_s is used rather than the compression-wave velocity V_p because water in rock fractures does not affect V_s, whereas it couples the seismic energy across joint openings, allowing much shorter travel times for P waves than if an air gap existed. Dynamic moduli are always higher than static moduli because the seismic pulse is of short duration and very low stress level, although the ratio of E_{static} to E_d will normally approach unity as rock-mass quality approaches sound, intact rock. E_d as determined from field testing is often referred to as E_{seis} and is correlated with other field and laboratory data to obtain a design modulus, as will be discussed.

Plate-Jack Test (In Situ Compression Test)

Performance

A load is applied with hydraulic jacks to a plate in contact with the rock mass using the roof of an adit as a reaction, as illustrated in Figure 2.71.

When the tests are set up, areas representative of rock-mass conditions are selected, and the size of the loaded area is scaled to the structural elements of the rock. (Depending on mass-defect spacings, the larger the loaded area the more representative are the results). Borehole extensometers (see Section 3.3.4) are grouted into the rock mass for

TABLE 2.28

Laboratory Tests to Determine Deformation Moduli of Intact Rock

Test Method	Parameters Measured	Apparatus	Performance	Comments
		Static Moduli		
Triaxial compression apparatus	E_i, E_{se}, E_t	See Figure 2.49 and Section 2.4.4. Rock testing equipment similar but larger. USBR machine can test cores of $D = 15$ cm and $L = 30$ cm under 800 tons axial load and 6000 kg/cm² lateral pressure	Specimen subjected to confining pressure and loaded axially to failure to obtain a stress–strain curve. Test repeated at various confining pressures to bracket *in situ* lateral pressure. Strain rate can be varied to suit field loading conditions	Modulus values controlled by orientation of applied load in rocks with bedding planes, foliations, schistosity, as well as major joints
Unconfined compression	E_i, E_{se}, E_t	See Section 2.4.4. Can use concrete testing apparatus	Specimen loaded to failure as stress–strain curve is recorded	Yields very conservative values for most practical uses. Often directly applicable for mine pillars
		Dynamic Moduli		
Resonant column device (sonic column)	E_d, G_d	Core specimen held fixed at the ends and subjected to vibrations while axial load applied equivalent to overburden (see also ASTM D2845)	Specimen subjected first to vibrations in the torsional mode, then in the longitudinal mode. End displacement is monitored for various frequencies	From specimen length and end displacement as a function of frequency. V_s is computed from the torsional test, and V_p ("rod" compression-wave velocity) from the longitudinal test

measurements of strain vs. depth beyond the upper and lower plates, and concrete facing is poured over the rock mass to provide flat-bearing surfaces. As load is applied, the extensometers sense compression strains that are recorded on electronic readouts.

See also ASTM D4394 and D4395 (Plate load tests).

Computing Moduli

Theoretical basis is the Boussinesq solution. The Young's modulus E_r and surface displacement y are related to the applied load P (Jaeger, 1972) as

$$y = P (1 - v)^2 / \pi E_r r \tag{2.59}$$

where r is the plate radius.

Moduli E_i, E_{se}, and E_t are determined from the stress–strain curve. A recovery modulus E_{sr} can be taken from the unload portion of the curve to provide a measure of rock elasticity. If $E_t = E_{sr}$ the material is perfectly elastic. A value of E_{se} much lower than E_t usually indicates the closure of fractures and plastic deformation.

Test Limitations

Rock-mass response is affected by disturbance during excavation and surface preparation for testing. It is also affected by the relatively small stressed zone, often to the degree that results are not representative of response under construction loading, although the results will usually be conservative (Rocha, 1970).

Radial Jacking and Pressure Tunnel Tests

Performance

Radial jacking test: The pressure to cause deformation of the wall of an adit excavated into the rock mass is applied mechanically. Displacements are measured with extensometers or other strain-measuring devices.

Pressure tunnel test: A portion of the adit is sealed with concrete and water is pumped in under high pressure to cause deformation. Displacements are measured with extensometers or some other strain-measuring device. Flow into the rock mass may cause large errors from pressure drop.

Computing Moduli

Analysis is based on the Boussinesq stress distribution and theories of stresses and deflections about a circular hole, or a hole cut in a plate (Wallace et al., 1970; Misterek, 1970).

Comments

Although both tests are relatively costly to perform, larger areas are stressed than is possible with plate-jack tests, which allows a better assessment of rock-mass deformation. Measurements of differential wall movements permit the assessment of the anisotropic properties of the rock mass. Excavation, however, results in straining of the rock mass, which affects its properties.

Flat-Jack Tests

Performance

A slot is cut into rock with a circular diamond saw or by a series of line-drilled holes. A steel flat jack Freyssinet jack (Figure 2.72) is grouted into the slot to ensure uniform

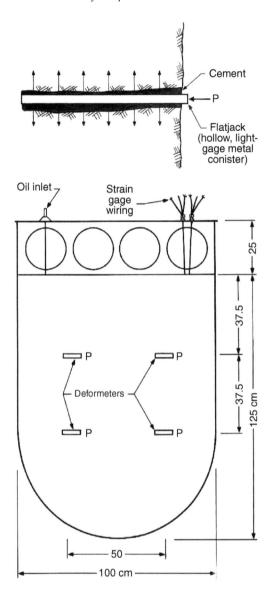

Cement

P

Flatjack
(hollow, light-
gage metal
conister)

Oil inlet

Strain
gage
wiring

25

37.5

37.5

125 cm

Deformeters

50

100 cm

FIGURE 2.72
The flat jack or Freyssinet jack used in slots sawed
into rock. (From Rocha, M., ASTM STP 477,
American Society for Testing and Materials,
Philadelphia, PA, 1970. Reprinted with permission
of the American Society for Testing and Materials.)

contact with the rock face. Jacks measuring 1 m in width by 1.25 m in depth and capable
of applying pressures of 100 kg/cm² (100 tsf) have been used (Rocha, 1970). Strain meters
measure the increase in distance between points on opposite sides of the slot as pressure
is applied (see also Section 3.4.4). Jacks at different orientations provide a measure of rock
anisotropy.

Computing Moduli

The increase in distance between points in the rock arranged symmetrically on opposite sides
of the slot can be related to the pressure in the jack by elastic theory. E_r is determined from

$$E_r = c(P/\delta) \tag{2.60}$$

where P is the pressure in the flat jack, c a laboratory constant for the jack and test geometry and δ the measured increase in distance between points.

The constant c depends on the position, shape, and size of the area under pressure as well as the location of the measured points, and has the dimension of length. It may be determined by model tests in large plaster blocks for a given jack and strain gage arrangement.

Borehole Jack Tests

Apparatus

The dilatometer (Figure 2.73) and the Goodman jack (Goodman et al., 1968), devices similar to the pressuremeter (see Section 2.5.4), are lowered into boreholes to measure moduli *in situ*.

Hydraulic pressure is applied between a metal jacket and a deformable rubber or metallic jacket in the apparatus which presses against the borehole walls. Linear variable differential transducers (LDVTs) in the instrument allow the measurement of four diameters, 45° apart, to account for lateral rock anisotropy. Designed to operate underwater in NX-size boreholes, Interfel dilatometers upto 6-in. in diameter are available. Dilatometers measure Young's modulus up to about 25 GPa and the Goodman jack up to about 50 GPa.

Computing Moduli

Rocha (1970) gives the following expression for computing E_r:

$$E_r = 2\,r(1 - v/\Delta)\,P \tag{2.61}$$

where $2\,r$ is the hole diameter, v the Poisson's ratio, and Δ the deformation when pressure P is applied

Methods Compared

A series of tests was performed in jointed granite for the Tokyo Electric Power Company using a number of methods (Hibino et al., 1977) including *in situ* triaxial testing, and the results were compared. The *in situ* triaxial compression test was performed on a 1m³ block of rock carved out of the rock mass but with its base remaining intact with the mass. The block was surrounded with 1500-ton-capacity flat jacks to provide confinement, the space between the rock and the jacks was backfilled with concrete, and loads were applied to the rock by the jacks.

Moduli values measured by the various tests are summarized in Figure 2.74. A wide range in values is apparent. The largest values were obtained from laboratory sonic column and uniaxial compression tests, which would not be expected to be representative of mass conditions. The larger plate-jack test resulted in less deflection (higher modulus) than did the smaller plate and would be considered as more representative of deformation characteristics, as would the water chamber test.

Modulus and Rock–Quality Relationships

Selection of Design Moduli

Deformation is related to rock-mass quality, which, because of fabric, discontinuities, and decomposition, can be extremely variable. Most practical problems involve poor-quality rock, which is the most difficult to assess.

The various test methods, in general, individually often do not yield representative values for E_r. All static methods disturb the rock surface during test setup, and most stress a

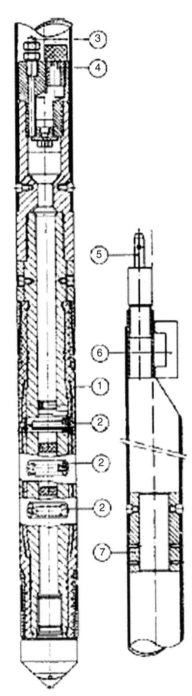

FIGURE 2.73
Simplified section of the dilatometer probe IF 096. (Courtesy of Interfels.)
1 — Reinforced rubber sleeve;
2 — Displacement transducers;
3 — Pressure line;
4 — Electric cable plug;
5 — Protection tube with; connection for string of setting rods;
6 — Hook for drilling rig rope;
7 — Connection.

relatively small area. Dynamic tests apply short-duration pulses at very low stress levels. Laboratory tests are performed on relatively intact specimens, seldom providing a representative model except for high-quality rock approaching an elastic material. Various investigators have studied relationships between rock quality and moduli as determined by various methods and correlative procedures in the attempt to provide bases for establishing design moduli.

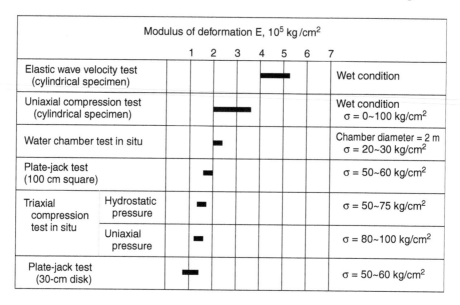

FIGURE 2.74

Deformation moduli obtained by various types of tests in a jointed granite. (From Hibino, S. et al., *Proceedings of the International Symposium on Field Measurements in Rock Mechanics*, Zurich, Vol. 2, 1977, pp. 935–948. With permission.)

Measuring Rock Quality

Rock-quality designation (RQD) as determined from core examination (see Section 1.4.5) provides an index to rock quality.

Modulus ratio E_{t50}/U_{ult} is defined by Deere and Miller (1966) as the ratio of the tangent modulus E_{t50} taken at a stress level of one half the ultimate strength, to the uniaxial compressive strength U_{ult} as obtained by laboratory test. It was selected as a basis for the engineering classification of intact rock because it is related to deformation and strength and provides a measure of material anisotropy. Low modulus values result from large deformations caused by the closure of foliation and bedding planes; high modulus values are representative of material with interlocking fabric and little or no anisotropy.

Modulus ratio E_r/E_{t50} is defined (Coon and Merritt, 1970) as the ratio of the *in situ* static modulus E_r to the intact static modulus E_{t50}.

Velocity ratio V_F/V_L is defined (Coon and Merritt, 1970) as the ratio of the field P-wave velocity, as determined by uphole or crosshole seismic methods or the 3-D velocity logger, to the laboratory P-wave velocity as determined by the resonant column device or ultrasonic testing (ASTM 2845). The velocity ratio approaches unity in high-quality massive rocks with few joints.

Velocity index (V_F/V_L) is defined (Coon and Merritt, 1970) as the square of the ratio of the field seismic velocity to the laboratory seismic velocity, or the square of the ratio of the *in situ* velocity to the intact velocity. The velocity ratio is squared to make the velocity index equivalent to the ratio of the dynamic moduli.

Modulus Reduction Factor β

Proposed by Deere et al. (1967), the modulus reduction factor β expresses the extent to which E_r is always lower than E_{seis} because of the short duration pulse and low stress level of the *in situ* seismic test. It is given as

$$\beta = E_r/E_{seis} \qquad (2.62)$$

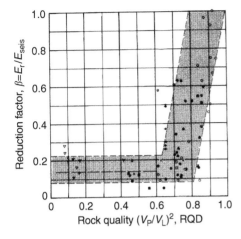

FIGURE 2.75
Variation in modulus reduction factor β with rock quality [E_r = static rock modulus (from load tests); E_{sets} = modulus from seismic velocity; V_p = field seismic velocity (compressional); V_L = laboratory sonic velocity of sound cores] (From Deere, D. U. et al., *Proceedings of the 8th Symposium on Rock Mechanics*, University of Minnesota, American Minerals and Metal Engineers, 1967, Chap. 11. With permission.)

β approaches unity as the rock quality approaches an elastic material. A relationship among β, RQD, and the velocity index is given in Figure 2.75. The reduction factor is applied to E_{seis} to arrive at a design value for E_r. Note that below an RQD = 60% (poor to very poor rock quality) the correction does not exceed 20% of E_{seis}. The application of such relationships requires substantial judgment and experience, and should be developed for a particular site on important projects.

2.5.4 Soil Deformation Measurements (Static)

Methods Summarized

Laboratory Testing

Undistorted specimens are statically tested in the laboratory in the triaxial compression apparatus to obtain measures of moduli (E_s or E_t, E_i, and E_{se}), and in the consolidometer to obtain measures of consolidation and expansive properties. The dynamic deformation moduli (E_d, G_d, and K) are usually measured in the cyclic triaxial, cyclic simple shear, or resonant column devices. Soils suitable for testing undisturbed samples include intact clays and cohesive granular soils containing a minimum of fine gravel and coarser particles.

Specimens of sand may be reconstituted at various values of D_R and tested in the consolidometer for static compression characteristics, or in the shaking table device or other devices for dynamic properties.

In Situ Testing

Materials in which undisturbed sampling is difficult or impossible, such as cohesionless sands and gravels, or materials with coarse particles such as some residual soils, decomposed rock, or glacial till, and fissured clays, are tested *in situ* to measure deformation characteristics. Tests include:

- Pressuremeter tests for compression modulus.
- Penetration tests (SPT) for correlations for estimating moduli.
- Penetration tests (CPT, DMT) for estimating consolidation coefficients.
- Full-scale load tests for the compression of sands, organic soils, and other materials.
- Plate-load test to measure the vertical modulus of subgrade reaction.
- Lateral pile-load tests to measure the horizontal modulus of subgrade reaction.
- Dynamic tests to measure shear modulus or peak particle velocity.

Consolidation Test for Clays

General

Purposes: The laboratory consolidation test is performed to obtain relationships for compression vs. load and compression vs. time for a given load.

Validity: The one-dimensional theory of consolidation of Terzaghi (1943) is valid for the assumptions that strains are small, the soil is saturated, and flow is laminar.

Apparatus: The consolidometer (oedometer), a one-dimensional test, includes a ring to contain the specimen, a small tank to permit sample submergence, a device for applying loads in stages, and a dial gage for deflection measurements as illustrated in Figure 2.76 and Figure 2.77. In the apparatus shown, loads are applied by adding weights to a loading arm. Some machines apply loads hydraulically through a bellows arrangement; these apparatus are often designed to permit application of back-pressure to the specimen to achieve saturation, or to permit the measurement of pore pressures. Compression and loads are monitored and recorded digitally with hydraulically loaded consolidometers.

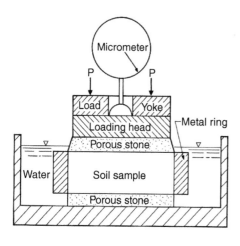

FIGURE 2.76
The floating-head-type consolidometer.

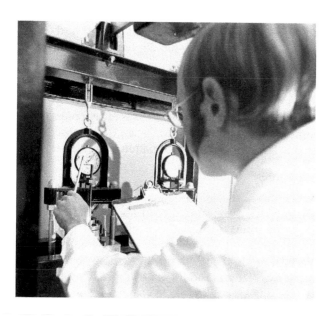

FIGURE 2.77
Reading deflection vs. time with the consolidometer.

Procedures (ASTM D2435-02, D4186-98)

A specimen from an undisturbed clay sample is trimmed carefully into a rigid ring (brass for most soils; Teflon for organic soils because of corrosion), and porous stones are placed on top and bottom to permit vertical drainage. (A variation to permit horizontal drainage uses a ring of porous stone and solid-end platens.)

The assembly is placed in a loading frame and subjected to a sequence of loads starting initially with very small loads. Then (normally) the load is increased by doubling until the test load significantly exceeds the anticipated field load, and the "virgin" compression curve and preconsolidation stress (p_c) have been defined (Figure 2.78). Usually, three loads beyond the range of p_c are required to define the virgin compression.

After the initial seating loading, the specimen is immersed in water to maintain saturation (unless the clay is expansive, as will be discussed). Each load remains until pore pressures are essentially dissipated and consolidation terminated; for most clays, 24 h is adequate for each load cycle. When first applied, the load is carried by the pore water. As water drains from the specimen, the voids close, the soil compresses, and the strength increases until it is sufficient to support the load and the extrusion of water ceases. At this point "primary compression" has terminated and the next load increment is applied (see the discussion of secondary compression below.) Measurements are made and recorded during the test of deflection vs. time for each load increment. (Plotting the data during the test shows the experienced technician when full consolidation under a given load has occurred, at which point, the next load is applied. This practice eliminates the need to wait 24 h between loadings. If pore-water pressures are measured, when they reach zero the next load increment is applied.)

Pressure vs. Void Ratio Relationships

The total compression that has occurred after the application of each load is plotted to yield a curve of either pressure vs. strain or pressure vs. void ratio (e–log p curve), where pressure is plotted on a log scale, and strain or void ratio on an arithmetic scale (Figure 2.78).

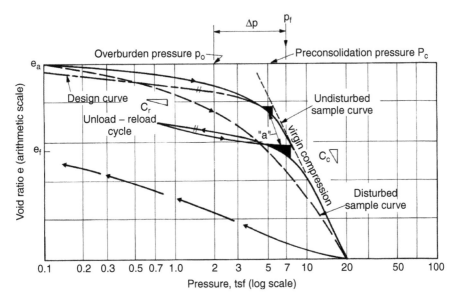

FIGURE 2.78
Consolidation test: pressure–void ratio curve (e–log p).

Curve shape significance: A truly undisturbed specimen will yield a curve with a relatively flat initial portion to the range of prestress, then a distinct change in slope near the maximum p_c, and thereafter a steep drop under loads causing virgin compression. Disturbed samples or samples with a large granular component yield a curve with a gradual change in slope and a poorly defined p_c as shown in Figure 2.78.

Methods to determine p_c: The Casagrande construction (Casagrande, 1936) is given in Figure 2.79 and the Burmister construction (Burmister, 1951b) in Figure 2.78. In the Burmister construction, a triangle from the unload–reload cycle is moved upward along the e–log p curve until the best fit is found, where the pressure is taken to be p_c. The method compensates for slight disturbances during sampling and laboratory handling, which cause the initial curve to be steeper than *in situ.*

Settlement analysis curve correction: Unload–reload cycles, usually made soon after p_c has been identified, provide data for the correction of the settlement curve to reduce conservatism caused by the slight sample expansion occurring during sampling and handling. The recompression portion of the unload–reload curve is extended back from the initial curve at p_c to form a new curve with a flatter initial slope.

Stress–Strain Relationships

When a series of increasing loads is applied to a clay soil, the amount of compression occurring up to the magnitude of the maximum past pressure or preconsolidation pressure p_c is relatively small to negligible. Stresses exceeding p_c enter the "virgin" portion of a stress–strain curve, and substantially greater compression occurs. It may be expressed by the *compression index* C_c, the slope of the virgin compression or void ratio vs. $\log_{10}$ time curve expressed as

$$C_c = (e_1 - e_2)/ \log(p_1/p_2) \tag{2.63}$$

C_c is useful for correlating data for normally consolidated clays of low to moderate sensitivity, or for computing settlements when data from a number of consolidation tests are available. It can be estimated from the expression (Terzaghi and Peck, 1967)

$$C_c \approx 0.009(LL - 10\%) \tag{2.64}$$

where LL is the liquid limit.

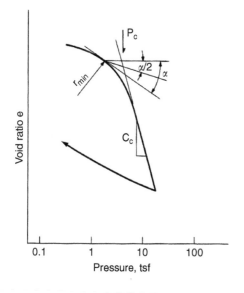

FIGURE 2.79
Casagrande construction for determining p_C. (From Casagrande, A., *Proceedings of the 1st Conference on Soil Mechanics and Foundation Engineering*, Cambridge, MA, Vol. 3, 1936, p. 60. With permission.)

Overconsolidation ratio (OCR) is defined as the ratio of the maximum past pressure p_c to the existing effective overburden pressure p'_o. In general, for NC clay, $p_c/p'_o = 0.8$ to 1.5 and, for OC clay, $p_c/p'_o > 1.5$ (Clemence and Finbarr, 1980). OCR is used to estimate consolidation in clays (Schmertmann, 1977) for correlation of strength properties (Terzaghi and Peck, 1967) and for estimating K_o in terms of PI (Brooker and Ireland, 1965). K_o as a function of OCR and PI is given in Figure 2.80.

The undrained shear strength s_u of NC clay normally falls within a limited fraction of the effective overburden stress, usually ranging from $s_u/p' = 0.16$ to 0.4 (Section 2.4.2). It can be estimated as an average value of

$$(s_u/p')NC = 0.33 \tag{2.65}$$

or in terms of PI (Terzaghi and Peck, 1967) as

$$s_u/p' = 0.11 + 0.0037PI \tag{2.66}$$

Typical e–log p curves for various soil types are given in Figure 2.81. They serve to illustrate that soils of various geologic origins have characteristic properties.

Compression vs. Time Relationships

Significance: The e–p curve provides an estimate of the compression occurring at 100% consolidation. In practice, it is important to estimate the amount of settlement that will occur under a given stress increment in some interval of time (end of construction, 15 years). The time rate of consolidation is analyzed from the compression vs. log time curve for a particular load increment (Figure 2.82).

Curve characteristics: The curve is divided into two portions for analysis: primary consolidation occurs while the excess pore pressures dissipate and consolidation proceeds in accordance with theory; secondary consolidation is a slow, continuing process of compression beyond primary consolidation after the excess pore pressures have been dissipated. The phenomenon is not clearly understood.

Primary consolidation: The time t to reach a given percent consolidation U is expressed as

$$t = (T_v/c_v)\,H^2 \tag{2.67}$$

where T_v is the theoretical time factor.

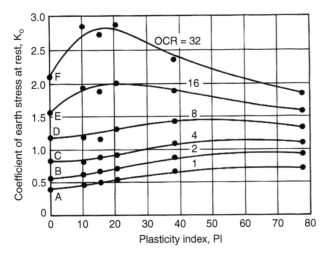

FIGURE 2.80
Relationship between K_0, OCR, and PI. (From Brooker, E. W. and Ireland, H. O., *Can. Geotech. J.* II, 1965. Reprinted with permission of the National Research Council of Canada.)

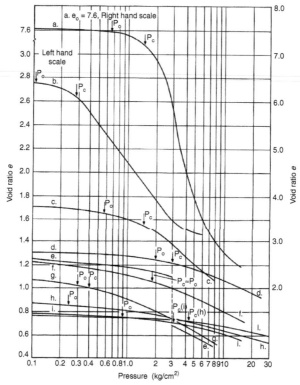

Material	γd	w	LL	PI	Material	γd	w	LL	PI
a. Soft silty clay (lacustrine, Mexico)	0.29	300	410	260	f. Medium sandy, silty clay (residual gneiss, Brazil)	1.29	38	40	16
b. Soft organic silty clay (alluvium, Brazil)	0.70	92			g. Soft silty clay (alluvium, Texas)		32	48	33
c. Soft organic silty clay (backswamp, Georgia)	0.96	65	76	31	h. Clayey silt, some fine sand (shallow alluvium, Georgia)	1.46	29	53	24
d. Stiff clay varve (glaciolacustrine, New York)		46	62	34	i. Stiff clay (Beaumont clay, Texas)	1.39	29	81	55
e. "Porous" clay (residual, Brazil)	1.05	32	43	16	j. Silt varve (glaciolacustrine, New York)				

FIGURE 2.81
Typical pressure–void ratio curves for various clay soils.

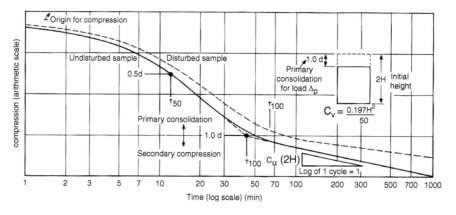

FIGURE 2.82
Compression vs. time for one load cycle.

H is the one half the stratum thickness if there is drainage at both interfaces, or the stratum thickness if there is a drainable layer at only one interface, and c_v the coefficient of consolidation.

The theoretical time factor T_v is a pure number that has been determined for all conditions of importance, given in terms of percent consolidation U. The values given in Figure 2.83 apply to the common cases of (1) a consolidating stratum free to drain through both its upper and lower boundaries, regardless of the distribution of the consolidation pressure, or (2) a uniform distribution of the consolidation pressure throughout a layer free to drain through only one surface. The cases of consolidation pressure increasing or decreasing through a consolidating stratum with an impervious boundary can be found in Terzaghi and Peck (1967, p. 181).

The coefficient of consolidation c_v is found from the compression–log time curve. Tangents are drawn to the primary section of the curve at its point of inflection and to the secondary sections of the curve to locate e_{100} of primary consolidation. The initial void ratio e_o is found by taking the amount of compression between 0.25 and 1.0 min and adding this value to the void ratio for 0.25 min. The void ratio corresponding to $U = 50\%$ is midway between e_o and e_{100} and the corresponding time t_{50} represents the time for 50% consolidation. The theoretical time factor for $U = 50\%$ is 0.197 (Figure 2.83), and c_v is found from

$$c_v = T_v h^2/t = 0.197\, h^2/t_{50}\ (\text{cm}^2/\text{sec, in}^2/\text{min, ft}^2/\text{day}) \tag{2.68}$$

where h is one half the thickness of the sample, since it has double drainage. The square root of time method is an alternative procedure (Lambe and Whitman, 1969).

Analysis of time rates: Consolidation is essentially complete at t_{90} (time factor = 0.848) since T, for U^{100} is infinity. Field time rates of consolidation can be quickly determined from c_v with the nomograph given as in Figure 2.84 or computed from Equation 2.67. The values are usually only approximations because field time rates are normally much higher than laboratory rates because stratification permits lateral drainage. Studies where time rates are critical employ piezometers to measure *in situ* pore pressures (see Section 3.4.2).

Computing coefficient of permeability k: The relationship between k and c_v may be expressed as

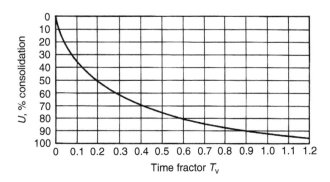

U	T_v	U	T_v
10	0.0077	60	0.286
20	0.0314	70	0.403
30	0.0707	80	0.567
40	0.126	90	0.848
50	0.197	100	Infinity

FIGURE 2.83
Theoretical time factor T_v vs. percent consolidation U.

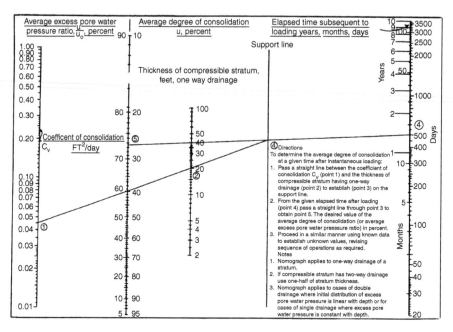

FIGURE 2.84
Nomograph for consolidation with vertical drainage ($1 \mathrm{cm}^2/\mathrm{sec} = 93 \mathrm{ft}^2/\mathrm{day}$). (From NAVFAC, 1982.)

$$c_v = k(1+e)/\gamma_w a_v = k/\gamma_w m_v \ (\mathrm{cm}^2/\mathrm{sec}) \tag{2.69}$$

where a_v is the compressibility coefficient, or the ratio between the change in void ratio and the change in vertical effective stress for the given increment, expressed as

$$a_v = -(e_o - e_1/p_1 - p_o) \ (\mathrm{cm}^2/\mathrm{g}) \tag{2.70}$$

and m_v is the coefficient of volume change, or the ratio of the change in vertical strain to the change in vertical stress, expressed as

$$m_v = \Delta e_v/\Delta \sigma_v = a_v/(1+e_o) \ \mathrm{cm}^2/\mathrm{g} \tag{2.71}$$

Settlement Analysis

Settlement from primary consolidation (S or ρ) is determined from the e–log p curve (Figure 2.78) from the expression

$$S = (\Delta e/1 + e_o)H \tag{2.72}$$

or, if representative values for C_c have been obtained from a number of tests on similar materials, from the expression

$$S = H(C_c/1 + e_o) \log(p_o - \Delta p/p_o) \tag{2.73}$$

where Δp is the average change in pressure resulting from the imposed stress (see Section 2.5.1).

Secondary compression can result in substantial compression in addition to primary consolidation in very soft clays and organic soils. It can be estimated from the compression vs. log time curve for a desired load. Since the relationship between compression and time on the semilog plot is essentially a straight line, it can be expressed (Perloff, 1975) as

$$\Delta e = - C_\alpha \log (t_1/t_2) \tag{2.74}$$

where C_α is the coefficient of secondary compression represented by the slope of the time curve, t_2 the time at which secondary compression is desired and t_1 the time at the start of secondary consolidation.

Secondary compression is a form of creep.

Lateral Strains

Lateral strains, prevented in the one-dimensional consolidation test, can be significant under conditions of relatively rapid loadings of soft clays. Analysis is based on triaxial test data of Lambe and Whitman (1969), Lambe (1964, 1973), Skempton and Bierrum (1957), Hansen (1967), and Davis and Poulos (1968).

Expansion of Clay Soils

General

Usually, clays with expansion potential are less than 100% saturated in the field; therefore, the consolidation theory does not apply. Tests are performed in the consolidometer or the California bearing ratio mold to obtain measures of percent swell or volume change under a given load, or the maximum swell pressure that may be anticipated.

Consolidometer Testing of Undisturbed Samples

Percent Swell Measurements:

- Place specimen in consolidometer at natural moisture (w) content and provide protection against changes in w.
- Add loads in the same manner as the consolidation test, although initial loads may be higher, measuring and recording compression until the final design foundation load is attained.
- Immerse specimen in water, permitting saturation, and measure the volume increase as a function of time, until movement ceases.

Maximum Swell Pressure:

- Place specimen in consolidometer under an initial seating load.
- Immerse in water and add loads as necessary to prevent specimen from swelling to determine the maximum swell pressure.

CBR Mold Testing of Compacted Samples

Procedures are similar to the percent swell test in the consolidometer.

Compression in Cohesionless Sands

Measurements

Cohesionless sands are seldom tested in the laboratory for compression characteristics since undisturbed samples cannot be obtained; they are occasionally tested in the consolidometer as reconstituted specimens placed at various values of D_R.

Evaluations of compression are normally based on *in situ* test data such as obtained with the pressure meter, SPT or CPT, or load test.

Load vs. Compression Characteristics

Under normal loading conditions, compression in quartz sands is essentially plastic and results from void closure. Compression of individual grains is insignificant except for sands composed of soft materials such as shell fragments, gypsum, or lightly cemented calcareous sands.

Compression magnitude in quartz sands is related to D_R, gradation characteristics, and the magnitude of applied static load or the characteristics of dynamic loadings. A family of curves representing pressure vs. void ratio for various values of D_R obtained by testing reconstituted samples in a consolidometer, is given in Figure 2.85.

Compression vs. Time

Compression under applied load results in the immediate closure of the voids as the grains compact, although in saturated silty soils some time delay will occur as pore pressures dissipate. The normal process of consolidation, however, does not occur. Swiger (1974) reports a case where primary settlements occurred within 1 h of load application during large-scale field load tests, but secondary compression, ultimately as large as primary compression, continued over a period of several years (see Section 2.5.5).

Pressuremeter Testing

General

Pressuremeters are used for *in situ* measurements of deformation moduli and strength. There are several types:

- Menard pressuremeter for soils and soft rock (ASTM D4719)
- Camkometer for soils
- Dynamometer and Goodman jack for rock (see Section 2.5.3)

Menard Pressuremeter

See Menard (1963, 1965, 1975), CGS (1978), and Terzaghi et al. (1996).

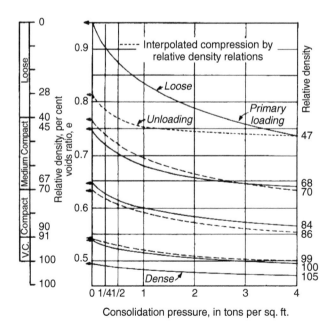

FIGURE 2.85
Pressure–void ratio curves and relative density relationships for a "coarse to fine sand, little silt." (From Burmister, D. M., ASTM, Vol. 48, Philadelphia, PA, 1948. Reprinted with permission of the American Society for Testing and Materials.)

Applicability: Although used in all soil types and soft rocks, the Menard pressuremeter is most useful in materials for which undisturbed sampling is difficult or not possible, such as sands, residual soils, glacial till, and soft rock (in a smooth borehole).

Procedure: The apparatus is illustrated in Figure 2.86. A cylindrical flexible probe is lowered into an NX-size borehole to the test depth, and increments of pressure are applied to the probe by gas, while radial expansion of the borehole is measured in terms of volume changes. The test can be performed up to failure of the surrounding materials; the limit of the radial pressure is in the range of 25 to 50 tsf.

Test Data: The volume of the expanded probe is plotted vs. the "corrected" pressure, which is equal to the gage pressure at the surface minus the probe inflation pressure plus the piezoelectric head between the probe and the gage, as shown in Figure 2.87. The quantities obtained from the test include:

- P_i in the initial or seating pressure at the beginning of the elastic stress stage, generally considered equal to P_o, the at-rest horizontal stress, P_f the creep pressure at the end of the elastic stress stage,
- P_L the limiting pressure, or ultimate pressure; the failure pressure,
- E_c the compression modulus; obtained from the slope of the compression curve between P_i and P_L Equation 2.73,
- E_r the rebound modulus, obtained from the slope of the rebound curve.

Computing the compression modulus E_c:

$$E_c = K \, (dp/dv) \ (\text{psi, kg/cm}^2) \tag{2.75}$$

where $K = 2(1 + v)(V_o + V_m)$ is a constant of the pressuremeter accounting for borehole diameter, probe size, and Poisson's ratio (usually taken as 0.3–0.4), V_o the initial hole vol-

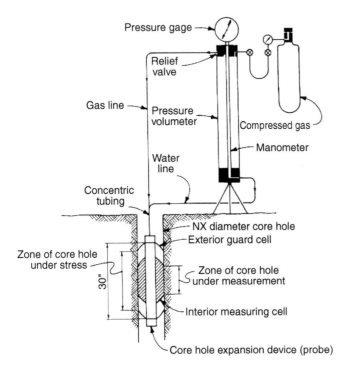

Pressure gage

Relief valve

Gas line

Pressure volumeter

Compressed gas

Manometer

Water line

Concentric tubing

Zone of core hole under stress

30"

NX diameter core hole

Exterior guard cell

Zone of core hole under measurement

Interior measuring cell

Core hole expansion device (probe)

FIGURE 2.86
Schematic of pressuremeter equipment. (From Dixon, S. J., *Determination of the In Situ Modulus of Deformation of Rock*, ASTM Special Technical Publication 477, American Society for Testing and Materials, Philadelphia, PA, 1970. Reprinted With permission of the American Society for Testing and Materials.)

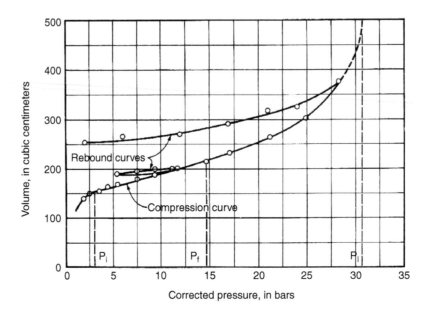

FIGURE 2.87
Typical pressuremeter test results. (From Dixon, S. J., *Determination of the In Situ Modulus of Deformation of Rock.*
ASTM Special Technical Publication 477, American Society for Testing and Materials, Philadelphia, PA, 1970.
Copyright ASTM International. Reprinted with permission.)

ume over length of cell (790 cm^3 for NX hole), and V_m the fluid volume injected into cell
for the average applied pressure. $K = 2700$ cm^3 typically, for an NX hole and $v = 0.33$.

Moduli and Strength Parameters

Typical values for E_c and P_L for various materials are given in Table 2.29.
  Estimating Young's modulus E from E_c: Menard (1965) suggests the use of a "rheological"
factor α to convert the compression modulus into Young's modulus, expressed as:

$$E = E_c/\alpha \qquad (2.76)$$

Menard's values for α are given in Table 2.30 in terms of the ratio E_c/P_L, which have been
found to be related to the amount of precompression in the material. There is some con-
troversy over the selection of the α factor (Silver et al., 1976), but most practitioners appear
to use Menard's values.

Data Applications:

- Estimating values for deformation modulus E_c.
- Estimating the undrained shear strength from the expression (Menard, 1975)

$$s_u = (P_L - P_o)/2K_b \qquad (2.77)$$

  where P_L is the limiting pressure, P_o the at-rest horizontal stress $\sigma'_v K_o$ and
  K_b the coefficient varying with E_c/P_L, typically 5.5 (Lukas and deBussy, 1976).

- Estimating allowable bearing value for foundations (CGS, 1978).

TABLE 2.29

Values of Compression Modulus E_c and Ultimate Pressure P_L from Menard Pressuremeter Testing[a]

Soil Type	E_c (tsf)	P_L (tsf)
Peat and very soft clays	2–15	0.2–1.5
Soft clays	5–30	0.5–3.0
Firm clays	30–80	3.0–8.0
Stiff clays	80–400	6–25
Loose silty sands	5–20	1–5
Silts	20–100	2–15
Sands and gravels	80–400	12–50
Till	75–400	10–50
Recent fill	5–50	0.5–3
Ancient fill	40–150	4.0–10
Glacial till (Ohio)	600–1500	Ward (1972)
Residual soil (schist)	20–160	Ward (1972)
Decomposed schist	200–500	Ward (1972)
Residual soils (schist and gneiss: SM, ML, saprolite)	50–2000	Martin (1977)

[a] After *Canadian Foundation Engineering Manual*, Part I 1978, unless otherwise noted. (Reprinted with permission of the National Research Council of Canada.)

TABLE 2.30

Rheological Factor α for Various Soil Conditions[a]

Material	Peat E_c/P_L	α	Clay E_c/P_L	α	Silt E_c/P_L	α	Sand E_c/P_L	α	Sand and Gravel E_c/P_L	α
Preconsolidated		1	>16	1	>14	0.67	>12	0.5	>10	0.33
Normally consolidated		1	9–16	0.67	8–14	0.5	7–12	0.33	6–10	0.25
Underconsolidated			7–9	0.5	5–8	0.5	5–7	0.50		0.25

[a] Menard, L.F., *Proceedings of the 6th International Conference on Soil Mechanics and Foundation Engineering*, Montreal, Vol. 2, 1965, pp. 295–299.

- Settlement analysis of shallow foundations (CGS, 1978; Menard, 1972).
- Analysis of deep foundations (CGS, 1978; Menard, 1972).
- Estimating horizontal subgrade reaction modulus k_h (Poulos, 1971):

$$k_h = 0.8 \, E_c/d \qquad (2.78)$$

where d is the pile diameter.

Limitations of Pressuremeter Test Data

The modulus is valid only for the linear portion of soil behavior (P_f in Figure 2.87). Stratified or otherwise anisotropic materials may have modulus values much lower in the vertical direction than in the horizontal direction. Since foundation stresses are applied vertically in most cases, values for E may be overestimated from pressuremeter tests.

Stiffness of the device may be significant compared with the compressibility of the tested material, as in very loose sands, soft clay, or organic soils. Borehole wall disturbance and irregularities greatly affect test results. Modulus values should be correlated with data obtained by other test methods.

Camkometer

Description: The Camkometer is a self-boring pressuremeter developed in the early 1970s by Cambridge University (Clough and Denby, 1980). The 80-mm-diameter device is covered by a rubber membrane that incorporates two very small cells for pore-pressure measurements. It is drilled into position and the membranes expanded; transducers permit the pressure response to be converted into electrical impulses. An effective stress–strain curve is plotted from the data.

Applicability: It is used primarily in soft clays and sands for *in situ* measurements of shear modulus, shear strength, pore-water pressure, and lateral stress K_o. Formerly, values for K_o were obtained only by empirical methods or laboratory tests.

Advantages over conventional pressuremeter: Lateral stress is measured directly because the instrument is self-boring and stress relief is not permitted. Records are obtained from precise electrical impulses, whereas conventional pressuremeters record total volume changes of hydraulic fluid within the flexible membrane and furnish only average values.

Disadvantages: As with all pressuremeters, soil anisotropy is not accounted for. Smear caused by drilling in soft clays reduces the true permeability and affects the pore-pressure measurements. To reduce drainage effects, undrained tests are performed at high strain rates.

Penetration Tests (SPT, CPT, and DMT)

General

SPT, CPT and FDT are discussed in Section 2.4.5. Intercorrelations between SPT and CPT test values are given in Figure 2.62. Data from the CPT, in particular point resistance q_c, have been correlated with various soil properties.

CPT Test Correlations

Estimating E_s: Values for the elastic modulus can be obtained from correlations with the cone resistance q_c and the normalized cone resistance g_t, and the shear wave velocity measured in the CPT containing an accelerometer (Section 1.3.4) (Robertson, 2000).

Consolidation parameters: c_h and c_v, and OCR can be estimated from CPTU tests (Section 1.3.4) (Robertson, 2000).

Flat Dilatometer Test (DMT)

Consolidation parameters c_h and OCR can be estimated from DMT tests (Section 2.4.5).

Plate-Load Test (ASTM D1196)

Purposes

Vertical modulus of subgrade reaction k_{sv} is obtained from plate-load tests. Results are used occasionally for estimating settlements in sands. In direct application, however, it must be considered that the depth of the stressed zone is usually much smaller than the zone that will be stressed by a larger footing, and the results will not necessarily be representative. This constraint is overcome by performing tests at various depths to stress the entire zone to be stressed by the footing or by performing full-scale load tests. Plate-load tests are not performed on clay soils for settlement measurement because of long-term consolidation effects.

Procedure

A 12- or 30-in.-diameter plate is jack-loaded against a reaction to twice the design load, and the plate deflection measured under each load increment. The test setup is shown in

Figures 2.88 and 2.89, and a plot of the test results in terms of load vs. deflection is given in Figure 2.90, from which the test modulus k_t is determined.

Modulus of Vertical Subgrade Reaction, k_{sv} and k_t

Determination: k_{sv} is defined as the unit pressure required to produce a unit of deflection expressed as

$$k_{sv} = p/y \text{ (force/length}^3 = \text{tons/ft}^3) \tag{2.79}$$

where p is the pressure per unit area between the contact surface of the loaded area and the supporting subgrade, and y the deflection produced by the load; length^3 = area $\times$ y in the denominator.

Values for p and y are taken at onehalf the yield point as estimated on a log plot of the field data (Figure 2.90) to determine subgrade modulus test value k_t. Values for k_{sv} depend on the dimensions of the loaded area as well as the elastic properties of the subgrade; therefore, the test value k_t always requires adjustment for design. Various expressions (Terzaghi and Peck, 1967) have been presented for converting k_t to k_{sv}.

Value ranges of k_t: For a given soil, the value of k_t decreases with increasing plate diameter, and the modulus for sands is reduced substantially below the water table. Typical value ranges for plates 1 ft square or beams 1 ft wide have been given for sands by Terzaghi (1955) as shown in Table 2.31 and for clays as shown in Table 2.32. In sands, the modulus increases with depth. In overconsolidated clays, the modulus remains more or

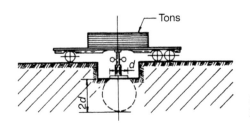

FIGURE 2.88
Reaction for plate-load test and the stressed zone.

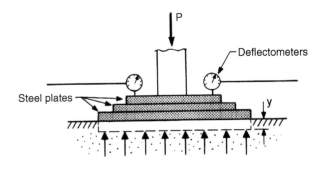

FIGURE 2.89
Load applied to plate in increments while deflection is measured.

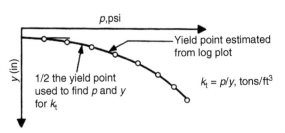

FIGURE 2.90
Load–deflection curve for load test on 1 ft^2 plate to find the subgrade modulus k_t.

TABLE 2.31

Values of Vertical Modulus of Subgrade Reaction (k_t) for Plates on Sand[a]

Relative Density of Sand	Loose[b]	Medium[b]	Dense[b]
Dry or moist sand, limiting values for k_t	20–60	60–300	300–1000
Dry or moist sand, proposed values for k_t	40	130	500
Submerged sand, proposed values for k_t	25	80	300

[a] From Terzaghi, K., *Geotechnique*, 4, 1955.
[b] k_t in ton/ft^3 for square plates, 1 × 1 ft, or beams 1 ft wide (1 ton/ft^3 = 0.032 kg/cm^3 = 32 ton/m^3).

TABLE 2.32

Values of Vertical Modulus of Subgrade Reaction (k_t) for Plates on Preconsolidated Clay[a]

Consistency of Clay	Stiff[b]	Very stiff[b]	Hard[b]
Values of U_c(ton/ft^2)	1–2	2–4	>4
Ranges for k_t, square plates	50–100	100–200	>200
Proposed values, square plates	75	150	300

[a] From Terzaghi, K., *Geotechnique*, 4, 1955.
[b] k_t in ton/ft^3 for square plates 1 × 1 ft, or long strips 1 ft wide (1 ton/ft^3 = 1.15 lb/in.^{3}).

less constant as long as the consistency remains constant. The test is not applicable to weaker clays because of the effect of consolidation.

Applications

The modulus of vertical subgrade reaction is used to determine shears and bending moments resulting from deflections of pavements, mat foundations, and continuous footings by employing the concept of a beam or plate on an elastic subgrade.

The modulus of horizontal subgrade reaction is determined by a pile-load test. It is used for the evaluation of shears and bending moments in flexible retaining structures and laterally loaded piles, employing the concept of a beam or plate on an elastic subgrade.

Full-Scale Load Tests

Footings

The test is performed in a manner similar to the plate-load test except that a poured concrete footing is used rather than a plate. Tests performed on various sizes and at various depths provide much useful information on the compressibility of sand subgrades, and offer the most reliable procedures for measuring settlements in these materials if representative sites are tested.

Pile-Load Tests

Vertical pile-load tests are performed to obtain measures of E and shaft friction in large-diameter piles, or end-bearing capacity and shaft friction on slender piles. A load test setup is illustrated in Figure 3.36.

Lateral load tests are performed on piles to obtain data on the horizontal modulus of subgrade reaction K_{sh}.

Embankment Tests

Embankments are constructed to obtain information on the compressibility of loose fills, or thick deposits of soft clays and weak organic soils, and are often designed to preload weak soils. An embankment with height and width, adequate to stress the weak soils to a substantial depth, is constructed with earth fill. Instrumentation is installed to monitor deflections and pore pressures as functions of time (see Section 3.5.2). Interpretation of the field data and correlations with laboratory test data provide information on the magnitude and time rate of settlements to be anticipated, as well as on the height limitations during actual construction required to avoid shear failures. This is the most reliable procedure for evaluating the characteristics of weak organic deposits.

2.5.5   Dynamic Deformation Moduli (Soils)

Methods Summarized

Laboratory Methods

Dynamic deformation moduli are measured in the laboratory in the cyclic triaxial and cyclic simple shear (see Section 2.4.4), and cyclic torsion, ultrasonic, and resonant column devices as summarized in Table 2.33.

   To date, the shaking table test has been used primarily in university research studies and is described in Novacs et al. (1971) and De Alba et al. (1976). Dynamic testing procedures are described in detail in USAEC (1972).

In Situ Methods

Dynamic moduli are measured in the field by seismic direct methods and steady-state vibration methods. Vibration monitors obtain data on ground motion (see Section 3.2.5).

Resonant Column Devices

Apparatus

Several types are in use as described in USAEC (1972).

Procedure

The specimen is placed in a chamber, subjected to a confining pressure stimulating overburden pressure, vibrated first in the torsional mode and then in the longitudinal mode, while end displacements are monitored.

   Shear-wave velocity is computed from the torsional test results and the compression-wave velocity from the longitudinal test results as functions of specimen and end displacements. Dynamic Young's modulus E_d and dynamic shear modulus G_d are computed from equations given in Table 2.27. When applicable, factors are applied to include the effects of damping and end conditions during test.

   See also ASTM D4015-92(2000).

Seismic Direct Methods

Purpose

As described in Section 1.3.2, seismic direct methods are used to obtain values of E_d and G_d, and have been used for estimating values of E_s in medium-dense to dense sands for settlement computations where small strains are critical.

TABLE 2.33

Laboratory Methods for Determining Dynamic Soil Properties[a]

Test Conditions	Test and Reference	Properties Measured[b]	Stress or Strain Conditions		Strain Amplitude
			Initial	Dynamic	
Low frequency	Cyclic triaxial (Seed and Chan, 1966; Castro, 1969)	E, D; stress vs. strain: strength	Axisymmetric consolidation	Pulsating axial or confining stress: constant-amplitude stress	10^{-4} to 10^{-1}
	Cyclic torsion (Zeevaert, 1967; Hardin and Drnevich, 1972a, b)	G, D, stress vs. strain	Axisymmetric consolidation	Pulsating shear stress, constant-amplitude stress or free vibration	10^{-4} to 10^{-2}
	Cyclic simple shear (Seed and Wilson, 1967)	G, D, stress vs. strain: strength	K_s consolidation	Pulsating shear stress: constant-amplitude stress (strain) or free vibration	10^{-4} to 3×10^{-2}
High frequency	Ultrasonic (Lawrence, 1965; Nacci and Taylor, 1967)	c_p or c_s	Axisymmetric consolidation	Dilation or shear, single pulse wave	10^{-5}
	Resonant column (Afifi, 1970)	c_p or c_s, E or G; D	Uniform or axisymmetric consolidation	Pulsating axial or shear (torsional) stress: constant-amplitude strain	10^{-6} to 10^{-2}

[a] From *Seismic Risk and Engineering Decisions*, Lomnitz and Rosenblueth, Eds., Elsevier, Amsterdam, 1976, ch. 4.

[b] E = dynamic Young's modulus, G = dynamic shear modulus, D = damping ratio, c_p = compression-wave velocity, c_s = shear-wave velocity.

Estimating Values of E_s

Moduli for strain levels in the order of 10^{-6} can be estimated from shear-wave velocities from crosshole or uphole seismic surveys (Swiger, 1974). Shear-wave velocities are used because they can be measured above and below the groundwater level, whereas compression-wave velocity can be measured only above groundwater level, since it is obscured by the compression-wave velocity for water.

Under loads of the order of 2 to 3 tsf, strains in dense sands are small, approximately 10^{-3}, but higher than the strains occurring during seismic testing that require adjustment for analysis. In granular soils, E_d and G_d have been found to decrease with increasing strain levels (Hardin and Drnevich, 1972). A relationship between shear strain and axial strain as a function of strain level is given in Figure 2.91. The ratios given on the abscissa are used to reduce the field shear and compression modulus for use in analysis.

Case Study

In a study reported by Swiger (1974), good agreement was found between settlements computed from seismic direct surveys and large-scale *in situ* load tests and the actual settlements measured on the structure for which the study was made. For a Poisson's ratio of 0.3, the values of E_s were of the order of 4×10^6 psf (2000 tsf). The primary settlements occurred within about 1 h of load application, but the magnitude of the secondary settlement appeared to approximate that of the primary and to continue over a period of some years. Approximately 25% occurred in the first year after load application (about 4–8 tsf foundation pressure).

Steady-State Vibration Methods

Purpose

Steady-state vibration methods are performed to obtain *in situ* values of E_d and G_d.

Principles

Ground oscillations are induced from the surface causing Rayleigh waves. The Rayleigh wave velocity V_r is used directly as the shear-wave velocity because, for Poisson's ratios of 0.35 to 0.45, $V_r = 0.935$ to $0.95\ V_s$, a difference which is of little engineering significance (Richart, 1975). E_d and G_d are then computed from V_r (for V_s) using equations given in Table 2.27.

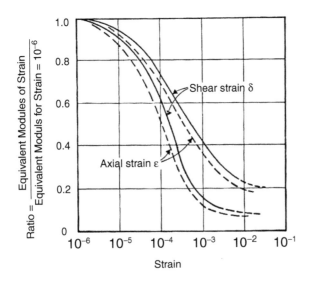

FIGURE 2.91
Strain modulus relations for sands. (After Seed, H. B., *Proceedings of the 7th International Conference on Soil Mechanics and Foundation Engineering*, Mexico City, 1969; from Swiger, W. F., *Proceedings of ASCE, Conference on Analysis and Design in Geotechnical Engineering*, University of Texas, Austin, Vol. II, 1974, pp. 79–92. With permission.)

Procedure

An electromagnetic vibrator located on the surface generates the Rayleigh waves. The wavelengths of these surface waves are determined by measuring the distance between two points vibrating in phase with the source. The generated vibrations are measured with a velocity or acceleration transducer and the wavelengths are measured by comparing the phase relationship of vibrations at various radii from the source with the vibrations of the source. Velocities are computed from the measured wavelength and the vibrating frequency.

2.6   Typical Values of Basic, Index, and Engineering Properties

2.6.1   Rock Masses

Relationships between the uniaxial compressive strength and the deformation modulus for various rock types are given in Figure 2.92.

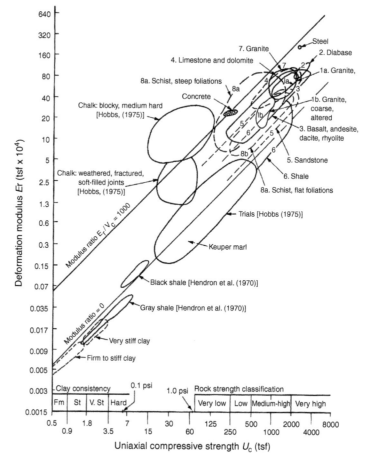

FIGURE 2.92

Relationships between uniaxial compressive strength and deformation modulus for various rock types and clays. (After Peck, R. B., *Proceedings of ASCE*, New York, Vol. 2, 1976, pp. 1–21; and Deere, D. U., *Rock Mechanics in Engineering Practice*, Stagg and Zienkie-Wicz, Eds., Wiley, New York, 1968, chap. 1.)

Typical permeability values for intact and *in situ* rocks are given in Table 2.12.

Typical strength values for rock mass discontinuities (joints and shear zones) for use in preliminary analysis are given in Table 2.34.

2.6.2 Weathered Rock and Residual Soil

Shear strength parameters for various rock types and in situ conditions are given in Table 2.35. Typical engineering properties of residual soils of basalt and gneiss are given in Table 7.5.

2.6.3 Cohesionless Soils

Common properties including relative density, dry density, void ratio, and strength as related to gradation and N are given in Table 2.36. Relationships between ϕ and D_R for various gradations are given in Figure 2.93.

Typical values for the permeability of soils are given in Tables 2.10– 2.14.

2.6.4 Clay Soils

Common properties, including relationships between consistency, strength, saturated weight and N, are given in Table 2.37. Correlation of N with U_c for cohesive soils of various plasticities are given in Figure 2.94.

TABLE 2.34

Typical Strength Values for Rock-Mass Discontinuties for Use in Preliminary Stability Analysis[a,b]

Defect	Condition	Strength Parameter[c]
Joint surface: smooth, unweathered rock[d,e]	Hard massive, well-interlocked or cemented(granite, basalt, gneiss, sandstone, limestone)	$\phi_u = 30\text{–}35°$
	Hard shaley or schistose (hard shale, slate, phyllite,mica-schist)	$\phi_u = 25\text{–}30°$
	Soft, laminated or schistose (clay shale, talc schist, chlorite schist)	$\phi_u = 20\text{–}25°$
Asperity angle j (add to ϕ_u)[e]	Very irregular surface	$j = 30\text{–}40°$
	Average surfaces	$j = 5\text{–}15°$
	Smooth, planar surface	$j = 0\text{–}2°$
Gouge or mylonite from fault or shear zones	Sandy material	$\phi_d = 24\text{–}32°$
	Intermediate material	$\phi_d = 16\text{–}23°$
	Clayey material	$\phi_d = 8\text{–}15°$

[a] After Deere, D.U., *Proceedings of ASCE, Rock Engineering for Foundations and Slopes, Speciality Conference*, Boulder, CO, Vol. II, 1976, pp. 55–85.

[b] For FS = 1.5 to 2.0, ϕ_u = basic friction angle, ϕ_d = drained test.

[c] Generally assume c = 0 for weak zones (weathered, altered, sheared), and in no case >1 tsf even (or hard) rock surfaces when tested.

[d] To ϕ_u add asperity angle j, or for analysis, $\bar{\phi} = \phi_u + j$. Joint roughness angle j defined in Figure 2.45.

[e] Reduce $\bar{\phi}, = \phi_u + j$ by 5 to 10° for weathered surfaces (Equation 2.59).

[f] Values for ϕ_r are generally 5 to 10° or more lower than values for, $\bar{\phi}$.

TABLE 2.35

Shear Strength Parameters of Residual Soil, Weathered Rocks, and Related Materials[a]

Rock Type	Weathering Degree	$c, \bar{c}$ (tsf)	$\phi, \bar{\phi}, \phi_r$ (deg)	Remarks
			Strength Parameters	
Igneous Rocks				
Granite	Decomposed	$C = 0$ $\phi_{avg} = 29$	$\phi = 27\text{--}31$ Dam	500 tests, Cherry Hill
	Quality index i	C	ϕ	*In situ* direct shear tests.
	15	1	41	Alto Rabagâo
	10	2	45–46	
	7	3	49–52	
	5	5	57	
	3	6–13	62–63	
	Weathered, zone IIB		$\phi_r = 26\text{--}33$	Lab direct shear tests.
	Partly weathered, zone IIB		$\phi_r = 27\text{--}31$	Alto Lindosa
	Relatively sound, zone III		$\phi_r = 29\text{--}32$	
	Red earth, zone IB		$\phi = 28$	
	Decomposed, zone IC		$\phi_{avg} = 35$	
	Decomposed, fine-grained	$c = 0$ if	$\phi = 25\text{--}34$	
	Decomposed, coarse-grained	saturated	$\phi = 36\text{--}38$	
	Decomposed, remolded		$\phi = 22\text{--}40$	
Quartz diorite	Decomposed; sandy, silty	$c = 0.1$	$\phi = 30 +$	Lab tests, UD samples
Diorite	Weathered	$c = 0.3$	$\phi = 22$	CU triaxial tests
Rhyolite	Decomposed		$\phi = 30$	
Metamorphic Rocks				
Gneiss (micaceous)	Zone IB	$c = 0.6$	$\phi = 23$	Direct shear tests
	Decomposed	$c = 0.3$	$\phi = 37$	
Gneiss	Decomposed, zone IC		$\phi = 18.5$	CU triaxial tests
	Decomposed (fault zone)	$c = 1.5$	$\phi = 27$	Direct shear tests on concrete-rock surfaces
	Much decomposed	$c = 4.0$	$\phi = 29$	
	Medium decomposed	$c = 6.5$	$\phi = 35$	
	Unweathered	$c = 12.5$	$\phi = 60$	
Schist	Weathered (mica schist soil)		$\phi = 24\frac{1}{2}$	From analysis of slides
	Partly weathered mica schists and phyllites (highly fractured)	$c = 0.7$	$\phi = 35$	Perpendicular to schistosity
	Weathered, intermediate zone IC	$\bar{c}, = 0.5$ $\bar{c}, = 0.7$	$\phi, \bar{\phi}, = 15$ $\underline{\phi} = 18$ $\bar{\phi}, = 21$	CU tests, $S = 50\%$ CU tests, $S = 100\%$
	Weathered		$\phi = 26\text{--}30$	Compacted rock fill, field direct shear tests
Phyllite	Residual soil, zone IC	$c = 0$	$\phi = 24$	Perpendicular to schistosity
		$c = 0$	$\phi = 18$	Parallel to schistosity (both from analysts of slides)
Sedimentary Rocks				
Keuper marl	Highly weathered	$\bar{c}, \leqslant 0.1$	$\bar{\phi} = 25\text{--}32;$ $\phi_r = 18\text{--}24$	2% carbonates
	Intermediately weathered	$\bar{c}, \leqslant 0.1$	$\bar{\phi} = 32\text{--}42$ $\phi_r = 22\text{--}29$	14% carbonates
	Unweathered	$\bar{c}, \leqslant 0.3$	$\bar{\phi} = 40$	20% carbonates (all triaxial tests, D and CU and cut planes)
			$\phi_r = 23\text{--}32$	

(Continued)

TABLE 2.35

Continued

Rock Type	Weathering Degree (see Table 6.15)	c, c̄ (tsf)	φ, φ̄, φr (deg)	Remarks
London clay	Weathered (brown)	$\bar{c}$, = 0.1–0.2	$\bar{\phi}$ = 19–22 ϕ_r = 14	
	Unweathered	$\bar{c}$, = 0.9–1.8	$\bar{\phi}$ = 23–30 ϕ_r = 15	
	Joint Filling			
"Black seams"	In zone IC		ϕ_r = 10.5 ϕ_r = 14.5	Seam with slickensides Seam without slicken sides (both CU tests)
	Shear Zones[b]			
Metamorphic rocks			ϕ_r = 15–25	Foliation shear
Shales			ϕ = 10–20	Mylonite seam
Fault gouge, general			ϕ_r = 15–30	

[a] From Deere, D.U. and Patton, *Proceedings of the 4th Panamerican Conference on Soil Mechanics and Foundation Engineering*, San Juan, Vol. I, 1971, pp. 87–100.

[b] From Deere, D.U., *Foundation for Dams*, ASCE, New York, 1974, pp. 417–424.

TABLE 2.36

Common Properties of Cohesionless Soils

Material	Compactness	D_R (%)	N[a]	γ dry (g/cm³)[b]	γ dry (pcf)[b]	Void ratio e	Strength[c] φ
GW:well-graded gravels,	Dense	75	90	2.21	138	0.22	40
gravel-sand mixtures	Medium dense	50	55	2.08	130	0.28	36
	Loose	25	<28	1.97	123	0.36	32
GP: poorly graded gravels,	Dense	75	70	2.04	127	0.33	38
gravel-sand mixtures	Medium dense	50	50	1.92	120	0.39	35
	Loose	25	<20	1.83	114	0.47	32
SW: well-graded sands,	Dense	75	65	1.89	118	0.43	37
gravelly sands	Medium dense	50	35	1.79	112	0.49	34
	Loose	25	<15	1.70	106	0.57	30
SP: poorly graded sands,	Dense	75	50	1.76	110	0.52	36
gravelly sands	Medium dense	50	30	1.67	104	0.60	33
	Loose	25	<10	1.59	99	0.65	29
SM: silty sands	Dense	75	45	1.65	103	0.62	35
	Medium dense	50	25	1.55	97	0.74	32
	Loose	25	<8	1.49	93	0.80	29
ML: inorganic silts, very fine	Dense	75	35	1.49	93	0.80	33
sands	Medium dense	50	20	1.41	88	0.90	31
	Loose	25	<4	1.35	84	1.0	27

[a] N is blows pet loot of penetration in the SPT. Adjustments for gradation are after Burmister (1962). See Table 2.23 for general relationships of D_R vs. N.

[b] Density given is for G_s = 2 65 (quartz grains).

[c] Friction angle φ depends on mineral type, normal stress, and grain angularity as well as D_R and gradation (see Figure 2.93).

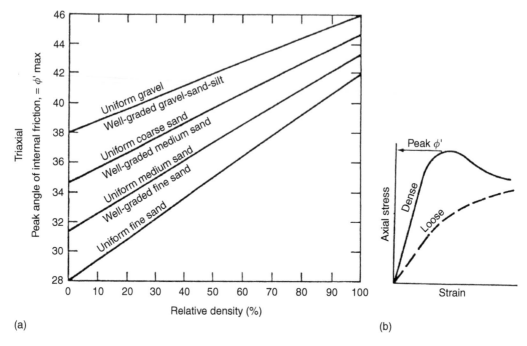

FIGURE 2.93
Friction angle and relative density relationships for granular soils: (a) chart for the approximate evaluation of the peak angle of internal friction from relative density — Schmertmann modification of Burmister (1948); (b) in problems where the sand may strain past the peak strength value before a general failure occurs, then a reduced value of ϕ must be used, particularly in the denser cohesionless soils. (*Note*: (1) For quartx sands. (2) Angular grains can increase ϕ by about 15% in the loose state and 30% in the dense state over rounded grains.) (Reprinted with permission of the Federal Highway Administration.)

TABLE 2.37

Common Properties of Clay Soils

Consistency	N	Hand Test	γ_{sat} (pcf)[a]	Strength[b] U_c (tsf)
Hard	>30	Difficult to indent	>125	>4.0
Very stiff	15–30	Indented by thumbnail	130–140	2.0–4.0
Stiff	8–15	Indented by thumb	120–130	1.0–2.0
Medium (firm)	4–8	Molded by strong pressure	110–120	0.5–1.0
Soft	2–4	Molded by slight pressure	100–110	0.25–0.5
Very soft	<2	Extrudes between fingers	90–100	0–0.25

[a] $\gamma_{sat} = \gamma_{dry} + \gamma_w \left(\dfrac{e}{1+e} \right)$.

[b] Unconfined compressive strength U_c is usually taken as equal to twice the cohesion c or the undrained shear strength s_u. For the drained strength condition, most clays also have the additional strength parameter ϕ, although for most normally consolidated clays $c = 0$ (Lambe and Whitman, 1969). Typical values for s_u and drained strength parameters are given in Table 2.38.

Typical properties of cohesive materials classified by geologic origin, including density, natural moisture contents, plasticity indices, and strength parameters, are given in Table 2.38.

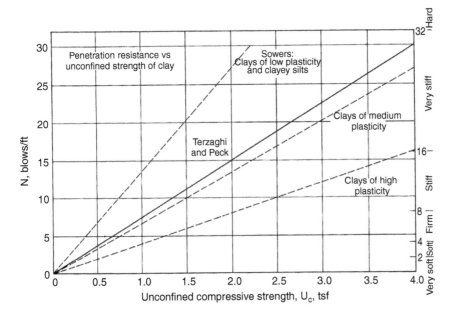

FIGURE 2.94
Correlations of SPT N values with U_C for cohesive soils of varying plasticities. (After NAVFAC, *Design Manual DM-7.1, Soil Mechanics, Foundations and Earth Structures*, Naval Facilities Engineering Command, Alexandria, VA, 1982.)

TABLE 2.38
Typical Properties of Formations of Cohesive Materials

Material	Type[a]	Locations	γd (pcf)	W (%)	LL (%)	PL (%)	s_u (tsf)	$\bar{c}$ (tsf)	$\bar{\phi}$	Remarks
Clay shales (Weathered)										
Carlisle (Cret.)	CH	Nebraska	92.4	18				0.5	45	ϕ extremely variable
Bearpaw (Cret.)	CH	Montana	89.9	32	130	90		0.35	15	
Pierre (Cret.)		South Dakota	91.7	28				0.9	12	
Cucaracha (Cret.)	CH	Panama Canal		12	80	45				$\phi_r = 10°$
Pepper (Cret.)	CH	Waco, Texas		17	80	58		0.4	17	$\phi_r = 7°$
Bear Paw (Cret.)	CH	Saskatchewan		32	116	92		0.4	20	$\phi_r = 8°$
Modeio (Tert.)	CH	Los Angeles	89.9	29	66	31		1.6		Intact specimen
Modelo (Tert.)	CH	Los Angeles	89.9	29	66	31		0.32	27	Shear zone
Martinez (Tert.)	CH	Los Angeles	103.6	22	62	38		0.25	26	Shear zone
(Eocene)	CH	Menlo Park, California	103.0	30	60	50		Free swell 100%; $P = 10$ tsf		
Residual Soils										
Gneiss	CL	Brazi; buried	80.5	38	40	16		0	40	$e_0 = 1.23$
Gneiss	ML	Brazil; slopes	83.6	22	40	8		0.39	19	c, ϕ
Gneiss	ML	Brazil; slopes	83.6		40	8		0.28	21	Unsoaked
Colluvium										
From shales	CL	West Virginia		28	48	25		0.28	28	$\phi_r = 16°$
From gneiss	CL	Brazil	68.6	36	40	16		0.2	31	$\phi_r = 12°$

(Continued)

TABLE 2.38

Typical Properties of Formations of Cohesive Materials (*Continued*)

Material	Type[a]	Locations	γd (pcf)	W (%)	LL (%)	PL (%)	s_u (tsf)	c (tsf)	φ	Remarks
Alluvium										
Back swamp	OH	Louisiana	35.6	140	120	85	0.15			
Back swamp	OH	Louisiana	62.4	60	85	50	0.1			
Hack swamp	MH	Georgia	59.9	54	61	22	0.3			$e_s = 1.7$
Lacustrine	CL	Great Salt Lake	48.7	50	45	20	0.34			
Lacustrine	CL	Canada	69.3	62	33	15	0.25			
Lacustrine (volcanic)	CH	Mexico City	18.1	300	410	260	0.4			$e_s=7, S_t=13$
Estuarine	CH	Thames River	48.7	90	115	85	0.15			
Estuarine	CH	Lake Maricaibo		65	73	50	0.25			
Estuarine	CH	Bangkok		130	118	75	0.05			
Estuarine	MH	Maine		DO	60	30	0.2			
Marine Soils (Other than Estuarine)										
Offshore	MH	Santa Barbara, California	52.0	80	83	44	0.15			$e_s = 2.28$
Offshore	CH	New Jersey		65	95	60	0.65			
Offshore	CH	San Diego	36.2	125	111	64	0.1			Depth = 2 m
Offshore	CH	Gulf of Maine	36.2	163	124	78	0.05			
Coastal Plain	CH	Texas (Beaumont)	86.7	29	81	55	10	0.2	16	$\phi_r=14, e_s=0.8$
Coastal Plain	CH	London	99.8	25	80	55	2.0			
Loess										
Silty	ML	Nebraska-Kansas	76.8	9	30	8		0.6	32	Natural $w\%$
Silty	ML	Nebraska Kansas	76.8	(35)	30	8		0	23	Prewetted
Clayey	CL	Nebraska-Kansas	78.0	9	37	17		2.0	30	Natural $w\%$
Glacial Soils										
Till	CL	Chicago	132.3	23	37	21	3.5			
Lacustrine (varved)	CL	Chicago	105.5	22	30	15	1.0			$e_s=0.6$ (OC)
Lacustrine (varved)	CL	Chicago		24	30	13	01			$e_s=1.2$ (NC)
Lacustrine (varved)	CH	Chicago	73.6	50	54	30	0.1			
Lacustrine (varved)	CH	Ohio	60.0	46	58	31	0.6			$S_t = 4$
Lacustrine) (varved)	CH	Detroit	74.9	46	55	30	08			$e_s=13$ (clay)
Lacustrine (varved)	CH	New York City		46	62	34	1.0			$e_s=1.25$ (clay)
Lacustrine (varved)	CL	Boston	84.2	38	50	26	0.8			$S_t = 3$
Lacustrine (varved)	CH	Seattle		30	55	22			30	$\phi_r = 13°$
Marine[b]	CH	Canada–Leda clay	55.5	80	60	32	0.5			$S_t = 128$
Marine[b]	CL	Norway	83.6	40	38	15	0.13			$S_t = 7$
Marine[b]	CL	Norway	80.5	43	28	15	0.05			$S_t = 75$

[a] See Figure 2.12.

[b] Marine clays strongly leached.

TABLE 2.39

Typical Properties of Compacted Soils[a]

Group Symbol	Soil Type	Range of Maximum Dry Unit Weight (pcf)	Range of Optimum Moisture (%)	Typical Value of Compression — Percent of Original Height		Typical Strength Characteristics				Typical Coefficient of Permeability (ft/min)	Range of CBR Values	Range of Subgrade Modulus k_s (lb/in.^{3})
				At 1.4 tsf (20 psi)	At 3.6 tsf (50 psi)	Cohesion as Compacted (psf)	Cohesion (Saturated) (psf)	Effective Stress Envelope ϕ (deg)	tan ϕ			
GW	Well-graded clean gravels, gravel–sand mixtures	125–135	11–8	0.3	0.6	0	0	>38	>0.79	5×10^{-2x}	40–80	300–500
GP	Poorly graded clean gravels, gravel–sand mix	115–125	14–11	0.4	0.9	0	0	>37	>0.74	10^{-1}	30–60	250–400
GM	Silty gravels, poorly graded gravel–sand silt	120–135	12–8	0.5	1.1	–	–	>34	>0.67	$>10^{-6}$	20–60	100–400
GC	Clayey gravels, poorly graded gravel–sand–clay	115–130	14–9	0.7	1.6	–	–	>31	>0.60	$>10^{-7}$	20–40	100–300
SW	Well-graded clean sands, gravelly sands	110–130	18–9	0.6	1.2	0	0	38	0.79	$>10^{-3}$	20–40	200–300
SP	Poorly-graded clean sands, sand–gravel mix	100–120	21–12	0.8	1.4	0	0	37	0.74	$>10^{-3}$	10–40	200–300
SM	Silty sands, poorly graded sand–silt mix	110–125	16–11	0.8	1.6	1050	420	34	0.67	5×10^{-5}	10–40	100–300
SM-SC	Sand–silt clay mix with slightly plastic fines	110–130	15–11	0.8	1.4	1050	300	33	0.66	2×10^{-6}	–	

(Continued)

TABLE 2.39

Typical Properties of Compacted Soils[a] (*Continued*)

Group Symbol	Soil Type	Range of Maximum Dry Unit Weight (pcf)	Range of Optimum Moisture (%)	Typical Value of Compression, Percent of Original Height		Typical Strength Characteristics				Typical Coefficient of Permeability (ft/min)	Range of CBR Values	Range of Subgrade Modulus k_s (lb/in.^{3})
				At 1.4 tsf (20 psi)	At 3.6 tsf (50 psi)	Cohesion as Compacted (psf)	Cohesion (Saturated) (psf)	Effective Stress Envelope φ (deg)	tan φ			
SC	Clayey sands, poorly graded sand–clay mix	105–125	19–11	1.1	2.2	1550	230	31	0.60	5×10^{-7}	5–20	100–300
ML	Inorganic silts and clayey silts	95–120	24–12	0.9	1.7	1400	190	32	0.62	10^{-5}	15 or less	100–200
ML-CL	Mixture of inorganic silt and clay	100–120	22–12	1.0	2.2	1350	480	32	0.62	5×10^{-7}	…	
CL	Inorganic clays of low to medium plasticity	95–120	24–12	1.3	2.5	1800	270	28	0.54	10^{-7}	15 or less	50–200
OL	Organic silts and silt-clays, low plasticity	80–100	33–21	–	–	–	–	–	–	–	5 or less	50–100
MH	Inorganic clayey silts, elastic silts	70–95	40–24	2.0	3.8	1500	420	25	0.47	5×10^{-7}	10 or less	50–100
CH	Inorganic clays of high plasticity	75–105	38–19	2.6	3.9	2150	230	19	0.35	10^{-7}	15 or less	50–150
OH	Organic clays and silty clays	65–100	45–21	–	–	–	–	–	–	–	5 or less	25–100

[a] From NAVFAC Manual DM 7 (1971). All properties are for condition of "standard Proctor" maximum density, except values of k and CBR which are for "modified Proctor" maximum density. Typical strength characteristics are for effective strength envelopes are obtained from USBR data. Compression values are for vertical loading with complete lateral confinement. (–) Indicates Inefficient data available for an estimate.

Relationships between the uniaxial compressive strength and the deformation modulus for clays are given in Figure 2.92.

2.6.5 Compacted Materials

Typical properties for compacted soils, including maximum dry weight, optimum moisture content, compression values, strength characteristics, coefficient of permeability, CBR values and range of subgrade modulus are given in Table 2.39.

References

Aas, C., "A Study of the Effect of Vane Shape and Rate of Strain on the Measured Values of In-Situ Shear Strength of Clays." Proceedings of the 6th International Conference Soil Mechanics on and Foundation. Engineering, Montreal, Vol. 1, 1965, pp. 141–145.

Alperstein, R. and Leifer, S. A., Site investigation with static cone penetrometer," *Proc. ASCE, J. Geotech. Eng. Div.*, 102, 539–555, 1976.

Arman, A., Engineering Classification of Organic Soils, HRB No. 30, Highway Research Board, Washington, DC., 1970, pp. 75–89.

Asphalt Institute, *Soils Manual for Design of Asphalt Pavement Structures*, The Asphalt Inst., Manual Series No. 1, College Park, MD, 1969.

Asphalt Institute, Thickness Design, Manual Series No. 1, College Park, MD, 1970.

Baligh, M. M., Vivatrat, V. and Ladd. C. C., Cone penetration in soil profiling, *Proc. ASCE J. Geotech. Eng. Div.*, 106, 447–461, 1980.

Bazaraa, A. R. S. Use of SPT for Estimating Settlements of Shallow Foundations on Sand, Ph.D. thesis, University of Illinois, Urbana, II, 1967.

Bieniawski. Z. T., Geomechanics Classification of Rock Masses and its Application to Tunneling, Proceedings of the 3rd International Congress for Rock Mechanics, International Society for Rock Mechanics, Vol. IIA, Denver, 1974, pp. 27–32.

Bishop, A. W. and Henkel, D. J., *The Triaxial Test*, Edward Arnold Ltd., London, 1962.

Bjerrum, L., Geotechnical properties of Norwegian marine clays, *Geotechnique*, 4, 49, 1954.

Bjerrum, L., Contribution to panel discussion, Session 5 "Stability of Slopes and Embankments," Proceedings of the 7th International Conference Soil Mechanics and Foundation Engineering, Mexico City, Vol. 3, 1969, pp. 412–413.

Bjerrum, L., Embankments on Soft Ground, State-of-the-Art Paper, Proceedings of ASCE Special Conference on Performance of Earth and Earth-Supported Structures, Purdue University, Lafayette, IN, 1972.

Bjerrum, L., Clausen, C. J. F. and Duncan, J. M., Earth Pressures on Flexible Structures: A State-of-the-Art Report, Proceedings of the 5th European Conference Soil Mechanics and Foundation Engineering, Madrid, Vol. II, 1972, pp. 169–196.

Bjerrum, L. and Landva, A., Direct simple shear tests on a Norwegian quick clay, *Geotechnique*, 16, 1–20, 1966.

Broch, E. and Franklin, J. A., The point load strength test, *Int. J. Rock Mech., Miner Sci.*, 8, 1972.

Brooker, E.W. and Ireland, H.O, Earth pressures at rest related to stress history, Can. Geotech. J., II,1965.

Brown, R. E., Drill Rod Influence on Standard Penetration Test, *Proc. ASCE J. Geotech. Eng. Div.*, 103, 1332–1336, 1977.

Burmister, D. M., The Importance and Practical Use of Relative Density in Soil Mechanics, ASTM, Vol. 48, Philadelphia, PA, 1948.

Burmister, D. M., Principles and Techniques of Soil Identification, Proceedings of the 29th Annual Meeting Highway Research Board, December, 1949.

Burmister, D. M., Identification and Classification of Soils — An Appraisal and Statement of Principles, ASTM Special Publication 113, American Society for Testing and Materials, Philadelphia, PA., 1951a pp. 3–24, 85–91.

Burmister, D. M., The Application of Controlled Test Methods in Consolidation Testing, ASTM Spec. Pub. No. 126, 1951b, pp. 83–97.

Burmister, D. M., The Place of the Direct Shear Test in Soil Mechanics, ASTM Special Technical Publication No. 131, 1953.

Burmister, D. M., Stress and Displacement Characteristics of a Two-Layer Rigid Base Soil System: Influence Diagrams and Practical Applications, Proceedings of HRB, Highway Research Board, Washington, DC, Vol. 35, 1956, pp. 773–814.

Burmister, D. M. Physical, Stress-Strain and Strength Responses of Granular Soils, *ASTM Special Technical Publication No. 322*, 1962a, pp. 67–97.

Burmister, D. M., Prototype Load-Bearing Tests for Foundations of Structures and Pavements, *Symposium on Field Testing of Soils*, STP No. 322, American Society for Testing and Materials, Philadelphia, PA, 1962b, pp. 98–119.

Casagrande, A., The Determination of the Preconsolidation Load and its Practical Significance, *Proceedings of 1st International Conference on Soil Mechanics and Foundation Engineering*, Cambridge, MA, Vol. 3, 1936, p. 60.

Casagrande, A., An Unsolved Problem of Embankment Stability on Soft Ground, *Proceedings of the 1st Panamerican Conference Soil Mechanics and Foundation Engineering*, Mexico City, Vol. II, 1959, pp. 721–746.

Casagrande, A. and Fadum, R. E., Notes on Soil Testing for Engineering Purposes, Harvard University Soil Mechanics Series No. 8, Bulletin 268, Cambridge, MA, 1940.

Cedergren, H. R., *Seepage, Drainage and Flow Nets*, Wiley, New York, 1967.

CGS, *Canadian Foundation Engineering Manual*, Canadian Geotech. Soc., Montreal, 1978.

Clemence, S. P. and Finbarr, A.O., Design Considerations and Evaluation Methods for Collapsible Soils, *Proceedings of ASCE*, Preprint 80–116, Apr., 1980

Clough, G. W. and Denby, G. M., Self-boring pressuremeter study of San Francisco Bay mud, *Proc. ASCE, J. Geotech. Eng. Div.*, 106, 45–64, 1980.

Coon, H. H. and Merritt, A. H., Predicting In Situ Modulus of Deformation Using Rock Quality Indexes, *Determination of the In-Situ Modulus of Deformation of Rock*, ASTM Specials Technical Publication 477, American Society for Testing and Materials, Philadelphia, PA, 1970, pp. 154–273.

Davis, E. H. and Poulos, H. G., The use of elastic theory for settlement prediction under three dimensional conditions, *Geotechnique*, XVII, 1968.

Davis, C. V. and Sorensen, K. E., *Handbook of Applied Hydraulics*, McGraw-Hill Book Co., New York, 1970.

De Alba, P., Seed, H. B. and Chan, C. K., Sand liquefaction in large scale simple shear tests, *Proc. ASCE. J. Geotech. Eng. Div.*, 102, 909–927, 1976.

Deere, D.U., Geological considerations, *Rock Mechanics in Engineering Practice*, Stagg, K.G. and Zienkiewicz, O.C., Eds., Wiley, New York, 1968, chap. 1.

Deere, D. U., Indexing Rock for Machine Tunneling, *Proceedings Tunnel and Shaft Conference*, Society of Mining Engineers of America, Minneapolis, May 1968, 1970, pp. 32–38.

Deere, D. U., Engineering geologist's responsibilities in dam foundation studies, *Foundations for Dams*, ASCE, New York, 1974, pp. 417–424.

Deere, D. U., Dams on Rock Foundations-Some Design Questions. *Proceedings of ASCE, Rock Engineering for Foundations and Slopes, Speciality Conference*, Boulder, CO, Vol. II, 1976, pp. 55–85.

Deere, D. U. and Miller, R. P., Classification and Index Properties for Intact Rock, Technical Report AFWL-TR-65-116, AF Special Weapons Center, Kirtland Air Force Base, New Mexico, 1966.

Deere, D. U., Hendron, A. J., Patton, F. D. and Cording, E. J., Design of Surface and Nearsurface Construction in Rock, *Proceedings of 8th Symposium on Rock Mechanics*, University of Minnesota, American Mineral and Metallurgy Engineers, chap. 11.

Deere, D. U. and Patton, F. D., Slope Stability in Residual Soils, *Proceedings of the 4th Panamerican Conference Soil Mechanics and Foundation Engineering*, San Juan, Vol. I, 1971, pp. 87–100.

Dick, R. C., *In Situ* measurement of rock permeability: influence of calibration errors on test results, *Bull. Assoc. Eng. Geol.*, XII, 1975.

Dixon, S. J., Pressuremeter Testing of Soft Bedrock, Determination of the *In Situ* Modulus of Deformation of Rock, ASTM Special Technical Publication 477, American Society for Testing and Materials, Philadelphia, PA, 1970.

Eide, O., Marine Soil Mechanics — Applications to North Sea Offshore Structures, Norwegian Geotechnical Institute Publication 103, Oslo, 1974, p. 19.

Faccioli, E. and Resendiz, D., Soil dynamics, in *Seismic Risk and Engineering Decisions*, Lomnitz, C. and Rosenblueth, E., Eds., Elsevier Pub. Co., Amsterdam, 1976, chap. 4.

Fahy, M. P. and Guccione, M. J., Estimating strength of sandstone using petrographic thin-section data, *Bull. Assoc. Eng. Geol.*, XVI, 467–486, 1979.

Farmer, I. W., *Engineering Properties of Rocks*, E. & F. N. Spon, London, 1968.

Foster, C. R., Field problems: compaction, In *Foundation Engineering*, G. A. Leonards, Ed., McGraw-Hill Book Co., New York, 1962, pp. 1000–1024.

Gaziev, E. G. and Erlikhman, S. A., Stresses and Strains in Anisotropic Rock Foundation (Model Studies). *Symposium International Society for Rock Mechanics*, Nancy, 1971.

Gibbs, H. J. and Holtz, W. G., Research on Determining the Density of Sands by Spoon Penetration Testing, *Proceedings of 4th International Conference on Soil Mechanics and Foundation Engineering*, London, Vol. I, 1957, pp. 35–39.

Goodman, R. E., Van Tran, D. and Heuze, F. E., The Measurement of Rock Deformability in Boreholes, *Symposium on Rock Mechanics*, University of Texas, Austin, 1968.

Handy, R.L., "Borehole Shear Test and Slope Stability," *Proceedings In Situ '86*, Geotechnical Division, ASCE, June 23–25, Blacksburg, VA, 1986.

Handy, R.L., Schmertmann, J.H. and Luttenegger, A.J., Borehole Shear Tests in a Shallow Marine Environment, Special Technical Testing Publication 883, ASTM, Philadelphia, PA, 1985.

Hansen, B. J., Refined Calculations of Foundation Movements. *Proceedings of 3rd International. Conference on Soil Mechanics and Foundation Engineering*, Caracas, Vol. III, 1967.

Hardin, B. M. and Dmevich, V. P., Shear Modulus and Damping in Soils: Measurement and Parameter Effects, *Proc. ASCE J. Soil Mech. Found. Eng. Div.*, 98, 1972.

Handy, R.L., Pitt, J.M., Engle, L.E. and Klockow, D.E., Rock Borehole Shear Test, *Proceedings of 17th U.S. Symposium on Rock Mechanics* Vol. 486, 1976. pp. 1–11

Haverland, M. L. and Slebir, E. J., Methods of performing and interpreting *in situ* shear tests, *Stability of Rock Slopes*, ASCE, New York, 1972, pp. 107–137.

Hendron, Jr., A. J., Mechanical properties of Rocks, in *Rock Mechanics in Engineering Practice*, Stagg, K.G. and Zienkiewicz, O.C., Eds., Wiley, New York, 1969, chap. 2.

Hibino, S.. Hayashi, M., Kanagawa, T. and Motojima, M., Forecast and Measurement of the Behavior of Rock Masses During Underground Excavation Works, *Proceedings of the International Symposium on Field Measurements in Rock Mechanics*, Zurich, Vol. 2, 1977, pp. 935–948.

Hoek, E. and Bray, J. W., *Rock Slope Engineering*, Institute of Mining and Metallurgy, London, 1977.

Hoek, E. and Brown, E. T., Empirical strength criterion for rock masses, *Proc. ASCE, J. Geotech. Eng. Div.*, 108, 1013–1035, 1980.

Hough, K. B., *Basic Soils Engineering*, The Ronald Press, New York, 1957.

Hunt, R.E., The significance of geology in determining soil profiles and stress histories, *Soils*, J. S. Ward and Assocs., Caldwell, NJ, 1967, chap. 1.

ISRM, Rock chatacterization and monitoring, in E.T. Brown, Ed., published for the *Comm. on Test Methods International Society for Rock Mechanics*, Pergamon Press, Oxford, 1981.

Jaeger, C., *Rock Mechanics and Engineering*, Cambridge University Press, England, 1972.

Jennings, J. E., An Approach to the Stability of Rock Slopes Based on the Theory of Limiting Equilibrium with a Material Exhibiting Anisotropic Shear Strength, *Stability of Rock Slopes, Proceedings of ASCE, 13th Symposium on Rock Mechanics*, University of Illinois, Urbana, 1972, pp. 269–302.

Kanji, M. A., Shear Strength of Soil Rock Interfaces, M.S. thesis, Department of Geology, University of Illinois, Urbana, 1970.

Kassif, G. and Baker, R., Swell Pressure Measured by Uni- and Triaxial Techniques, *Proceedings of 7th International Conference on Soil Mechanics and Foundation Engineering*, Mexico City, Vol. 1, 1969, pp. 215–218.

Kavazaniian, E., Jr. and Mitchell, J. K., Time-dependent deformation of clays. *Proc. ASCE, J. Geotech. Eng. Div.*, 106, 593–610, 1980.

Kayen, R.E., Mitchell, J.K., Seed, R.B. and Coutinho, R., Evaluation of SPT-CPT, and shear wave-based methods for liquefaction potential assessment using Loma Prieta data, *Proceedings of the 4th Japan–US Workshop on Earthquake Resistant Design of Lifeline Fac. and Countermeasures for Soil Liquefaction*, Vol. 1, 1992, pp. 177–204.

Kierstad and Lunne, *International Conference on Behavior of Offshore Structures*, paper, London, (from Construction Industry International, Dec. 1979).

Kjellman, W., Testing the shear strength of clay in Sweden, *Geotechnique*, 2, 225–235, 1951.

Krynine, D. and Judd, W. R., *Principles of Engineering Geology and Geotechnics*, McGraw-Hill Book Co., New York, 1957.

Lacroix, Y. and Horn, H. M., Direct Determination and Indirect Evaluation of Relative Density and Its Use on Earthwork Construction Projects, Evaluation of Relative Density and Its Role in Geotechnical Projects Involving Cohesionless Soils, ASTM STP 523, American Society for Testing and Materials, 1973, pp. 251–280.

Ladd, C. C. and Foott, R., New design procedure for stability of soft clays, *Proc. ASCE, J. Geotech. Eng. Div.*, 100, 753–786, 1974.

Lambe, T. W., *Soil Testing for Engineers*, Wiley, New York, 1951.

Lambe, T. W., Pore pressures in a foundation clay, *Proc. ASCE J. Soil Mech. Found. Eng. Div.*, 88, 19–47, 1962.

Lambe, T. W., Predictions in soil engineering, *Geotechnique*, 23, 1973.

Lambe. T. W., Methods of estimating settlements, *Proc. ASCE. J. Soil Mech. Found. Eng. Div.*, 90, 47–71, 1964.

Lambe, T. W. and Whitman, R. V., *Soil Mechanics*, Wiley, New York, 1969.

Leet, L. D., *Vibrations from Blasting Rock*, Harvard University Press, Cambridge, MA, 1960.

Leonards, G. A., *Foundation Engineering*, McGraw-Hill, New York, 1962, chap. 2.

Liao, S. and Whitman, R.V., Overburden correction factors for SPT in sand, *J. Geotech. Eng*, ASCE, 112, 373–377, 1986.

Lo, K.Y. and Lee, C.F., Analysis of progressive failure in clay slopes, *Proceedings of the 8th International Conference on Soil Mechanics Foundation Engineering*, 1, 251–258, 1973.

Lowe, III, J. and Zaccheo, P. F., Subsurface explorations and sampling, in *Foundation Engineering Handbook*, Winterkorn, H.F. and Fang, H.-Y., Eds., Van Nostrand Reinhold, New York, 1975, chap. 1.

Lukas, R. G. and deBussy, B. L., Pressuremeter and laboratory test correlations for Clays, *Proc. ASCE J. Geotech. Eng. Div.*, 102, 945–962, 1976.

Manchetti, S., *In situ tests by flat dilatometer, ASCE, J. Geotech. Div.*, GT3, 299–321, 1980.

Manchetti, S., Monaco, P., Totani, G. and Calabrese, M., The Flat Dilatometer Test in Soil Investigations. A Report by the ISSMGE Committee TC16, *IN SITU 2001 International Conference on In Situ Measurment of Soil Properties*, Bali, Indonesia, 2001.

Marcuson, W. F. and Bieganousky, W. A., Laboratory penetration tests on fine sands, *Proc. ASCE J. Geotech. Eng. Div.*, 103, 565–588, 1977a.

Marcuson, W. F. and Bieganousky, W. A., SPT and relative density in coarse sands, *Proc. ASCE J. Geotech. Eng. Div.*, 102, 1295–1309, 1977b.

Martin. R. E., Estimating settlements in residual soils, *Proc. ASCE J. Geotech. Eng. Div.*,103, 197–212, 1977.

Mather, K., Cement-Aggregete Reaction: What is the Problem?, Paper prepared for presentation, Panel Discussion on Cement Aggregate Reaction, Annual Meeting, American Institute of Mining and Metallurgical Engineers, New York, Feb. 21, 1956, pp. 83–85.

Menard, L. F., Calculations of the Bearing Capacity of Foundations Based on Pressuremeter Results, in *Sols-Soils*, Paris, Vol. 2, No. 5, June, and Vol. 2, No. 6, September.

Menard. L. F., Rules for the Computation of Bearing Capacity and Foundation Settlement Based on Pressuremeter Results, *Proceedings of the 6th International Conference on Soil Mechanics and Foundation Engineering*, Montreal, Vol. 2, 1965, pp. 295–299.

Menard, L. F., Rules for the Calculation of Bearing Capacity and Foundation Settlement Based on Pressuremeter Tests, Draft Translation 159, U.S. Army Corps of Engineers, Cold Regions Research and Engineering Lab, 1972.

Menard, L. F., Interpretation and Application of Pressuremeter Test Results, in *Sols-Soils*, Paris, Vol. 26, 1975, pp. 1–43.

Meyerhof, G. G., Penetration tests and bearing capacity of cohesionless soils, *Proc. ASCE, J. Soil Mech. Found. Eng. Div.*, 82, 866, 1956.

Miller, R. P., Engineering Classification and Index Properties for Intact Rock, Ph.D. thesis, University of Illinois, Urbana.

Misterek, D. L., Analysis of Data from Radial Jacking Tests. Determination of the In Situ Modulus of Deformation for Rock, ASTM STP 477, American Society for Testing and Materials, Philadelphia, PA, 1970.

Mitchell, J. K. and Lunne, T. A., Cone resistance as a measure of sand strength, *Proc. ASCE J. Geotech. Eng. Div.*, 104, 995–1012, 1978.

Murphy, D. J., Soils and Rocks: Composition, Confining Level and Strength, Ph.D. thesis, Department of Civil Engineering, Duke University, Durham, NC, 1970.

NAVFAC, *Design Manual DM-7.1, Soil Mechanics, Foundations and Earth Structures*, Naval Facilities Engineering Command, Alexandria, VA, 1982.

NCE, No Known Cure for Jersey Dam, NCE International, Institute of Civil Engineers, London, July, 1980, pp 38–39.

Novacs, W. D., Seed. H. B. and Chan. C. K., Dynamic moduli and damping ratios for a soft clay, *Proc. ASCE. J. Soil Mech. Found. Eng. Div.*, 97, 59–75, 1971.

Patton, F.D., Multiple modes of shear failure in rock, *Proceedings of the Ist International Congress of Rock Mechanic* Lisbon, Vol. 1, 1966, pp. 509–513.

Patton, F. D. and Hendron, A. J., Jr., General Report on Mass Movements, *Proceedings of the 2nd International Congress International Association Engineering of Geology*, Sao Paulo, 1974, p. V-GR 1.

Peck, R. B., Stability of Natural Slopes, Proceedings of ASCE, Stability and Performance of Slopes Embankments, Berkeley, CA, Aug. 1969, pp. 437–451.

Peck, R.B., Rock Foundations for Structures, Rock Engineering for Foundations and Slopes, *Proceedings of the ASCE*, New York, Vol. 2, 1976, pp. 1–21.

Peck, R. B., Hanson, W. E. and Thornburn, T. H., *Foundation Engineering*, 2nd ed., Wiley, New York, 1973.

Pells, P. J. N., Stress Ratio Effects on Construction Pore Pressures, *Proceedings of the 8th International Conference on Soil Mechanics and Foundation Engineering*, Moscow, Vol. 1, 1973, pp. 327–332.

Perloff, W. H., Pressure distribution and settlement, in *Foundation Engineering Handbook*, Winterkorn, H.F. and Fang, H.-Y., Eds., Van Nostrand Reinhold, New York, 1975, chap. 4.

Poulos, H. G., Behavior of laterally loaded piles: I-single piles, *Proc. ASCE, J. Soil Mech. Found. Eng. Div.*, 97, 722, 1971.

Richart, Jr., P. E., Foundation vibrations, in *Foundation Engineering Handbook*, Winterkorn, H.F. and Fang, H.-Y., Eds., Van Nostrand Reinhold, New York, 1975, chap. 24.

Robertson, P.K., *Cone Penetration Testing: Geotechnical Applications Guide:* produced by ConeTec Inc & Gregg In Situ, Inc., Vancouver, 2002.

Robertson, P.K and Campanella, R.G., Interpretation of cone penetration test: Part I, Sand, *Can. Geotech. J.*, 20, 718–733, 1983.

Robertson, P.K. and Wride, C.E., Evaluating cyclic liquefaction potential using the cone pentetration test, *Can. Geotech. J.*, 35, 442–459, 1998.

Rocha, M., New Techniques in Deformability Testing of in Situ Rock Masses, in *Determination of the In-Situ Modulus of Deformation of Rock*, ASTM STP 477, American Society for Testing and Materials, Philadelphia,PA, 1970.

Roscoe, K. H., An Apparatus for the Application of Simple Shear to Soil Samples, *Proceedings of the 3rd International Conference on Soil Mechanics and Foundation Engineering*, Switzerland, Vol. 1, 1953, pp. 186–291.

Sanglerat, G., *The Penetrometer and Soil Exploration*, Elsevier, Amsterdam, 1972.

Schmertmann, J. H., Static cone to compute static settlement over sand, *Proc. ASCE, J. Soil Mech. Found. Eng. Div.*, 96, 1011–1043, 1970.

Schmertmann, J. H., Guidelines for CPT Performance and Design, Pub. No. FHWA-TB 78-209, Federal Highway Administration, Washington, DC, 1977.

Schmertmann, J. H., Statics of SPT, *Proc. ASCE J. Geotech. Eng. Div.*, 105, 655–670, 1979.

Seed, H. B., Influence of Local Soil Conditions in Earthquake Damage, Speciality Session 2, Soil Dynamics, *Proceedings of the 7th International Conference on Soil Mechanics and Foundation Engineering*, Mexico City, 1969.

Seed, H. B., Evaluation of Soil Liquefaction Effects on Level Ground During Earthquakes, in *Liquefaction Problems in Geotechnical Engineering*, Preprint 2752, ASCE National Convention, Philadelphia, PA, 1976, pp. 1–104.

Seed, H. B., Soil liquefaction and cyclic mobility evaluation for level ground during earthquakes, *Proc. ASCE J. Geotech. Eng. Div.*, Vol. 105, 201–255, 1979.

Serafim, J. L., Influence of interstitial water on the behavior of rock masses, in *Rock Mechanics Engineering Practice*, Stagg and Zienkiewicz, Eds., Wiley, New York, 1969, chap. 3

Silver, V. A., Clemence, S. P. and Stephenson, R. W., Predicting deformations in the Fort Union formation, in *Rock Engineering for Foundations and Slopes, Proc. ASCE*, I, 13–33, 1976.

Simons, N. E., Prediction of Settlements of Structures on Granular Soils, *Ground Eng.*, 5, 1972.

Skempton, A. W., The Colloidal Activity of Clays, *Proceedings of the 3rd International Conference Soil Mechanics and Foundation Engineering*, Switzerland, Vol. I, 1953, pp. 57–61.

Skempton, A.K., Standard penetration test procedures and the effects of sands on overburden pressure, relative density, particle size, aging and overconsolidation, *Geotechnique*, 36, 425–447, 1986.

Skempton, A. W. and Bierrum, L., A contribution to the settlement analysis of foundation on clay, *Geotechnique*, VII, 1957.

Skogland, G. R., Marcuson, III, W. F. and Cunny, R. W., Evaluation of resonant column test device, *Proc. ASCE J. Geotech. Eng. Div.*, 102, 1147–1158, 1976.

Stagg, K. G., In-situ tests in the rock mass, in *Rock Mechanics in Engineering Practice*, Stagg and Zienkiewicz, Eds., Wiley, New York, 1969, chap. 5.

Swiger, W. F., Evaluation of Soil Moduli, *Proceedings of ASCE Conference on Analysis and Design* in *Geotechnical Engineering*, University of Texas, Austin, Vol. II, 1974, pp. 79–92.

Tarkoy. P. J., A Study of Rock Properties and Tunnel Boring Machine Advance Rates in Two Mica Schist Formations, Applications of Rock Mechanics, *Proceedings of 15th Symposium on Rock Mechanics*, Custer State Park, South Dakota, September 1973, ASCE, New York, pp. 415–447.

Taylor, D. W., *Fundamentals of Soil Mechanics*, Wiley, New York, 1948.

Terzaghi, K., *Theoretical Soil Mechanics*, Wiley, New York, 1943.

Terzaghi, K., Evaluation of the modulus of subgrade reaction, *Geotechnique*, 4, 1955.

Terzaghi, K. and Peck, R. B., *Soil Mechanics in Engineering Practice*, Wiley, New York, 1948.

Terzaghi, K. and Peck, R. B., *Soil Mechanics in Engineering Practice*, 2nd ed., Wiley, New York, 1967.

Terzaghi, K., Peck, R.B., and Mesri, G., Soil Mechanics in Engineering Practice, Wiley, New York, 1996, 549 pp.

Totani, G., Manchetti, S., Monaco, P. and Calabrese, M., Use of Flat Dilatometer Test (DMT) in geotechnical design, *IN SITU 2001 International Conference on In Situ Measurement of Soil Properties*, Bali, Indonesia, 2001.

Trofimenkov, J. G., Penetration Testing in USSR, *State-of-the-Art Report, European Symposium Penetration Testing*, Stockholm, Vol. 1, 1974.

Turnbull, W.J., Compaction and Strength Tests on Soil, paper presented at Annual ASCE Meeting, Jan. 1950.

USAEC, Soil Behavior under Earthquake Loading Conditions, National Technical Information Service Pub. TID-25953, U.S. Dept. of Commerce, Oak Ridge National Laboratory, Oak Ridge, TN, Jan., 1972.

USBR, *Earth Manual*, U.S. Bureau of Reclamation, Denver, Co, 1974.

USBM Borehole Device for Determining In-place Strength of Rock or Coal, Bureau of Mines, U.S. Dept. of Interior, Technology News No. 122, 1981.

Vesic, A. S., A Study of Bearing Capacity of Deep Foundations, Final Report, Project B-189, Georgia Institute of Technology, March 1967, p. 170.

Vijavergiya, V. N. and Ghazzaly, O. I. Prediction of Swelling Potential for Natural Clays, *Proceedings of the 3rd International Conference on Expansive Soils*, Haifa, Vol. 1, 1973.

Vucetic, M., Emerging Trends in Dynamic Simpler Shear Testing. International Workshop on Uncertainties in Nonlinear Soil Properties and Their Impact on Modeling Dynamic Soil Response. PEER-Lifelines Program, University of California, Berkeley, Mar. 18–19, 2004.

Ward, J. S., In situ pressuremeter testing on two recent projects, *Soils*, J. S. Ward and Assocs., Caldwell, NJ, Nov. 1972, pp. 5–7.

Wallace, B., Slebir, E. J. and Anderson, F. A., In Situ Methods for Determining Deformation Modulus Used by the Bureau of Reclamation, Determining the In-Situ Modulus of Deformation of Rock, STP 477, American Society for Testing and Materials, Philadelphia, PA, 1970, pp. 3–26.

Wineland, Borehole Shear Device, *Proceedings Conference on In Situ Measurement of Soil Properties,* Geotechnical Engineering Division, ASCE, North Carolina State University, June 1–4, 1975.

Wu, T. H., *Soil Mechanics*, Allyn and Bacon, Inc., Boston, 1966.

Further Reading

ASTM, Testing Techniques for Rock Mechanics, ASTM Special Technical Publication No. 402, American Society for Testing and Materials, Philadelphia, PA, 1965.

Barton, N., Estimating Shear Strength on Rock Joints, Proceedings of the 3rd International Congress on Rock Mechanics, International Society for Rock Mechanics, Denver, Vol. llA, 1974, pp. 219–221.

Deklotz, E. J. and Boisen, B., Development of Equipment for Determination of Deformation Modulus and In-Situ Stress by Large Flatjacks, in *Determination of In-Situ Modulus of Deformation of Rock*, ASTM Special Technical Publication 477, American Society For Testing and Materials, Philadelphia, PA, 1970.

DeMello, V. F. B., The Standard Penetration Test, State-of-the-Art Session 1, *Proceedings of the 4th Panamerican Conference Soil Mechanics and Foundation Engineering*, San Juan, Puerto Rico, Vol. 1, 1971, pp. 1–86.

Goodman, R. E., The Mechanical Properties of Joints, *Proceedings 3rd International Congress on Rock Mechanics, International Society for Rock Mechanics*, Denver, Vol. lA, 1974, pp. 127–140.

Grubbs, B. R. and Nottingham, L. C., Applications of Cone Penetrometer Tests to Gulf Coast Foundation Studies, presented at Meeting of Texas Section, ASCE, Corpus Christi, Apr., 1978.

Heuze, F. E., Sources of errors in rock mechanics field measurements and related solutions, *Intl. J. Rock Mech., Miner. Sci.,* 8, 1971.

Hilf, J. W., Compacted fill, *Foundation Engineering Handbook*, Winterkorn, H.F. and Fang, H.-Y., Eds., Van Nostrand Reinhold Co., New York, 1975, chap. 7.

Mitchell, J. K., *Fundamentals of Soil Behavior*, Wiley, New York, 1976.

NCE, Instruments-Cambridge Camkometer Goes Commercial, NCE International (The New Civil Engineer), Institute of Civil Engineers, London, Oct. 1978, p. 53.

Nelson, J. D. and Thompson, E. G., A theory of creep failure in overconsolidated clay, *Proc. ASCE, J. Geotech. Eng. Div.,* 102, 1281–1294, 1977.

Obert, L. and Duval, W. F., *Rock Mechanics and the Design of Structures in Rock*, Wiley, New York, 1967.

Poulos, H. G., deAmbrosis, L. P. and Davis, E. H., Method of calculating long-term creep settlements, *Proc. ASCE J. Geotech. Eng. Div.,* 102, 787–804, 1976.

Rapheal, J. M. and Goodman, R. E., Strength and deformability of highly fractured rock, *Proc. ASCE J. Geotech. Eng. Div.,* Vol. 105, 1285–1300, 1979.

Richart, Jr., F. E., Foundation vibrations, *Proc. ASCE J. Soil Mech. Found. Eng. Div.,* 88, 1960.

Richart, Jr.,F. E., Hall, Jr., J. R., and Woods, R. D., *Vibrations of Soils and Foundations*, Prentice-Hall, Englewood Cliffs, NJ, 1970.

Sharp, J. C., Maini, Y. N. and Brekke, T. L., Evaluation of the Hydraulic Properties of Rock Masses. *New Horizons in Rock Mechanics, Proceedings of the 14th Symposium on Rock Mechanics, ASCE,* University Park, PA, 1973, pp. 481–500.

Woodward. R. J., Gardner, W. S. and Greer, D. M., *Drilled Pier Foundations*, McGraw-Hill Book, New York, 1972.

3

Field Instrumentation

3.1 Introduction

3.1.1 Methods and Instruments Summarized

Monitoring methods and some instruments available are summarized for general reference in the following tables:

- Table 3.1 — method or instrument vs. condition to be monitored
- Table 3.2 — applications of simple vs. complex instruments
- Table 3.3 — methods or instrument summarized on basis of category of application

Comprehensive references on instrumentation have been prepared by Cording et al. (1975) and Dunicliff (1988). Catalogs available from Boart Longyear Interfels GmbH, Geokon, Soil Instrumentation Ltd., and Slope Indicator Co. are very informative.

3.1.2 Objectives

Instrumentation is installed to measure and monitor field conditions that are subject to changes including:

- Surface movements
- Subsurface deformations
- *In situ* earth and pore pressures
- Stresses on structural members

3.1.3 Applications

Foundation Design Studies

Measurements are made of deflections and stresses during load tests and preloading operations.

Construction Operations

During construction of buildings, embankments, retaining structures, open excavations, tunnels, caverns, and large dams, instrumentation is installed to monitor loads, stresses, and deformations to confirm design assumptions and determine the need for changes or remedial measures.

TABLE 3.1

Field Instrumentation: Methods and Devices

Condition to be Monitored	Surface Movements											Subsurface Deformation							Loads and Stresses								
	GPS	Survey nets	Water-level device	Settlement plates	Vertical extensometer	Tiltmeters	Inverse pendulum	Convergence meters	Strain meter	Terrestrial photography	Vibration monitors	Settlement points	Inclinometer	Deflectometer	Shear-strip indicator	Borehole extensometer	Electric strain meter	Acoustic emissions	Piezometers	Pressure cells	Load cells	Tell tales	Strain gages	Strain meters	Stress meters	Shallow-rock stress	Deep-rock stress
Settlements																											
Steel and masonry structures	X	X	X	X	X	X	X			X		X							X								
Fill embankments	X	X		X	X							X							X								
Dam embankments	X	X		X	X				X			X					X		X								
Ground subsidence	X	X			X	X			X			X					X		X								
Stability																											
Soil slopes	X	X							X				X		X			X	X								
Rock slopes		X				X			X				X		X			X	X					X			
Embankment and dam foundation		X						X		X			X	X		X		X	X	X							
Retaining structures										X									X	X	X						
Underground openings in rock		X				X		X	X					X		X	X	X	X						X	X	X
Tunnel linings		X						X						X		X				X	X						
Deformations																											
Subsurface																											
Soils, vertical		X		X	X							X			X				X								
Soils, lateral		X							X				X	X	X				X								
Rock		X											X	X	X	X			X						X	X	X
Faults		X						X	X		X						X										
Piles during load test																				X	X	X	X	X			
Structural elements																				X	X	X	X	X			
Pressures																											
Pore water																			X								
Embankments																				X							
Against walls, beneath foundations																				X							
Loads: on structural elements																					X		X	X			
Residual stresses: rock masses																							X	X	X		
Vibrations: seismic, man induced											X												X	X	X	X	

TABLE 3.2

Instrumentation: Simple vs. Complex Methods and Devices

Applications	Simple[a]	Complex[b]
Surface movements	Optical surveys of monuments, settlement plates	Global positioning system (GPS)
	Water-level device	Electrical tiltmeter
	Simple strain meter	Electrical strain meter
	Wire extensometer	Vertical extensometer
Subsurface deformations	Settlement points	Inclinometer
	Borros points	Deflectometer
	Rock bolt and rod-type MPBX extensometers	Wire-type MPBX (extensometer)
	Shear-strip indicator	Acoustical emissions
In situ pressures and stresses	Tell tales (pile load tests)	Pneumatic and electric piezometers
	Open-system piezometers	Strain gages
		Pressure cells
		Vibrating-wire stress meter
Residual rock stresses	Strain meter on rock surface	Borehole devices
	Flat jacks	

[a] Simple types are read optically or with dial gages. They are less costly to install and less subject to malfunction.

[b] Complex devices are attached to remote readouts, and many can be set up to monitor and record changes continuously. The device and installation are more costly, and reliability will usually be less than that of the simpler types. In many cases, however, there is no alternative but to use the more complex type.

TABLE 3.3

Instrumentation of Movements, Deformations, and Stresses

Method/Instrument	Applications
Surface Movements	
Survey nets	Vertical and horizontal movements of slopes, walls; settlements by precise leveling, theodolite, or laser geodimeter. Requires stable bench mark. Slow except with laser. GPS to monitor slope movements
Water-level device	To monitor building settlements optically
Settlement plates	Installed at base of fills and read optically for settlement monitoring
Remote settlement monitor (settlement extensometer)	Installed at base of fills, over tunnels in soft ground, adjacent to excavations to monitor vertical deflections
	Reference point installed below in rock or strong soils
	Instrument connected electrically to readout or recorder
Tiltmeters	Measure rotational component of deflection electronically
	Used in buildings, on walls, and rock-cut benches
Pendulums	Monitor tilt in buildings adjacent to excavations
Convergence meters	Measure convergence in tunnels, between excavation walls, and down cut slopes in rock
Surface extensometers or strain meters	Measure linear strains downslope or across faults. Small strain meters for joints
Terrestrial stereophotography	Monitor movements of slopes, retaining structures, and buildings. Less accurate than optical systems
Vibration monitoring	Monitor vibrations caused by blasting, pile driving, traffic, etc.
Subsurface Deformations	
Vertical rod extensometer	Settlement points installed at various depths allow measurement of increments of vertical deflection. Borros points, cross-arm device, and the remote settlement monitor also are used

(Continued)

TABLE 3.3

Instrumentation of Movements, Deformations, and Stresses (*Continued*)

Method/Instrument	Applications
Subsurface Deformations	
Inclinometer	Measure lateral deflections. Used behind walls, in lateral pile-load tests, for measuring deflections beneath loaded areas over soft soils, and to locate the failure surface in a slope and monitor slope movements
Deflectometers	Used in rock as permanent installation to monitor movements perpendicular to the borehole in rock slopes, open-pit mines, and fault zones
Shear-strip indicators	Used to locate failure surface in earth mass and to send an alarm when failure occurs
Borehole extensometers	Installed singly or in series (MPBX) in boreholes to monitor deflections occurring parallel to hole. Used to monitor slopes in rock, tunnels, and caverns. Installed in any orientation
Subsurface Deformations	
Electrical strain meters	Installed below the surface in earth dams to monitor longitudinal strains between embankment and abutment and to locate transverse cracks
Acoustical emissions device	Detect and monitor subaudible noise in soil and rock resulting from distress caused by slope movements and mine collapse, and along faults. Also used to locate leakage paths in dams
In Situ Pressures and Stresses	
Piezometers	Monitor pore-water pressures in slopes, dewatered excavations, beneath embankments, in dams, beneath buildings, and during preloading. Various systems available. Application depends on soil or rock conditions, response time required, and necessity for remote readout and recording
Stress or pressure cells	Measure stresses behind walls, in tunnel linings, beneath foundations during load test, and in embankments
Load cells	Measure loads in anchors, wall braces, and tunnel lining
Tell tales	Measure deflections at various depths in a pile during load test. Used to compute side friction and end bearing
Strain gages	Measure strains in piles during load test, bracing for retaining structures, earth and rock anchors, and steel storage tank walls during hydrostatic testing
Strain meters	Purposes similar to tiltmeters above, but meters tire encased so as not to be susceptible to short circuits, and are welded to the structure so as to not be subject to long-term creep of a cementing agent
Stress meters	Installed in borehole to measure stress changes during tunneling and mining operations
Residual Rock Stresses	
Shallow-Depth Methods	
Strain meters or rosettes	Stresses a short distance behind the wall remain unknown and excavation for test relieves some residual stress
Flat Jacks	Relatively low costs. Used in good-quality rock
Deep Methods	Permits deep measurement of residual stresses by borehole
Borehole devices	overcoring techniques. Any borehole orientation is possible,
Deformation gage	but installation and overcoring are difficult operations.
Inclusion stress meter Strain gage	Practical depth limit is about 10 to 15 m
Hydraulic fracturing	Allows very deep measurement, about 300 to 1500 m
	Boreholes are limited to vertical or near-vertical. Technique is in development stages

Postconstruction

After construction has been completed, stresses and deformations may be monitored to provide an early warning against possible failure of slopes, retaining structures, earth dams, or concrete dams with rock foundations and abutments.

Existing Structures

Structures undergoing settlement are monitored to determine if and when remedial measures will be required and how stability will be achieved.

Instrumentation is installed to monitor the effects on existing structures of (1) dynamic loadings from earthquake forces, vibrating machinery, nearby blasting or other construction operations; and (2) deformations resulting from nearby tunneling or open excavations.

Slopes

Instrumentation is installed to monitor movements of rock and soil slopes, to locate failure surfaces and monitor their deflections, and to monitor groundwater conditions.

Mineral Extraction

The extraction of oil, gas, water, coal, or other minerals from the subsurface can result in surface deformations that require monitoring.

Tectonic Movements

In areas of crustal activity, movements and stresses related to surface warping and faulting are monitored.

3.1.4   Program Elements

General

In planning an instrumentation program, the initial steps involve determining what quantities should be measured and selecting the instrument or instruments. The choice of instruments requires consideration of a number of factors, followed by a design layout of the instrumentation arrays for field installation.

Execution requires installation and calibration, data collection, recording, processing, and interpretation.

Instrumentation Selection

Conditions to be Monitored

Instrumentation types and methods are generally grouped by the condition that has to be monitored, i.e., surface movements, subsurface deformation, and *in situ* pressures and stresses.

Application

Instruments vary in precision, sensitivity, reliability, and durability. The relative importance of these factors varies with the application or purpose of the program, such as collecting design data or monitoring construction and postconstruction works, existing structures, or the effects of mineral extraction or tectonic activity.

Instrumentation Characteristics

Precision refers to the degree of measurement agreement required in relation to true or accepted reference values of the quantity measured.

Sensitivity refers to the smallest unit detectable on the instrument scale, which generally decreases with variations in the instrument measurement range.

Repeatability refers to obtaining sequential readings with similar precision. It can be the most significant feature, since it indicates the trend in change of the measurement quantity and is often more important than sensitivity or precision.

Drift refers to the instrument error progression with time and affects precision.

Reliability refers to the resistance to large deformations beyond the desired range, high pressures, corrosive elements, temperature extremes, construction activities, dirty environment, humidity and water, erratic power supply, accessibility for maintenance, etc. For a particular application, the less complicated the instrument the greater is its reliability. Thus, mechanical instruments are preferred to electrical; stationary parts are preferred to moving parts. Electrical instruments should be battery-powered for greater reliability.

Service life required can vary from a few weeks, to months, to a year, or even several years and is influenced by the reliability factors.

Calibration

Instrument Precision

Precision is verified upon the receipt of an instrument from the manufacturer by checking it against a standard and corrections are considered for temperature and drift. After installation, precision is verified, then checked periodically during monitoring and upon project completion.

Bench Marks or Other Reference Data

Reference points such as bench marks are installed or established and protected to ensure reliability against all possible changes that may occur during the project life. Care should be taken to ensure that the bench mark does not move. The best bench marks are supported on sound rock; Borros points (Section 3.3.2) set in hard soil are an option where the rock is very deep. "Temporary" reference points that require transfer to permanent references should be avoided.

Installation and Maintenance

Some systems are relatively simple to install and maintain, since they are portable and require only surface reference points; others require boreholes or excavations and are therefore difficult to install and maintain. These factors substantially affect costs.

Operations

Data Collection and Recording

Data collection should take place on a planned basis. Data may be collected and recorded manually or automatically, directly or remotely, and on a periodic or continuous basis. Automatic collection and remote recording can be connected to alarm systems and are useful in dangerous situations, or in situations where long periods of inactivity may be followed by some significant occurrence such as an earthquake or heavy rainfall. Readout systems measure and display the measured quantity.

Frequency of Observations

Construction progress, data trends, and interpretation requirements influence the frequency of the observations. Periodic monitoring is always required. The frequency of readings increases when conditions are critical.

Data Processing and Interpretation

Data are recorded, plotted, reviewed, and interpreted on a programmed basis, either periodic or continuous. Failures have occurred during monitoring programs when data were recorded, collected, and filed, but not plotted and interpreted.

3.1.5 Transducers

General

Transducers, important elements of complex instruments, are devices actuated by energy from one system to supply energy, in the same or some other form, to a second system. They function generally on the membrane principle, the electrical resistance gage principle, or the vibrating-wire principle. Occasionally they function on the linear-displacement principle.

Membrane Principle

Pressure against a membrane is measured either hydraulically (e.g., by hydraulic piezometers) or pneumatically (e.g., by the Gloetzl pressure cell, the pneumatic piezometer, or geophones).

Resistance Strain Gages

Applications

Resistance strain gages are cemented to structural members or used as sensors in load cells, piezometers, extensometers, inclinometers, etc.

Principle

The straining of a wire changes its cross-sectional area and consequently its electrical resistance. When a strain-gage wire is attached to a structural member, either externally or as the component of an instrument, measurements of changes in electrical resistance are used to determine strains in the structural element. There are several types of gages.

Bonded Gages

The most common form is the bonded gage, which consists of a thin wire filament or metal foil formed into a pattern and bonded to a backing of paper or thin plastic (Figure 3.1). The backing is then cemented to the surface where the measurements are to be made.

Gage lengths range from 1/16 to 6 in., and strain sensitivity is usually of the order of 2 to 4 microstrains, with ranges up to 20,000 to 50,000 microstrains (2–5% strain) for normal gages. High elongation gages ranging up to strains of 10 to 20% are available.

Encapsulated Gages

Bonded gages mounted and sealed in the factory into a stainless steel or brass envelope are termed as encapsulated gages. They are welded to the measurement surface or embedded in concrete. Encapsulated gages provide better protection against moisture than unbonded gages.

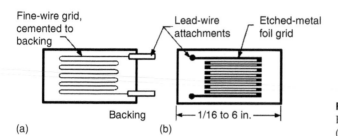

FIGURE 3.1
Bonded resistance strain gages:
(a) wire grid, (b) foil grid.

Unbonded, Encapsulated Gages

A fine wire is strung under tension over ceramic insulators mounted on a flexible metal frame to form a resistance coil in unbonded, encapsulated gages (Figure 3.2). Usually two coils are used and arranged so that one contracts while the other expands when the frame is strained. The coils and the frame are factory-sealed into a tubular metal cover to provide moisture protection. Mounting is achieved by bolting the gage to saddle brackets previously tack-welded or bolted to the measuring surface. The Carlson strain meter is one version of the unbonded, encapsulated resistance gage.

Readouts

Strain indicators measure the resistance changes of the gages. They consist of a power supply, various fixed and variable resistors, a galvanometer, and a bridge circuit that can be switched to connect the strain gages, resistors, and galvanometers in various configurations.

Temperature Compensation

Temperature compensation is always required and can be achieved by a dummy gage mounted so that it responds only to temperature-induced resistance changes or by the use of self-compensating gages.

Vibrating-Wire Devices

Applications

Vibrating-wire devices are used for measurements of:

- Strain with the transducer mounted on steel or embedded in concrete
- Displacement and deformation with extensometers and joint meters
- Tensile and compressive forces
- Pore and joint water pressures
- Changes in rock stresses
- Changes in surface or subsurface inclinations
- Temperature changes

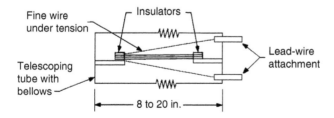

FIGURE 3.2
Unbonded resistance strain gage. (From Cording, E.J. et al., University of Illinois, Urbana, Vols. 1 and 2, 1975. With permission.)

Principle

A vibrating steel wire is attached to the object on which a measurement has to be made. A change in the quantity to be measured causes a change in stress in the measuring wire and consequently a change in its frequency. The measuring wire oscillates within a magnetic field, inducing in a coil an electrical oscillation of the same frequency, which is transmitted by cable to a receiving instrument.

Types

Intermittent vibrating-wire systems, used for static and semistatic measuring systems, consist of a transducer containing the measuring wire and an electromagnet, and a receiver. *Continuous vibrating-wire systems* contain the measuring wire, an oscillator, and two electromagnets in the transducer, and are used for static and dynamic measuring devices and for alarm and control systems (Dreyer, 1977).

Characteristics

Gage lengths generally range from 4 to 14 in., and strain sensitivity has typical maximum ranges from 600 to 7000 microstrains.

Linear-Displacement Transducers

Applications

Linear-displacement transducers are generally used in extensometers.

Linear Potentiometer

Resistance devices, linear potentiometers, consist of a mandrel wound with fine wire or conductive film. A wiper attached to a shaft rides along the mandrel and divides the mandrel resistance into two parts. The resistance ratio of these parts is measured with a Wheatstone bridge circuit to determine the displacement of the wiper and shaft.

Movement ranges are generally about 0.5 to 24 in. with an average sensitivity of 0.01 to 0.001. Linear potentiometers are extremely sensitive to moisture and require absolute sealing.

Linear Variable Differential Transformer (LVDT)

The LVDT converts a displacement into a voltage change by varying the reluctance path between a primary coil and two or more secondary coils when an excitation voltage is applied to the primary coil. Variations in the output signal are calibrated to displacements of the LVDT core.

Movement ranges are generally about 0.1 ft to several feet; sensitivity is about 10 microstrains. LVDTs are much less sensitive to moisture and less affected by temperature than are linear potentiometers.

3.2 Surface Movements

3.2.1 Forms and Significance

Vertical Displacements

Settlement or heave is a vertical surface displacement that results in detrimental distortion of structures, especially if differential, when significant magnitudes are reached (see also Section 3.3.2).

Tilt

Tilt can indicate differential settlement of a structure, impending failure of walls and slopes, or major tectonic movements.

Lateral Displacement

Lateral surface displacements can signify slope failure or fault activity. The surfaces of tunnels and other underground openings undergo convergence from stress relief, which may lead to failure.

Vibrations

Construction activity or earthquakes can induce vibrations in structures, resulting in over-stress and failure.

Summary

Forms of surface movement, their occurrence, and applicable monitoring methods are summarized in Table 3.4.

TABLE 3.4

Surface Movements: Forms, Occurrence, and Monitors

Movement	Occurrence	Monitoring Method	Reference
Settlement or subsidence	Fills, surcharge: buildings	Optical survey nets with monuments or settlement plates	Section 3.2.2
	Over tunnels, behind walls	Settlement extensometers	Section 3.3.2
Heave or rebound	Fills, surcharge: buildings	Optical survey nets	Section 3.2.2
		Settlement extensometers	Section 3.3.2
Differential settlement	Within buildings	Optical survey nets	Section 3.2.2
		Water-level device	Section 3.2.2
Tilt	Retaining structures, buildings, rock-slope benches	Portable electric tiltmeter	Section 3.2.3
		Pendulums	Section 3.2.3
		Terrestrial stereophotography	Section 3.2.2
	Ground surface	Mercury pools set in piers	Section 3.2.3
Lateral displacement	Slopes	Survey nets and GPS	Section 3.2.2
		Terrestrial photography	Section 3.2.2
		Wire extensometer	Section 3.2.4
		Precision electric strain meter	Section 3.2.4
		Sliding wire contact	Section 3.2.4
	Tension cracks, joints	Simple strain meter with pins	Section 3.2.4
	Faults	Survey nets	Section 3.2.2
		Precision electric strain meter	Section 3.2.4
Convergence	Tunnels	Wire extensometer	Section 3.2.4
Vibrations	Blasting, pile driving, etc.	Vibration monitors	Section 3.2.5

3.2.2   Surveying

GPS (Global Positioning System)

Applications

GPS is one of the most important instrumentation systems to have been developed in the past 20 years. Its applications in geotechnical engineering will increase as the system sensitivity increases. At present it is used to monitor slope movements and surface movements in seismically active areas.

Method

GPS is a worldwide radio-navigation system at present made up of a constellation of 24 satellites and their ground stations. GPS receivers have been miniaturized to just a few integrated circuits. Positioned on a surface, the antennas measure the travel time of radio signals from at least three of the 24 satellites. Antenna locations are determined by triangulation and by applying complex geometric relationships and very accurate timing to the radio signals from the satellites. Antenna positions are calculated accurately to a matter of meters with standard equipment, and with advanced forms of GPS the positions can be calculated to about 1 cm.

The GPS receiver locations and elevations can be read when the data are received, or the data can be sent by a radio modem and antenna to a personal computer where a graphical output of vertical and horizontal movements are stored and can be displayed.

Survey Nets

Applications

Survey nets making use of optical systems are used to monitor deflections of slopes, walls, buildings, and other structures, as well as ground subsidence and heave.

Methods

Ranges, accuracy, advantages, limitations, and reliability of various surveying methods are summarized in Table 3.5.

Procedures

Reference points are installed to provide a fixed framework. They include ground monuments, pins in structures, and immovable bench marks.

Close-distance surveys are usually performed with a first-order theodolite, and critical distances are measured by chaining with Invar or steel tapes, held under a standard tension.

Long-distance surveys, such as for slope movements, are more commonly performed with the geodimeter (electronic distance-measuring unit or EDM) because of a significant reduction in measurement time and increased accuracy for long ranges. Distances are determined by measuring the phase difference between transmitted and reflected light beams, using the laser principle. Calculation and data display can be automated by using an electronic geodimeter with a punched-tape recorder. The accuracy for slope monitoring is about 30 mm (as compared with 300 mm for the theodolite) (Blackwell et al., 1975), depending on the sight distance. An example of a scheme for geodetic control of a landslide is given in Figure 3.3. Sequential plotting of the data reveals the direction and velocity of the movements.

TABLE 3.5

Surveying Methods Summarized[a]

Method	Range	Accuracy	Advantages	Limitations and Precautions	Reliability
Chaining	Variable	$\pm\dfrac{1}{5000}$ to $\pm\dfrac{1}{10,000}$ Ordinary (third order) survey $\pm\dfrac{1}{20,000}$ to $\pm\dfrac{1}{200,000}$ Precise (first-order) survey	Simple, direct observation Inexpensive	Requires clear, relatively flat surface between points and stable reference monuments Corrections for temperature and slope should be applied and a standard tension used	Excellent
Electronic distance measuring	50 to 10,000 ft	$\pm\dfrac{1}{50,000}$ to $\pm\dfrac{1}{300,000}$ of distance	Precise, long range, fast Usable over rough terrain	Accuracy influenced by atmospheric conditions Accuracy at short ranges (under 100–300 ft) is limited	Good
Optical leveling Ordinary–second and third order		±0.01 to ±0.02 ft	Simple, fast (particularly with self-leveling instruments)	Limited precision. Requires good bench mark nearby	Excellent
Precise parallel-plate micrometer attachment, special rod, first-order technique		±0.002 to ±0.004 ft	More precise	Requires good bench mark and reference points, and careful adherence to standard procedures	Excellent
Offsets from a baseline Theodolite and scale	0 to 5 ft off-baseline	±0.002 to ±0.005 ft	Simple direct observation	Requires baseline unaffected by movements and good monuments.	Excellent

(Continued)

TABLE 3.5

(Continued)

Method	Range	Accuracy	Advantages	Limitations and Precautions	Reliability
				Accuracy can be improved by using a target with a vernier and by repeating the sight from the opposite end of the baseline	
Laser and photocell detector	0 to 5 ft off-baseline	±0.005 ft	Faster than transit	Seriously affected by atmospheric conditions	Good
Triangulation		±0.002 to 0.04 ft	Usable when direct measurements are not possible. Good for tying into points outside of construction area	Requires precise measurement of base distance and angles. Good reference monuments are required	Good
Photogrammetric methods		$\pm\dfrac{1}{5000}$ to $\pm\dfrac{1}{50,000}$	Can record hundreds of potential movements at one time for determination of overall displacement pattern	Weather conditions can limit use	Good

Note: GPS becoming standard surveying practice (Section 3.2.2). Current precision about 1 cm.

[a] From Cording, E. J. et al., Department of Civil Engineering, University of Illinois, Urbana, Vols. 1 and 2, 1975.

Water-Level (Hose-Level) Device

Application

Floor, wall, or column deflections inside structures undergoing settlements or heave may be monitored with the water-level device. Its accuracy ranges from ±0.001 to 0.5 in.

Procedure

Monument pins are set at a number of locations on walls, columns, and floors. Elevations relative to some fixed point are measured periodically with the device, which consists of two water-level gages connected by a long hose. In operation, one gage is set up at a bench-mark elevation and the other is moved about from pin to pin and the elevation read

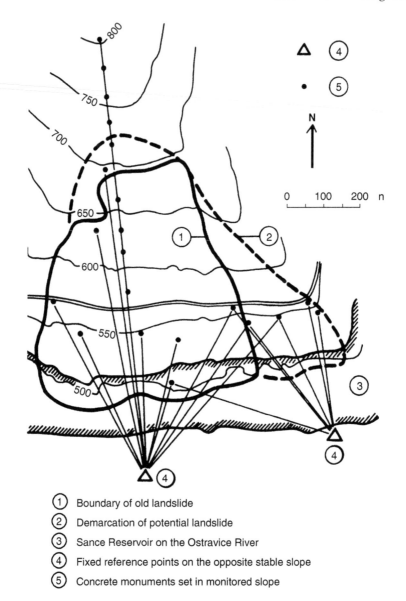

① Boundary of old landslide
② Demarcation of potential landslide
③ Sance Reservoir on the Ostravice River
④ Fixed reference points on the opposite stable slope
⑤ Concrete monuments set in monitored slope

FIGURE 3.3
Scheme of geodetic control of the Recice landslide. Note that a laser geodimeter was used in the survey because of significant reduction in measurement time and increased accuracy. (From Novosad, S., *Bull. Int. Assoc. Eng. Geol.*, 17, 71–73, 1978. With permission.)

at the control gage. The principle is similar to that of the theodolite, except that the water-level device is quicker and simpler for working around walls and columns inside buildings. Temperature differential, as might be caused by radiators or sunshine affecting the hose, may introduce considerable error.

Data Reduction

Plots of movements measured periodically reveal the rate and magnitude of differential settlement or heave within the structure. Determination of total movements requires an immovable reference point such as a bench mark set into rock.

Settlement Plates

Application

Ground surface settlements under fills and surcharges are monitored with settlement plates.

Procedure

Plates, at least 3 ft (90 cm) square, made of heavy plywood or steel, are placed carefully on prepared ground. A 1 in. rod is attached to the plate and its elevation is determined by survey.

As fill is placed, readings of attachments are taken, and when necessary, additional rod sections are attached. A relatively large number of plates should be installed on a given site, not only to provide data on differential movements, but also because some will invariably be disturbed or destroyed during filling operations.

Data Reduction

Plots of movements provide data on the rate and magnitude of settlement when referenced to an immovable bench mark.

Terrestrial Photography

Applications

Slopes, retaining structures, buildings, and other structures experiencing movements may be monitored on a broad basis with terrestrial photography.

Procedure

A bench-mark reference point is established and photographs are taken periodically from the same location, elevation, and orientation with a high-resolution camera and film. After each photo, the camera is moved a short distance laterally (about 1 m) to obtain at least a 60% overlap of adjacent photos. Sequential photos provide pairs suitable for stereoscopic viewing when enlarged. Phototheodolites provide a sophisticated system.

Data Obtained

The direction and relative magnitude of movements are detected by comparing sequential photos. The precision is not equal to that of the standard survey techniques, but it may be adequate for detecting changes in the locations of survey reference points and providing for approximate measurements if a reference scale is included in the photo. Stereo-terrestrial photographs are extremely useful for mapping joints and other discontinuities on rock slopes.

3.2.3 Tiltmeters

Electrical Tiltmeter

Application

Tilting or rotational movements of retaining walls, structures, and rock benches are monitored with electrical tiltmeters.

Instrument

The portable tiltmeters utilize a closed-loop force-balanced servo-accelerator to measure tilt. Accuracy is reported to be equivalent to a surface displacement of 0.06 in. of a rock mass rotating on an axis 100 ft beneath the surface, or a displacement of 200 μin. over the 4-in. length of the instrument.

Procedure

Ceramic plates are cemented to the surface at locations to be monitored. The plates contain pegs that which serve as reference points for aligning the tiltmeter and measuring the angular deflection. Changes in tilt are found by comparing the current reading to the initial reading.

Pendulums

Application

Pendulums are used to measure the horizontal distance between two points at different elevations. They are installed in the highest elevations of structures adjacent to open excavations areas or over tunneling operations, where ground subsidence may cause differential settlement and tilting.

Instrument

The two general pendulum types, direct and inverted, are illustrated in Figure 3.4.

Mercury Pools Set in Piers

Application

Mercury pools set in piers were installed by National Oceanic and Atmospheric Administration's (NOAA) Earthquake Mechanism Laboratory, Stone Canyon, California, adjacent to the San Andreas Fault to monitor ground surface movements over a large area (Bufe, 1972).

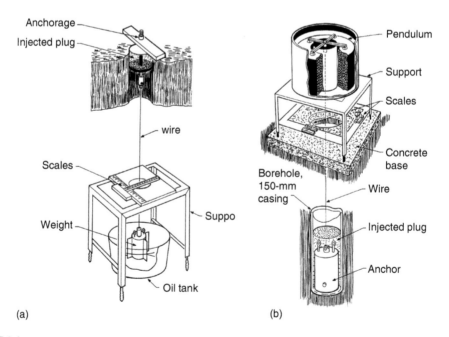

(a)   (b)

FIGURE 3.4
Pendulums of Soil Instruments Ltd: (a) direct pendulum, (b) inverted pendulum. (From Franklin, J.A., *Int. J. Rock Mech. Miner. Sci. Geomech. Abstr.*, 14, 163–192, 1977. With permission.)

Instrument

Similar in principle to the water-level device, mercury pools were set in piers 20 m apart in an interconnected array and attached to Invar rods driven to resistance in the ground.

3.2.4 Extensometers

Simple Strain Meter

Application

Joint or tension crack deformations in rock masses may be monitored with simple strain meters; they are particularly useful on slopes.

Installation

Pins are installed in a triangular array adjacent to the crack and initial reference measurements are made. Periodic measurements monitor vertical, lateral, and shear displacements along the crack.

Simple Wire Extensometer

Application

Slope movements are monitored with simple moving wire contacts that can give warning of large deformations, a particularly useful feature for open-pit mine and other rock slopes.

Installation

A stake is driven into the ground, upslope of the crack, and is attached to a tensioned wire that becomes the measurement station. The wire is attached to a stake driven into an unstable area. The wire passes over guides in the measurement station to connect to a weight to maintain wire tension. At the measurement station, trip blocks are attached to the wire that passes along a steel scale. Displacements are measured periodically from the locations of the trip blocks along the scale. A trip block can be placed on the scale at the location of the desired maximum movement and can be connected to an alarm (see Figure 3.5).

Electrical Crack Gage

Application

An electrical crackmeter continuously monitors rock slope joint movements when connected to a data logger. The initial reading is used as a datum. Subsequent readings provide data on magnitude, rate, and acceleration of movements across the joint.

Installation

A displacement transducer is connected to anchors or steel pins set in the rock on each side of the joint. A change in distance across the joint causes a change in the frequency signal produced by the transducer when excited by the readout or data logger. The readout device processes the signal, applies calibration factors, and displays a reading in millimeters or inches.

Wire Extensometer (Convergence Meter)

Applications

Wire extensometers are used to measure the movements of soil- or rock-cut benches in the downslope direction as indicators of potential failure or to monitor convergence in tunnels, particularly along shear zones where movements are likely to be relatively large. In tunnels they provide data based on the need for additional supports (Hartmann, 1966).

Instrument

An Invar steel wire is attached to a tensioning device and dial indicator as shown in Figure 3.6. Reference pins are grouted into tunnel walls or mounted on monuments on slopes. One end of the wire is connected to a pin (Figure 3.6) and the instrument is attached to the opposite pin (Figure 3.7). The wire is adjusted to a calibrated tension and the distance between pins is measured with the dial gage. Well-made instruments, such as those of Interfels, used properly, have an accuracy of 1×10^{-5} times the measurement length (Silveira, 1976).

Slope Monitoring

As shown in Figure 3.8, pins are set in monuments on slope benches in a series of parallel lines running up- and down-slope, with each line starting well behind the slope crest and

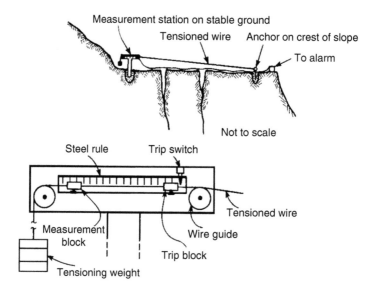

FIGURE 3.5
Wire extensometer on crest of rock slope. (From FHWA, USDOT, FHA, Pub. No. FHWA-TS-89-045, September 1989. With permission.)

FIGURE 3.6
Tensioning device and dial indicator used to measure convergence.

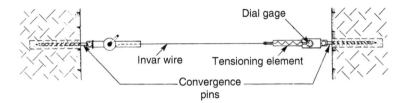

FIGURE 3.7
Convergence meter. (From Silveira, J.F.A., *Proceedings of the 1st Congress*, Brasileira de Geologia de Engenharia, Rio de Janeiro, August, Vol. 1, 1976, pp. 131–154. With permission.)

FIGURE 3.8
Installation of convergence monuments on a cut slope in a mine.

any potential failure surface, and extending to lower bench levels (Figure 3.41). Either positive or negative movements will be measured between any two adjacent pins. Interpretation of where the movements are occurring along the slope requires plotting the readings between any two pins on a section showing all the pins.

Tunnel Convergence Monitoring

There are two basic layouts for positioning the pins in the tunnel walls: one provides for measurements across the diameter but has the often serious disadvantage of interfering with operations, and the other provides for measurements around the perimeter with less interference in tunneling operations.

For making measurements across the diameters of tunnels in rock, the pins are installed in arrays positioned on the basis of the geologic structure. In Figure 3.9, an array for horizontally bedded rock provides a concentration around the roof arch, since roof deflections are likely to be the most significant movements. The base pins at points I and II are reference pins. In Figure 3.10, the pin array is positioned for measurements when the rock structure is dipping. The pins are concentrated where the structure dips out of the tunnel wall, since failure is most likely at this location. These arrays provide relatively precise measures of tunnel closure.

Stringing the wire through a series of rollers attached to pins located around the tunnel wall, as in Figure 3.11a, leaves the tunnel unobstructed for construction activities. Measurements of the net change in tunnel diameter are made but the locations of maximum

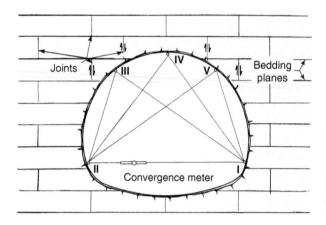

FIGURE 3.9
Convergence meter pin array for
horizontally bedded rock in a tunnel. (From
Silvera, J.F.A., *Proceedings of the 1st Congress*,
Brasileira de Geologia de Engenharia, Rio
de Janeiro, August, Vol. 1, 1976, pp.
131–154. With permission.)

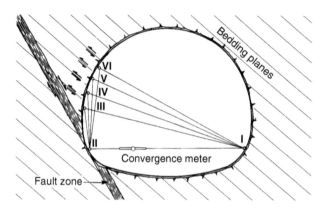

FIGURE 3.10
Convergence meter pin array for dipping
rock structure in a tunnel. (From Silvera,
J.F.A., *Proceedings of the 1st Congress*,
Brasileira de Geologia de Engenharia, Rio
de Janeiro, August, Vol. 1, 1976, pp.
131–154. With permission.)

deflection are not known. If creep occurs in the wire, accuracy reduces to values of about 0.2
to 0.8 in. (Silveira, 1976).

The Bassett convergence system is a more sophisticated system. Tilt sensors are used to
monitor the position of reference pins installed around a tunnel section (Figure 3.11b). The
sensors are linked to pins by a system of arms. Spatial displacement of the pins results in
changed tilt readings that are recorded on a data logger. A computer calculates displacement data and generates a graphic display of tunnel displacements. Deformations as small
as 0.02 mm can be detected.

3.2.5   Vibration Monitoring

Purpose

Ground motion is monitored to obtain data for the evaluation of vibration problems associated with blasting, road and rail traffic, pile driving, the operation of heavy equipment,
and rotating or reciprocating machinery.

Equipment

Small portable seismographs (particle-velocity seismographs) are used to determine frequency, acceleration, and displacement of surface particles in three rotational directions of
vibration motion. They also record sound pressure levels.

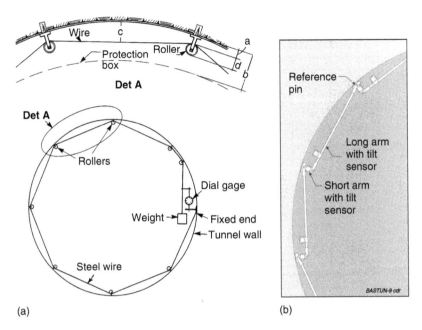

FIGURE 3.11
Monitoring tunnel wall movements (a) Stringing an Invar wire through a series of rollers attached to pins around the tunnel wall leaves the tunnel unobstructed for construction activity. (From Silvera, J.F.A., *Proceedings of the 1st Congress*, Brasileira de Geologia de Engenharia, Rio de Janeiro, August, Vol. 1, 1976, pp. 131–154. With permission.) (b) The Bassett system uses tilt sensors to monitor changes in the position of reference pins. A data-logger automatically records tilt readings. (Courtesy of Slope Indicator Co.)

Principles

Frequency and amplitude are the basic elements of the waveform of harmonic motion from which acceleration, force, particle velocity, and kinetic energy can be determined. Acceleration or displacement, when combined with the measured frequency, can be used to control or measure vibration damage by providing a measure of the energy transmitted by the vibration source. Some instruments directly measure the peak particle velocity. The energy transmitted is directly proportional to the square of the peak velocity.

Ground motion is the result of the induced vibrations. A scale of vibration damage to structures and the limits of human perception in terms of amplitude and frequency is given in Figure 3.12.

3.3 Subsurface Deformations

3.3.1 Forms and Significance

Vertical Displacements

Compression or consolidation of strata under applied stress results in surface settlements as described in Section 3.2. In some cases, it is important to determine which strata are contributing most significantly to the vertical displacements.

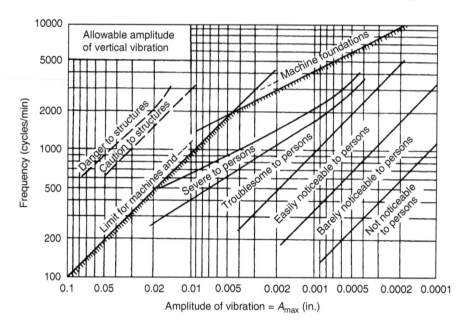

FIGURE 3.12
Scale of vibration damage to structures and limits of human perception. (From Richart, F.E., Jr., *Proc. ASCE J. Soil Mech. and Found. Eng. Div.*, 86, 1960. With permission.)

Lateral Displacements

Slopes, retaining structures, pile foundations, soft ground beneath embankment loads, and faults all undergo lateral deformations. Earth pressures against retaining walls cause tilting or other wall deflections. Slope failures often occur along a zone or surface, resulting in lateral displacements. Surface loads can cause lateral displacement of buried, weak soil strata. Foundation piles deflect laterally during load tests and while in service.

Strain Gradients

Differential deflection of earth dam embankments and reinforced earth walls, the closure of excavations in rock masses, and the imposition of high loads in rock masses result in strain gradients.

Internal Erosion

Seepage forces through, around, or beneath earth embankments or beneath or around concrete dams or cofferdams cause internal erosion.

Summary

The monitoring methods are summarized in terms of the type of ground movement and its occurrence in Table 3.6.

3.3.2   Vertical Displacement

Borehole Extensometers

Various types of borehole extensometers are available to provide settlement profiles. They are commonly based on lowering a probe down the borehole that can contain flush-coupled

TABLE 3.6

Subsurface Deformations: Forms, Occurrence, and Monitors

Movement	Occurrence	Monitoring Method	Reference
Vertical displacement	Ground compression at various depths	Vertical-rod extensometer	Section 3.3.2
		Borros points	
		Cross-arm device	
	Mine roof deflection	Acoustical emissions device	Section 3.3.5
Lateral displacement	Slope failure zone (*a*, *b*, *c*, *d*)	(*a*) Inclinometer for soil and rock masses	Section 3.3.3
	Wall or pile deflection (*a*)	(*b*) Deflectometers for rock	
		(*c*) Shear strip indicators	
	Weak soils under stress (*a*)	(*d*) Slope failure sensor	
	Fault movements (*a*, *b*, *d*)	(*e*) Acoustical emission device	
Linear strain gradients	Dam foundations or abutments	Rock bolt extensometers	Section 3.3.4
	Open or closed rock excavations	Multiple position borehole extensometers (rod or wire)	
	Differential settlement of dam bodies and foundations	Electrical strain meters	Section 3.3.4
Piping erosion	Through, around, or beneath dams	Acoustical emission device	Section 3.3.5

pipe, corrugated plastic pipe, or inclinometer casing. The probes record depths by sensing magnets or other sensor rings set in the borehole, or by a hook-type probe as described in the Slope Indicator Catalog.

Rod Extensometers (Settlement Reference Points)

Applications

The vertical rod extensometer is a simple device for measuring deformations between the surface and some depths in most soil types, except for very soft materials.

Instrument

An inflatable bag is set into an HX-size borehole at a depth above which the deformations are to be monitored. The bag is connected to the surface with a 1-in.-galvanized pipe installed inside a 3-in.-diameter pipe, which acts as a sleeve permitting freedom of movement of the inner pipe. A dial gage is mounted to read relative displacement of the 1-in.-diameter pipe and a reference point on the surface.

Installation

An HX-size hole is drilled to test depth and the 3 in. pipe installed as shown in Figure 3.13a. The 1 in. pipe and bag are lowered to test depth, with the pipe centered with rubber spacers free to move in the 3 in. pipe. Cement grout is injected into the inflatable bag through the 1 in. pipe under pressures of about 50 psi to expand the bag and secure it against the sides of the hole, below the casing. An expanded bag under surface test is illustrated in Figure 3.13b. When the bag and pipe are filled, the grout returns to the surface. This return can be signaled by the bursting of a short length of rubber tubing installed in the top of the grout pipe. A 10-in.-diameter steel pipe set in a concrete pad at the ground surface, on which a dial gage is mounted, serves as the reference measuring component (Figure 3.13c).

Piezometers are also installed (see Section 3.4.2) to monitor porewater pressures during the settlement observations.

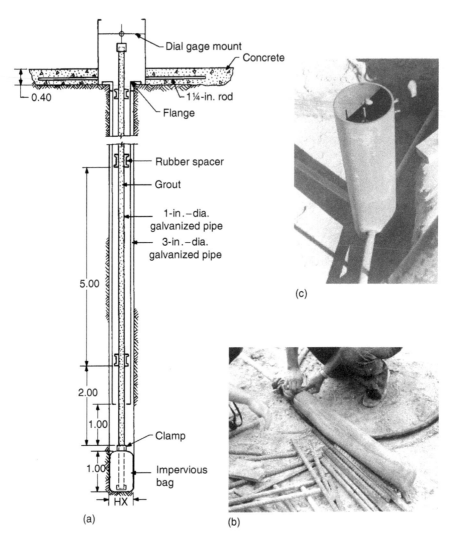

FIGURE 3.13
Settlement point installation: (a) schematic of the settlement point (depth are in meters); (b) clamp and bag being tested on the surface; (c) dial gage mount and protection.

Borros Points

Application

Borros points are simple and inexpensive devices used to monitor settlements from the compression of soft soils.

Instrument and Installation

The Borros point consists of an anchor, with prongs folded to the sides of a driving point. It is attached to a ¼-in. rod contained within a 1-in. pipe that is pushed or driven through a 2½-in. casing to the observation depth (Figure 3.14). The 1 in. pipe is held fixed while the ¼ in. pipe is advanced about 1 in., extending the prongs into the soil. The 2½-in. casing is removed and the 1-in. pipe withdrawn a short distance. The ¼-in. rod, attached to the points, is free to move within the 1-in. pipe. Settlements are determined by periodically surveying the elevation of a cap on the pipe.

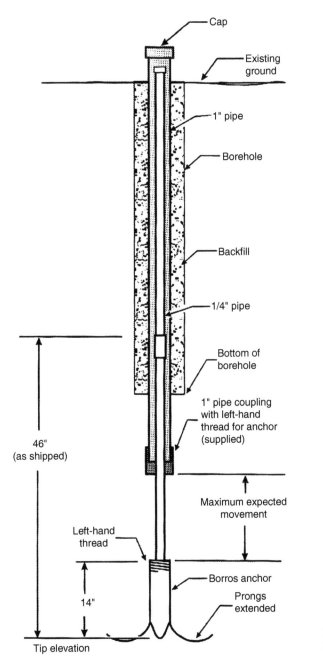

FIGURE 3.14
Borros point.

Cross-Arm Device

Application

Cross-arm devices are installed at various elevations in earth dam embankments to provide measurements of embankment compression (USBR, 1974).

Instrument

Cross arms, fabricated of 3-in. metal channels, are installed at 10-ft intervals (or some other spacing) and connected by metal pipe as the embankment is raised. The final assembly is illustrated in Figure 3.15. The bottom section is grouted into the dam foundation.

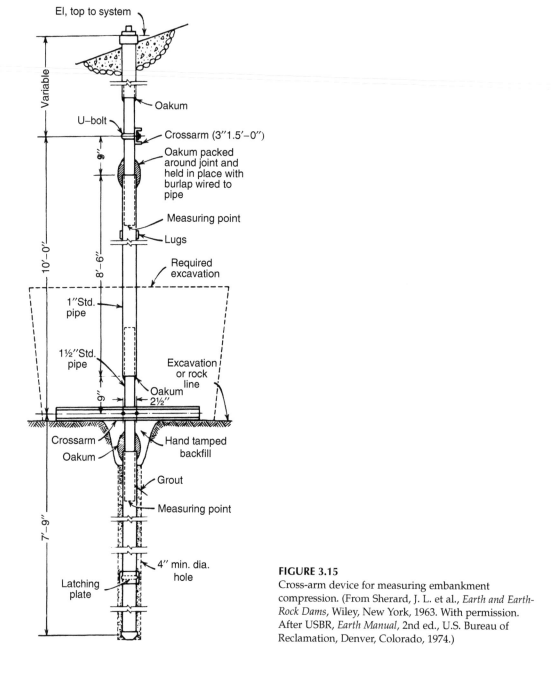

FIGURE 3.15
Cross-arm device for measuring embankment
compression. (From Sherard, J. L. et al., *Earth and Earth-
Rock Dams*, Wiley, New York, 1963. With permission.
After USBR, *Earth Manual*, 2nd ed., U.S. Bureau of
Reclamation, Denver, Colorado, 1974.)

Installation and Measurements

The metal channels, 6 ft long, are placed one by one in excavated trenches as the embankment is constructed. At each channel a 1½ in. standard pipe is attached, over which a 2-in. pipe that is free to slide is placed, forming a telescoping member. The lower tip of the 1½-in. pipe provides a measuring point. The vertical settlement of each segment is measured with a torpedo device lowered into the pipe on a steel surveyor's tape. At each measurement point the torpedo is raised until collapsible fins catch on the end of the 1½ in. pipe. The fins are collapsed for final retrieval. Probes are also lowered for water-level readings.

The pipe must be protected from disturbance during construction. Plant growth and corrosion within the pipe may in time inhibit torpedo movement.

3.3.3   Lateral Displacement

Inclinometer

Application

Continuous measurements of lateral deflections are made from the surface with the inclinometer. Commonly used in slope studies, this device may also be used to measure deformations in soft soils beneath fills, or installed behind retaining structures or in pile foundations.

Instruments

The inclinometer contains servo-accelerometers that can detect lateral movements of the order of ± 0.0001 ft per 2 ft of casing (the normal depth increment at which readings are taken). Since the voltage output is proportional to the sine of the angle of inclination of the long axis of the sensor, from the vertical, it can be used to measure true deviations from verticality.

Installation and Operation

The inclinometer is lowered and raised in specially grooved casing (3.38 or 2.79 in. O.D.) installed in a borehole and extended to a depth below the anticipated movement zone. The casing, the servo-type inclinometer, cable, and digital readout are shown in Figure 3.16. In soil formations, it is advisable to install 6-in. casing to allow for packing sand between the casing and the grooved casing, as the 6-in. casing is withdrawn to provide a sure contact between the inclinometer casing and the borehole wall. The casing bottom is often grouted into place to assure fixity. For monitoring pile and wall movements, the casing is attached directly to the structural member.

The casing is grooved at 90° intervals to guide the inclinometer wheels and to allow measurements along two axes. The relatively flexible casing deflects freely during ground movements. From readings taken at regular depth intervals, a profile of the casing is constructed. Repeating the measurements periodically provides data on location, magnitude, direction, and rate of movement. At some amount of casing deflection the inclinometer will not pass the bend. Slope Indicator Co. give the shortest radius at which the probe can be retrieved as 10 ft and the shortest radius curve at which the probe will pass for reading as 17.6 ft.

Figure 3.17 is an example of inclinometer readings of a failure zone in a slope. Constructing a vector from the readings at 90° provides the resultant of movement direction and amount. Plots of deflection vs. time are prepared; movement acceleration is an indication of approaching failure.

In-place Inclinometer Sensors

Slope Indicator Co. provide an in-place inclinometer system consisting of a string of inclinometer sensors permanently deployed in the casing (Figure 3.18). The sensors are set to span the zone where movement measurements are desired. The sensors can be connected to a data acquisition system that continuously monitors movements and can trigger an alarm when it detects a rate of change in deflection that exceeds a preset value.

FIGURE 3.16
Slop indicator inclinometer, grooved casing, cable, and digital readout.

Deflectometer

Application

Deflectometers are used as permanent installations in boreholes to measure movements normal to the hole axis such as that occuring along a fault zone, or along some other weakness plane in a rock mass.

Operation

The elements of the deflectometer are illustrated in Figure 3.19. Any movement in the mass perpendicular to the axis of the hole causes an angular distortion between consecutive rods or wires, which is monitored by transducers. Chains of deflectometers can be made with eight transducers to depths of about 60 m (200 ft). The system is connected to a remote readout and can be attached to an early warning system. Precision is about 0.025 mm (0.01 in.) for a 5 m distance between the transducers.

Shear-Strip Indicator

Application

Shear-strip indicators are used to locate the failure surface in a moving earth mass.

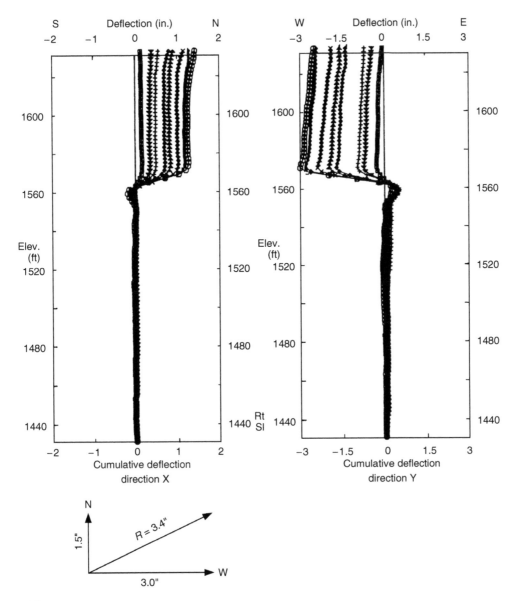

FIGURE 3.17
Inclinometer readings of failure zone in slope taken over a period of 11 months. (Courtesy of Woodward-Clyde Consultants.)

Instrument

A row of electrical transducers is wired in parallel and mounted on a waterproofed brittle backing strip about 15 cm (6 in.) apart.

Installation and Operation

The strip is mounted on a PVC pipe placed in a borehole drilled downward into a slope to intercept a potential failure surface. When shear displacement at any point exceeds 2 to 3 mm, the electrical circuit is broken. The failure surface is determined by measuring resistances at the top and bottom of the strip (Broms, 1975). The strip can be connected to an automatic recording and alarm system.

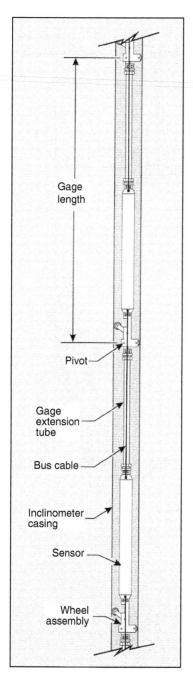

FIGURE 3.18
In-place inclinometer sensors. (Courtesy of Slope Indicator Co.)

Slope Failure Plane Sensor

Application and Instrument

A simple, low-cost instrument, a 2-m-long rod on a steel fishing line, is used to locate the failure surface in a moving mass (Brawner, 1975).

Installation and Operation

A borehole is made through the sliding mass and a 2-m-long rod on a steel fishing line is lowered into the borehole. The rod is raised each day until eventually an elevation is

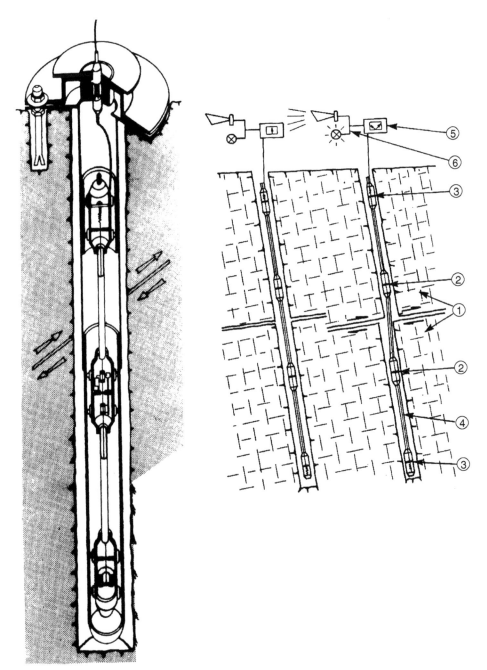

FIGURE 3.19
Schematic of the function of the Interfels deflectometer. *Note*: 1. shear zone; 2. transducer; 3. deflectometer head with tension; system from the steel wire; 4. steel wire; 5. electrical readout; 6. optical or acoustical alarm. (From Silvera, J.F.A., *Proceedings of the 1st Congress*, Brasileira de Geologia de Engenharia, Rio de Janeiro, August, Vol. 1, 1976, pp. 131–154. With permission.)

reached where the rod stops during lifting, indicating the bottom of the failure surface. Another rod is lowered from the ground surface until it stops, indicating the location of the top of the failure zone.

3.3.4 Linear Strain Gradients

Occurrence and Measurements

Rock Masses

Linear strains occur around the openings for tunnels, caverns, open-pit mines, or other rock slopes as a result of excavation, or in the abutments or foundations of concrete dams as a result of excavations and imposed loads.

In tunnels and caverns, borehole extensometers are installed to measure the tension arch or zone of relaxation which determines the depth to which support is required, the type of support, and the excavation methods. Deformations are also monitored during and after construction in critical zones characterized by faulting, intense fracturing, or swelling ground (in which horizontal stresses can be vertical stresses many times).

Earth Dams and Reinforced Earth Walls

Linear strains occurring in earth dams and reinforced earth walls are measured with electrical strain meters. In earth dams, the meters are installed in trenches at various depths to monitor strains and detect and locate cracks resulting from the differential movement of the dam foundation or the embankment itself, and between the embankment and the abutment (see Figure 3.38). In reinforced earth walls the meters are installed to monitor horizontal movements in the body of the backfill.

Wire Extensometers

Wire extensometers to monitor slope movements and deformations around a tunnel opening are discussed in Section 3.2.4.

Rock-Bolt Borehole Extensometer

Application

Rock bolts are installed to monitor deflections occurring parallel to a borehole in a rock mass. They can be installed in any orientation, and are rugged and low in cost because they are relatively simple and can be installed rapidly. Precision is of the order of 0.001 in.

Installation

A small-diameter borehole is drilled to a certain depth and a 1-in.-diameter steel rod is grouted into a fixed position as shown in Figure 3.20. A Teflon collar is fitted on the rod as a spacer near the tunnel face and a head is fixed to the wall, where a strain gage is mounted to measure deflection readings. The bolts are installed at various depths to measure differential

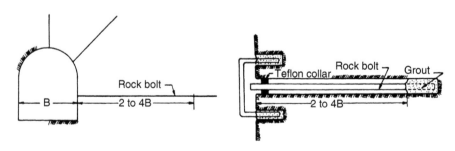

FIGURE 3.20
Simple rock bolt extensometer.

strains and to determine the extent of the zone in which significant strains occur, which is a function of the tunnel diameter.

Multiple-Position Borehole Extensometer (MPBX): Rod Type

Application

The rod-type MPBX is installed to monitor deflections occurring parallel to a borehole at a number of positions in a single hole. Vertical overhead installation is difficult and lengths are limited to about 50 ft. Sensitivity is of the order of 0.001 in.

Installation

The ends of up to six metal rods are cemented into a borehole at various distances from the excavation face and the rods are attached to a measuring platform fixed to the excavation wall as shown in Figure 3.21a.

Multiple-Position Borehole Extensometer (MPBX): Wire Type

Application

The wire-type MPBX is used like the rod type but can be installed to far greater depths.

Installation

Up to eight stainless-steel tension wires are installed in a borehole and anchored by grouting or by some other means, to the side of the hole at various positions along the hole as shown in Figure 3.21b. A measuring head is threaded over the wires and installed in the collar of the hole, and each wire is connected to a measuring element in the head. The measuring element can be a steel rod running in a track, pulling against a tension spring or pushing against a compression spring, which is read by dial gages; it can be an electric sensor, such as an LVDT, linear potentiometer, or strain gage connected to an electrical readout.

Calibration is difficult and allowances are made for wire stretch, friction between the wires and the borehole, and temperature. Wire-type MPBXs can be read to the nearest 0.001 in., but because of calibration difficulties, the deeper the installation, the greater should be the changes in readings before they are judged as significant.

Horizontal Strain Meters and Inclinometers

Application

Linear strains occurring during slope movements or fault displacements can be monitored with electrical strain meters or inclinometers. When used in dam embankments, they provide a plot of the vertical deformation that is occurring. They can be used in nets, connected to a data acquisition system that continuously monitors movements, and set as an early warning system when a preset value is exceeded.

Instrument

Linear potentiometers are mounted in a slip-jointed polyvinyl chloride (PVC) pipe. The Slope Indicator horizontal in-place inclinometer consists of a string of inclinometer sensors deployed in inclinometer casing as discussed in Section 3.3.3.

Performance

Installed in shallow trenches, the inclinometers are read and continuously recorded at remote receiving stations, and can be set in series to extend over long distances. Changes of several inches can be measured over gage lengths of several feet with a sensitivity of 10^{-3} times the movement range.

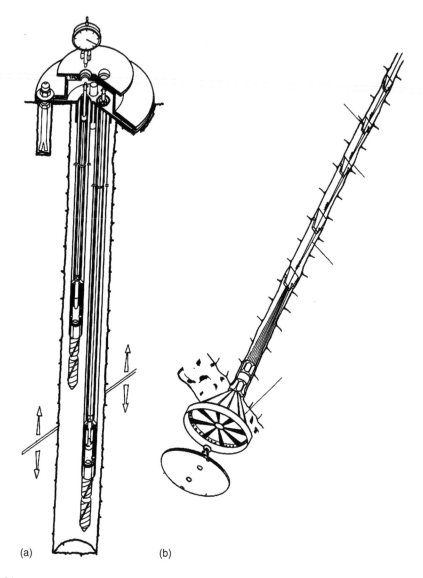

FIGURE 3.21
Schematics of the multiple-position borehole extensometer: (a) rod type, (b) wire type. (From Silvera, J.F.A., *Proceedings of the 1st Congress*, Brasileira de Geologia de Engenharia, Rio de Janeiro, August, Vol. 1, 1976, pp. 131–154. With permission.)

3.3.5   Acoustical Emissions

Description

Acoustical emissions refer to subaudible noises resulting from distress in soil and rock masses. They are also termed microseismic activity, microseisms, seismic–acoustic activity, stress-wave activity, and rock noise.

Applications

Monitoring acoustical emissions provides an aid to the anticipation of failure by rupture or internal erosion. The method was first used in the underground mining industry to detect the instability of the mine roof, face, or pillar rock.

Some of the applications include:

- Monitoring natural slopes, open-pit mines, and other cut slopes; excavations for tunnels, caverns and underground storage facilities and pressure chambers; and the stability of dams and embankments
- Locating leakage paths in dams, reservoirs, pressure pipelines, and caverns
- Monitoring fault zones
- Inspecting steel and concrete structures under test loading

The Phenomenon

General

A material subject to stress emits elastic waves as it deforms, which are termed acoustic emissions.

Rock Masses

Testing has shown that both the amplitude and number of emissions increase continuously as macroscopic cracks initiate and propagate first in a stable manner, then in an unstable manner. Near rupture, friction along crack surfaces, as well as crack propagation and coalescence, contributes to the acoustical emission activity. Mineralogical and lithological differences, moisture content, and stress conditions affect the emissions (Scholz, 1968).

Soils

Frictional contact in well-graded soils produces the greatest amount of "noise" during stress, and activity increases with the confining pressure. Clays exhibit a different form of response than sands (Figure 3.22), and the emission amplitude for sands can be 400 times that of clay (Koerner et al., 1977).

Detection

General

Sensors detect microseismic activity at a specific location in an earth mass by monitoring displacements, velocities, or accelerations generated by the associated stress waves. At times, the sounds are audible, but usually they are subaudible because of either low magnitude or high frequency, or both. In application, if after all extraneous environmental noise is filtered and there are no emissions, the mass can usually be considered as stable. If, however, emissions are observed, a nonequilibrium condition exists, which may eventually lead to failure.

Instrumentation

Accelerometer or transducer sensors are used to detect acoustic emissions by converting mechanical energy associated with the microseisms into an electrical signal proportional to the amplitude of sound or vibration that is detected. The detected signal is then passed through a preamplifier, amplifier, and filter, and finally to a display on a cathode-ray oscilloscope or into a recorder as shown in Figure 3.23.

The system components are selected for a specific study to provide suitable frequency response, signal-to-noise ratio, amplification, and data-recording capacity. The design, construction, and calibration of equipment are discussed in Hardy and Leighton (1977).

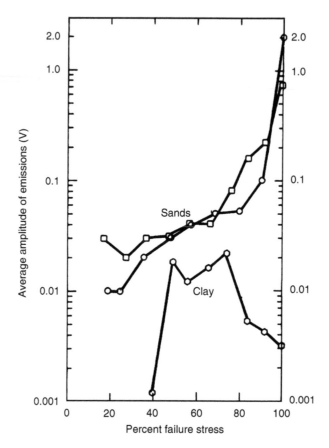

FIGURE 3.22
Average acoustical emissions amplitude (peak signal voltage output) vs. percentage of failure stress in triaxial creep (5-psi confining pressure). *Note*: (□) concrete sand (soil 3); (○) Ottawa sand (soil 2); (◐) Kaolinite clay (soil 6). (From Koerner, R. M. et al., *Proc. ASCE, J. Geotech. Eng. Div.* 103, 837–850, 1977. With permission.)

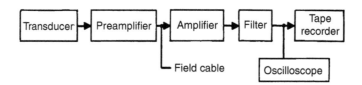

FIGURE 3.23
Simplified diagram of typical microseismic field-monitoring system. (From Hardy, H. R., Jr. and Mowrey, G. L., *Proceedings of the 4th International Conference*, Zurich, April, A. A. Balkema, Rotterdam, pp. 75–72. With permission.)

Sensors

Geophones can detect signals in the frequency range of 2 to 1000 Hz, the normal seismic energy range for large geologic structures. They are generally used as sensors to monitor mines and tunnels because of their greater sensitivity in the lower frequency ranges (Hardy and Mowrey, 1977). Accelerometers are employed for high-frequency components ($f > 1000$ Hz), and displacement gages are employed for very low-frequency ranges ($f < 1$ Hz).

Installation

Geophones can be installed at shallow depths below the surface, cemented into boreholes, or mounted on a rock face in an excavation as shown in Figure 3.24. The best results for rock studies are obtained when the transducer is located in the rock mass.

Case Studies

Cut-Slope Failure (Koerner et al., 1978)

Operations: The toe excavation of a large fill was monitored by an acoustic emissions device. Five cuts were required to cause failure over a 21-day period. After each cut, the

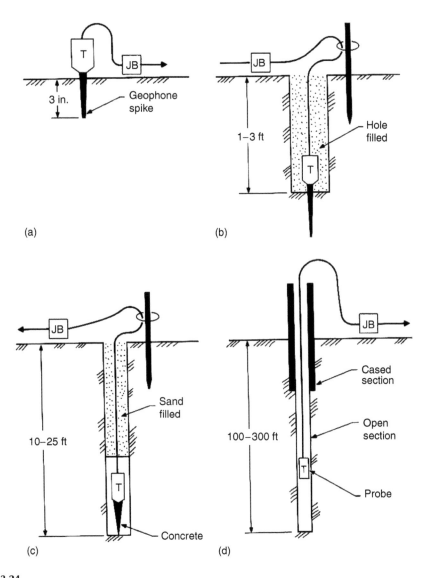

FIGURE 3.24
Types of microseismic transducer mounting techniques: (a) surface mounting; (b) shallow burial; (c) deep burial; (d) borehole probe. (From Hardy, H. R., Jr. and Mowrey, G. L., *Proceedings of the 4th International Conference*, Zurich, April, A. A. Balkema, Rotterdam, pp. 75–72. With permission.)

"acoustic counts" were high, then tapered off. During one operation rainfall caused an increase in acoustic counts. During the fifth cut, the emission rate followed the trend for the previous cuts, but 30 min after the cut was made, the rate increased rapidly and reached its maximum when a large portion of the fill slid downslope.

Acoustic count: For the equipment used, acoustic count was defined as being registered each time the amplitude of the electric signal exceeded the threshold level of 0.025 V as recorded on the electronic counter. From several studies (Koerner et al., 1978) it was concluded that:

1. Soil masses with no emissions are stable.
2. Granular soil masses generating moderate levels of emissions (from 10 to 100 counts/min) are considered as marginally stable.
3. High-emission levels (100 to 500 counts/min) indicate that the soil mass is deforming substantially and that immediate remedial measures are required.
4. At above 500 counts/min the mass is in the failure state.

Active Slide (Novosad, 1978)

The inclinometer and the acoustic emissions device were compared during the study of an active landslide. The emissions identified the shear zone after only three measurements, even though movements were very slow, of the order of 5 to 10 mm/year. Novosad concluded that the emissions method had the advantages of a simple, common borehole casing, smaller casing diameter (40 cm I.D. is sufficient), and a deeper reach. The disadvantage was the qualitative character of the measurements.

Open-Pit-Mine Slopes (Kennedy, 1972)

Rock noises emanating from an unstable zone in an open-pit-mine slope are identifiable by their characteristic envelope shape, a frequency between 6 and 9 Hz, and a characteristic irregular amplitude. Experience is required to distinguish extraneous noise from drills, trucks, shovels, locomotives, etc. operating in the pit. Earthquakes are easily identifiable because they have a very definite waveform envelope.

In application, records of the number of rock noises for a given period (counts) and of the frequency, duration, and equivalent earth motion of each seismic disturbance are maintained. As the instability increases, so does the number of rock noises released during a given period. Plotting the cumulative number of rock noises against time may give a useful prediction chart similar in shape to those obtained from plotting displacement vs. time.

Advantages of the method include its extreme sensitivity and continuous recording capabilities. Early warnings of impending failure allow the removal of workers and equipment from the threatened area.

3.4   *In Situ* Pressures and Stresses

3.4.1   General

Significance

Deformations occur as the result of response to changes in stress conditions *in situ*, which can be measured under certain conditions. The data obtained by instrumentation are used as a design basis, as a basis for checking design assumptions and determining the need for additional support, or as a warning of impending failure.

Categories

Pressure and stress conditions measured *in situ* include:
1. Groundwater levels and pore pressures
2. Stresses in embankments, beneath loaded areas, or in structural members
3. Earth pressures against retaining structures or tunnel linings and supports
4. Residual stresses in rock masses

Summary

In situ pressures and stresses, their occurrence, and methods of measurement are summarized in Table 3.7.

3.4.2   Pore-Water Pressures

Applications and Summary

Observation wells are installed to measure changes in the static water table. Piezometers are used to monitor water levels (static, perched, artesian) in excavations, slopes, and dam embankments, to measure excess hydrostatic pressures beneath dams and embankments, and to aid in control of preloading operations and placement of fill over soft ground.

The types, applications, advantages, and disadvantages of piezometers are summarized in Table 3.8. Piezometers are divided into open systems, in which measurements are made from the surface and the water level is generally below the surface, and closed systems, in which measurements are made remotely and the water level may be at any location.

TABLE 3.7

In Situ Pressures and Stresses: Occurrence and Measurements

Phenomenon	Occurrence	Measurement	Reference
Pore-water pressures	Natural static GWL	Standpipe piezometer	Section 3.4.2
	Artesian conditions	Casagrande piezometer	
	Excess pore pressures under static or dynamic loads	Hydraulic piezometer	
		Pneumatic piezometer	
		Electrical piezometer	
Loads and stresses	Beneath foundations	Pressure cells	Section 3.4.3
	In embankments	Load cells	Section 3.4.3
	In structural members	Tell tales	Section 3.4.3
	Against retaining walls	Strain gages	Section 3.4.3
	Tunnel linings and supports	Acoustical-emissions devices	Section 3.3.5
		Vibrating-wire stress meter	Section 3.4.3
Residual rock stresses	In closed and open excavations in rock masses	Flat-jack test	Section 3.4.4
		Strain meters and overcoring	Section 3.4.4
		Borehole methods	Section 3.4.4
		Deformation meter	
		Inclusion stress plug	
		Strain gage (Leeman doorstopper)	
		Hydraulic fracturing	

TABLE 3.8

Piezometer Types and Applications

Type	Application	Advantages	Disadvantages	Figure
Open Systems (Vertical Readout; Water Level below Ground Surface)				
Single-tube open system (Standpipe)	Coarse-grained granular soils, free-draining rock masses	Simple, rugged, inexpensive	Indicates average head, relatively insensitive, time lag in impervious soils	3.25
Casagrande type (single tube or double tube)	Coarse-grained to silty soils and rock masses	Measures response at a particular depth	Time lag in relatively impervious soils	3.26
		Decreased time lag	Low sensitivity	
		Double tube allows flushing entrapped air or gas from lines	Single-tube system subject to clogging from entrapped air or gas	
Closed Systems (Remote Readout: Water Level any Location)				
Pneumatic (diaphragm principle)	Fine-grained soils, slow-draining rock masses	Negligible time lag	Relatively costly device and installation	3.27
		Increased sensitivity		
		Durable and reliable	Requires protection against pinching of lines	
		Lines readily purged and extended to avoid construction		
Hydraulic	Earth dams with soils of low to medium permeability	System can be flushed to remove air and gas	Slow response time	3.28
		Remote sensing of pore pressure to leave construction area free	Long tubing lines require careful flushing	
			Fittings may be subject to leaks	
			Requires protection against pinching	
Electric (diaphragm principle)	Fine-grained soils and slow-draining rock masses	Extreme sensitivity	Relatively costly	3.29
		Fast response		
		Continuous recording possible (only type with this capability)	Decreased durability and reliability over other closed systems because of electrical circuitry	

Single-Tube Open Piezometer

The simplest piezometer consists of a tube or pipe which connects a tip or sensor (porous stone, well point, or slotted pipe) (Figure 3.25) to the surface and which is installed in a borehole as the casing is withdrawn.

Readings are made by plumbing with a chalked tape or with an electric probe contacting the water in the standpipe. The static head measured is the average head existing over the depth of the inflow part of the borehole below the water table. The head may be higher

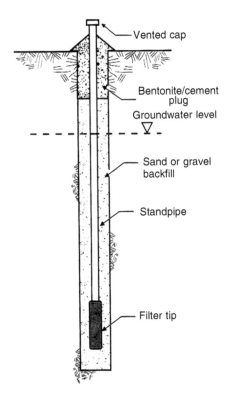

Vented cap

Bentonite/cement
plug

Groundwater level

Sand or gravel
backfill

Standpipe

Filter tip

FIGURE 3.25
Observation well. (Courtesy of Slope Indicator Co.)

or lower than the free water table, and in moderately impervious to impervious soils is subject to time lag.

Casagrande Open-Tube Piezometer

Casagrande-type piezometers are similar to the open-tube system, except that the tip is surrounded by a bulb of clean sand and a clay seal is placed above the tip as shown in Figure 3.26 to confine response to a particular depth interval. Before the clay seal is placed, the annular space around the tip is filled with gravel or sand to prevent soil migration and system clogging. The hole is backfilled above the clay seal. The casing is withdrawn slowly as the clean sand, clay seal, and backfill are placed.

In granular soils, large-diameter (1 in.) tubes adequately reflect changes in pore pressure, but small tubes are required for soils of low permeabilities. Water levels are read in the same manner as for simple systems, except that in small tubes readings are made with electric probes with a 1/8 in. O.D.

Double-tube systems are a variation, which permits flushing to remove entrapped air or gas that clogs lines, a common occurrence in single-tube systems of small diameter. In cold weather, tubes are filled with alcohol or antifreeze.

Pneumatic-Type Piezometers

In pneumatic piezometers the sensor is a sealed porous tip that contains a gas- or fluid-operated diaphragm and valve, which are connected by two lines to a pressure supply and outlet system on the surface as shown in Figure 3.27. When air pressure applied to the connecting line is equal to the pore-water pressure acting on the diaphragm, the valve closes. This pressure is assumed equal to the pore-water pressure in soil, or the cleft-water pressure in rock masses.

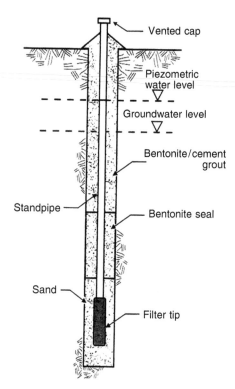

Vented cap

Piezometric
water level

Groundwater level

Bentonite/cement
grout

Standpipe

Bentonite seal

Sand

Filter tip

FIGURE 3.26
Casagrande-type open-tube piezometer. (Courtesy of Slope
Indicator Co.)

Other types of pneumatic transducers include the Warlam piezometer, which operates with air, and the Glotzl piezometer, which employs a hydraulic fluid. In all cases, the lines can be extended through an embankment to a readout so that they do not interfere with construction operations. The lines must be protected against pinching.

Hydraulic Piezometers

Commonly installed in earth dams, the hydraulic piezometer used by the U.S. Bureau of Reclamation (USBR, 1974) uses a porous ceramic disk as a sensor. The disk is connected by two plastic tubes to a pressure gage near the downstream face of the embankment (Figure 3.28). The pressure gage and its housing are located at an elevation slightly higher than the piezometer tip.

The system requires long tubing lines. Deairing by water circulation must be done carefully, and all fittings must be well made to avoid leaks. The lines require protection against pinching. To operate, fluid is pumped into the system until a desired balancing pressure is obtained as read on the gage; then an inlet valve to the piezometer tip is opened slightly and the response pressure observed.

Electrical Piezometers

The tip of an electrical piezometer contains a diaphragm that is deflected by pore pressure against one face. The deflection is proportional to the pressure and is measured by means of various types of electrical transducers. The system components are illustrated in Figure 3.29. Electrical piezometers are extremely sensitive and have negligible time lag and are used with impervious materials, where measurements are critical and continuous monitoring and recording are required.

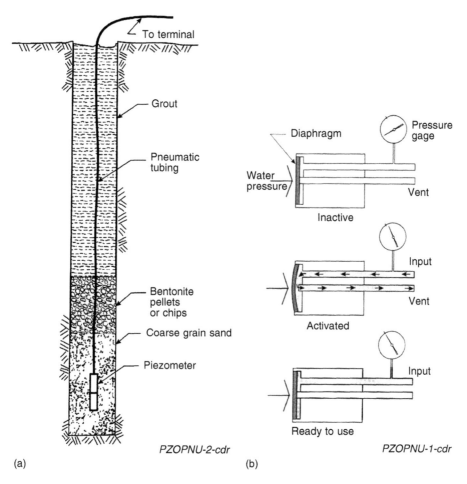

FIGURE 3.27
Pneumatic piezometer: (a) piezometer installed in borehole; (b) operating principler. (Courtesy of Slope Indicator Co.)

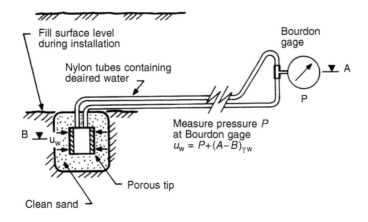

FIGURE 3.28
Hydraulic-type piezometer (installed in an earth dam).

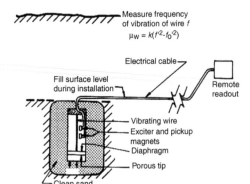

FIGURE 3.29
Electrical-transducer-type piezometer.

3.4.3   Loads and Stresses

Applications and Summary

Loads and stresses are measured to check design assumptions, provide early warning against failure, and to permit the installation of remedial measures to provide additional support wherever necessary. Measurements are made primarily during construction. The various devices and their applications are summarized in Table 3.9.

Strain Gages

Purpose

Strain gages are used to measure deformations of structural members to provide the basis for estimating stresses and loads. There are two general types: electrical (see Section 3.1.4) and mechanical.

Resistance Strain Gages (see Figures 3.1 and 3.2)

Bonded gages are used primarily for short-term surface measurements. Installation requires meticulous care and skill during the preparation of the receiving surface and the cementing of the gage in place. Moisture sensitivity is high and adequate field waterproofing is

TABLE 3.9

Load or Stress vs. Device Suitable for Measurements

Condition	Load or Stress to be Measured	Device Suitable
Soil formations	Stress distribution beneath foundations or in embankments	Pressure cell
	Earth pressure distributions behind retaining structures	Pressure cell
Rock formations	Stress distribution beneath foundations or in abutment walls, during or after construction	Carlson strain meter
		Borehole stress meter
Structural members	Stress distribution in piles to differentiate between end bearing and shaft friction during load test	Tell tale
		Strain gage
	Stresses on retaining structure bracing, earth and rock anchors, steel storage-tank walls	Strain gage
	Stresses on tunnel linings	Pressure cell
	Loads on tunnel lining elements, earth and rock anchors, pile tips	Load cell

difficult to achieve. Reliability is generally low but may be partially compensated for by the installation of a relatively large number of gages on a given member.

Encapsulated gages are factory waterproofed and should have a service life of about a year or more if installed properly. Installation is simple, but as with normal bonded gages, lead wires require protection.

Unbonded, encapsulated gages are simple to install and have a reliability much higher than other types. Carlson strain meters installed in concrete dams have performed successfully for periods of over 20 years. They are also left inserted in boreholes to measure rock stresses. Deflection ranges, however, are somewhat less than that of the other types.

Vibrating-Wire Strain Gage

Installation is simple and the gages are recoverable, remote-reading with excellent long-term stability, strong, and waterproof. The gage, however, is a large instrument and its measurement range is more limited than that of the resistance-gage types. Its cost is generally higher than that of most resistance gages.

Mechanical Strain Gages

Mechanical strain gages are used to measure deformations of accessible metal surfaces. They have two pointed arms that fit into conical holes (gage points) drilled or punched into the surface to be measured, or into studs that are set into or on the surface. The change in distance between gage holes is measured by determining the distance between arms when inserted into the gage points. Measurement is done with a dial gage. Wittier gages are a popular type. Gage lengths of 2 to 80 in. are available. 10-in.-long gage is used for most engineering studies; sensitivity is of the order of 10 microstrains, and gages are read usually to the nearest 0.0001 in.

Mechanical gages are relatively simple and reliable. Temperature corrections are easily made and gage calibration is easily checked. However, they are suited only for measuring surface strains, cannot be read remotely, and have a lower sensitivity than the electrical types. Repeatability depends on the skill and experience of the operator.

Pressure Cells

Purpose

Dunnicliff (1988) divides pressure cells into two general categories: (1) embedment earth pressure cells that are installed in a soil mass, and (2) contact earth pressure cells that are installed between a structure and a soil mass. Pressure cells are used to measure pressures against retaining walls or tunnel linings, or stresses beneath foundations or in embankments.

Device

The embedment earth pressure cell consists of a circular (or rectangular) double-wall metal pad, and is extremely thin compared to its diameter (ratio of about 1.1 to 23 cm). A common type is the Glotzl cell, which is filled with oil or antifreeze liquid and functions hydraulically as shown in Figure 3.30. Pressurized fluid is delivered to the small pressure diaphragm by a pump or compressor until the pressure equals the resisting pressure on the outside of the diaphragm. The diaphragm deflects a slight amount and opens a bypass orifice that permits the fluid in the system to return to the reservoir. The pressure creating a balance is read on gages that provide a measure of the external pressures acting on the cell.

One type of contact earth pressure cell used to measure loads at a pile tip or against sheet pile wall, for example, includes a hydraulic cell with a pneumatic transducer.

Load Cells

Purpose

Load cells are used to measure loads on earth or rock anchors, in the various elements of tunnel linings, in struts in braced cuts, at the bottom of piles (total pile load minus tip load equals shaft friction), and at the top of piles in pile load tests.

Device

Cell function is often based on the electric transducer principle. The cells are mounted within variously shaped containers. Various capacities are available.

Calibrated hydraulic jacks are used for the application of load to rock bolts, tieback anchors, piles, drilled shafts, and cross-lot struts.

Tell Tales

Purpose

Tell tales are installed in piles at various depths for strain measurements during load testing, from which shaft friction and end bearing are calculated.

Device and Installation

A small-diameter rod or pipe is fixed at the bottom of the concrete or steel pipe. It is extended up through the pile top and encased in a slightly larger pipe to permit free movement. A dial gage at the top reads deflections. A tell-tale set in the pile bottom provides a measure of both base movement and shaft shortening when compared with the total pile deflection measured at the top by separate instruments.

A group of rods is usually installed at the same depth in large-diameter piles, and often at various depths in long piles. They are distributed as a function of the soil profile to measure shaft friction for the various soil types.

Analysis

The average load P_s from shaft friction, carried by a segment of the pile ΔL from the measurement of two tell tales, is expressed as

$$P_s = a_s E_p ((R_1 - R_2)/\Delta L) \tag{3.1}$$

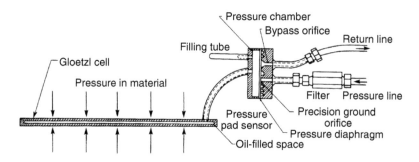

FIGURE 3.30
Schematic of the hydraulic pressure cell. (From Hartman, B. E., *Terrametrics*, Wheatridge, Colorado, 1966. With permission.)

where a_s is the section area, E_p the modules of concrete (or steel, or steel and concrete) composing the pile section and R_1 and R_2 are the deflection readings of tell tales at depths 1 and 2, respectively.

Subtracting the values for shaft friction from the total applied load provides a measure of the load reaching the pile tip.

Vibrating-Wire Stress Meter

Purpose

Changes in *in situ* stresses occurring in rock masses during mining and tunneling operations have been measured with vibrating-wire stress meters (Sellers, 1977).

Device

A transducer consisting of a hollow steel cylinder (proving ring), across which a steel wire is tensioned, is wedged into a 1.5- to 3-in.-diameter borehole. Rock stress release flexes the proving ring, changing the wire's tension and natural frequency or period of vibration. An electromagnet excites the vibrations in the wire and their frequency is measured as excavation proceeds in the rock. The frequency provides a measure of the proving-ring strain and, therefore, the stress change.

3.4.4   Residual Rock Stresses

General

Occurrence

The geologic conditions causing overstress or residual stresses in rock masses in excess of overburden stresses are significant in open excavations and tunnels where their effects are mitigated with rock bolts or bracing systems.

Measurements

Measurements are made at the rock surface where disturbance and stress relief from excavation may affect values obtained with strain meters, strain rosettes, or flat jacks. Alternatively, measurements are made in boreholes. Although they are difficult to perform, borehole measurements provide more representative data. Borehole devices require overcoring. Hydraulic fracturing is also performed in boreholes.

Strain Meters or Strain Rosettes

SR-4 strain gages (bonded gages) are attached to the rock wall, a zero reading is taken, slots are cored around the gage to relieve rock stresses as shown in Figure 3.31, and the gages are read again. The principal stresses can be calculated with a suitable array of gages. Results are difficult to interpret unless rock modulus values have been measured.

Flat Jacks

Flat jack tests have been described in Section 2.5.3. Reference pins or strain gages are installed on the rock face and distance measurements are accurately made (or strains are measured by gages mounted on the jack). A slot is drilled, and the jack inserted and grouted (Figure 3.32) and then expanded under pressure to restore the original strain or reference-point measurements. The restoration pressure is taken as equal to the residual rock stress. Flat jacks as well as strain meters are effective only in tightly jointed high-quality rock.

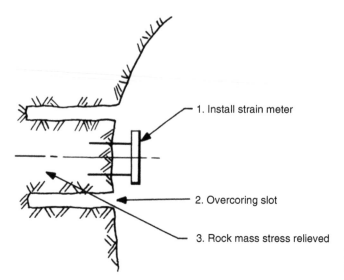

1. Install strain meter

2. Overcoring slot

3. Rock mass stress relieved

FIGURE 3.31
Strain meter for residual rock stresses.

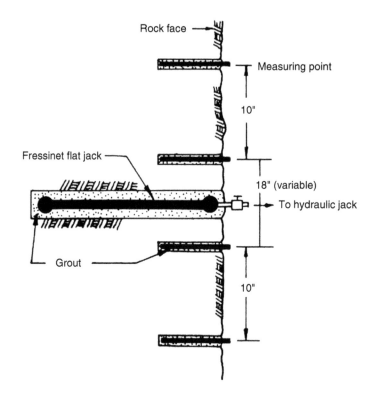

Rock face

Measuring point

10"

Fressinet flat jack

18" (variable)

To hydraulic jack

Grout

10"

FIGURE 3.32
Flat jack test to measure *in situ* rock stresses.

Borehole Devices

Devices

Residual stresses can be measured in boreholes with the borehole deformation meter (Figure 3.33a), the high-modulus stress plug or inclusion stress meter (Figure 3.33b), or the Leeman "door-stopper" strain gage (Figure 3.33c).

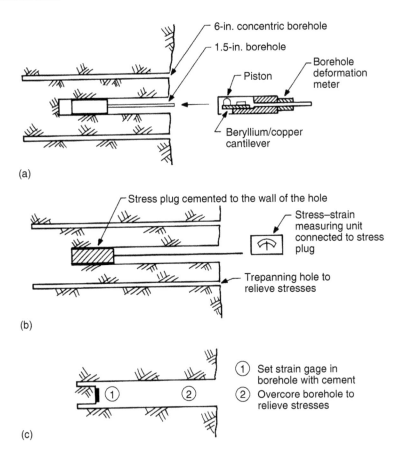

FIGURE 3.33
Measurements of *in situ* stresses by stress relief in boreholes (a) borehole deformation meter; (b) high-modulus stress plug or inclusion stress meter (a rigid or near-rigid device calibrated directly in terms of stress); (c) Leeman "doorstopper" strain gage.

Procedure

The device is inserted into a small-diameter borehole (NX) and the stresses are relieved by overcoring. These stresses are read directly with the inclusion stress meter, or computed from strains measured with the deformation meter or strain gages. To compute stresses when only strains are measured requires either a measurement or an assumption for the rock modulus. Installation and interpretation are described by Roberts (1969). Maximum applicable depths are of the order of 10 to 15 m because of the difficulties of making accurate overcores, especially in holes that are not vertical.

Hydraulic Fracturing

Application

Hydraulic fracturing has been used to measure stresses in deep boreholes (Haimson, 1977) and is still in the development stages.

Technique

A section of borehole is sealed off at the depth to be tested by means of two inflatable packers. The section is pressurized hydraulically with drilling fluid until the surrounding rock mass ruptures in tension. The pressure required to initiate failure and the subsequent pressure required to maintain an open hole after failure are used to calculate the magnitude of the *in situ* stress.

3.5   Instrumentation Arrays for Typical Problems

3.5.1   Importance

Soil Formations

Design Data Procurement

Design data are usually based on properties measured in the laboratory or *in situ*. Load tests on piles or other types of foundations, and preloading with embankments, provide additional design information.

Construction Control

Instrumentation is required where analytical limitations result in marginal safety factors, if excessive deformations are anticipated or if remedial measures are to be invoked. Such conditions exist for retaining structures constructed in soft ground, structures on shallow foundations, embankments on soft ground, and earth dams and cut slopes.

Rock Masses

Design Data Procurement

Design is usually based on assumed properties because of the cost and limitations of the validity of *in situ* tests; therefore, careful monitoring of construction is required on important projects. *In situ* load tests and stress measurements require instrumentation.

Construction Control

Instrumentation is used to monitor the changes in *in situ* conditions caused by construction and to provide an early warning of impending failures. It is also used to assess the effectiveness of the installation of supports or other changes to improve the stability of tunnels, caverns, mines, cut slopes, and other excavations.

3.5.2   Settlement of Structures

Case 1: Structure Undergoing Differential Settlement

Problem

The structure in Figure 3.34a is undergoing excessive differential settlements resulting in wall cracks, distortions of doors and windows, misaligned machinery, and floor warping, as a result of the consolidation of a weak soil stratum.

Objectives

It is desired to determine the magnitude and anticipated rate of the remaining settlements of the floors and structural frame. This information will provide the basis for determining the remedial measures.

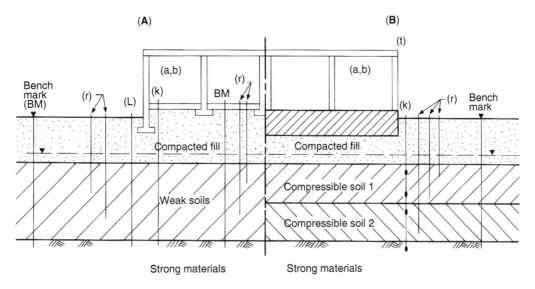

FIGURE 3.34
Instrumentation for building settlements: (A) spread footing-supported light structure; (B) mat-supported heavy structure. Legend: (a) optical survey; (b) water level device; (k) settlement points; (L) inclinometer; (r) piezometer; (t) tiltmeter.

Instrumentation

The amounts and time rates of settlement at various locations in the structure and pore pressures at various depths are monitored with the devices shown in Figure 3.34a. The inclinometer is installed to measure possible lateral deflections. The data obtained are analyzed with laboratory test data to determine the magnitude and time rates of the remaining settlement.

Case 2: Structure Undergoing Tilt

Problem

The relatively rigid, mat-supported structure in Figure 3.34b is undergoing tilt from differential settlement, which is affecting the balance of turbines.

Objectives

It is desired to determine which strata are contributing to the settlements, to estimate the magnitude of the remaining settlement, and to judge the time required for its essential completion in order to arrive at remedial treatments.

Instrumentation

As shown in the figure, building deflections are monitored, as are the compressions occurring in each stratum and the pore pressures.

Case 3: Construction over Soft Ground and Preloading

Problem

An embankment (Figure 3.35a) or steel storage tank (Figure 3.35b) is constructed over soft ground. Preloading is achieved by adding fill or loading the tank with water.

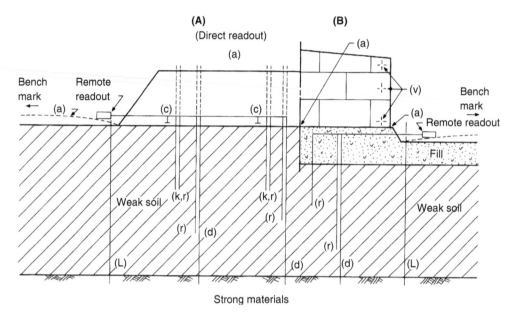

FIGURE 3.35
Instrumentation for embankment or steel storage tank over soft ground: (A) embankment over soft ground,
(B) steel storage tank over soft ground. Legend: (a) optical survey; (c) settlement plates; (d) borehole
extensometer; (k) borros points; (L) inclinometers; (r) piezometers; (v) strain gages.

Objectives

Loading rates must be controlled to prevent foundation failure. Data on the magnitudes,
rates of settlement, and on the pore pressures provide the basis for determining when the
surcharge may be removed. (After the storage tank is surcharged, it is raised, releveled,
and used for storing products lighter than water.)

Instrumentation

Precise optical surveys monitor settlements as a function of time as well as heave occurring beyond loaded areas. Settlement plates, extensometers, and borros points are monitored at the fill surface. Remotely read piezometers and inclinometers may be installed in
the fill and beneath the tank. (With respect to tank rupture, the critical differential settlements are between the center and the bottom and along the perimeter.) Stresses in the tank
wall are monitored with strain gages; mechanical types are preferred.

  The piezometers, inclinometer, and heave reference points are necessary to provide control against failure by foundation rupture. Pore-pressure data are needed to determine
when the preload may be removed. Bench marks must be installed beyond any possible
ground movements.

Case 4: Pile Load Testing

Objectives

A pile load test may be performed (1) to determine the capacity of a pile as required to
satisfy a building code, (2) to determine the proportion of the total load carried in end
bearing and that carried in shaft friction during vertical load testing, and (3) to measure
horizontal deflections during lateral load testing for the determination of the horizontal
modulus of subgrade reaction.

The information is particularly important for the design of long, high-capacity piles penetrating several formations where design analysis based on laboratory test data is often not reliable, especially in formations which are difficult to sample undisturbed.

Instrumentation

Deflections at the pile top shown in Figure 3.36 are monitored by dial gages and optical surveys. Deflections at the pile tip and at various locations along the pile's length are monitored with tell tales and strain gages. A load or pressure cell installed at the base provides data on the end bearing pressure.

During horizontal testing, dial gages and optical surveys monitor deflections at the top and the inclinometer monitors deflection along the length.

Dynamic Load Testing

The Case–Goble method of dynamic pile testing (Goble and Rausche, 1970; Rausche et al., 1971; La Fond, 1977) utilizes transducers and accelerometers attached near the pile top. For each hammer blow, a "Pile Driving Analyzer" processes the signals to calculate the hammer energy, pile stresses, and pile bearing capacity. Instrumentation removes the need to evaluate the unknown factors of energy loss in the driving system and allows the indirect elimination of the dynamic pile characteristics. It thus provides a more accurate measurement of driving resistance. The data are input to the wave equation method of dynamic pile analysis.

3.5.3 Excavation Retention

Objectives

A retaining structure in an open excavation must be designed to provide adequate support for the excavation sides and to restrict wall deflection to a minimum in urban areas. As the

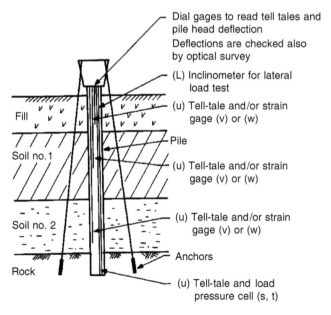

FIGURE 3.36
Instrumentation for pile load test.

wall deflects, the ground surface behind the wall subsides, and significant movements can result in detrimental settlements of adjacent structures. The problem is particularly serious in soft to firm clays and in clays underlain by sands or silts under water pressure.

Wall Design Criteria

The design of a braced excavation in soft ground is based on calculations of earth pressure magnitudes and distributions that can be determined only by empirical or semiempirical methods. Calculations of wall deflections and backslope subsidence cannot be made with the classical theories, although the finite element method is used in current analysis for such predictions. As a rule, however, the engineer cannot be certain of the real pressures and deflections that will be encountered, especially in soft ground. The engineer also may have little control over construction techniques that will influence wall performance.

A wall may become unnecessarily costly if the design provides for all contingencies. Because the nature of wall construction provides the opportunity for strengthening by placing additional braces, anchors, or other supports, initial design can be based on relatively low safety factors but with provisions for contingency measures. This approach to design and construction is feasible only if construction is monitored with early warning systems.

Construction Monitoring

Elements to be monitored during construction include:

- Movements of the backslope area and adjacent structures, heave of the excavation bottom, and movements in the support system
- Groundwater levels and the quality and quantity of water obtained from the dewatering system
- Strains and loads in the support system
- Vibrations from blasting, pile driving, and traffic

Instrumentation

As illustrated in Figure 3.37, vertical deflections of the wall, excavation bottom, backslope, and adjacent buildings are monitored by precise leveling (a). In important structures, measurements should be made with tiltmeters (e) or pendulums (f). Inclinometers (L) are installed immediately behind the wall to measure its lateral deflections. Excavation closure can be measured with convergence meters (g) and backslope subsidence with vertical extensometers (k) or shallow buried strain meters (h).

Loads in anchors and braces are monitored with strain gages (v or w) or load cells (t), or checked periodically by jacks. Strain gages must be protected against temperature changes, or the temperatures must be monitored, to provide useful data. Pressure cells (s) installed behind the wall have the advantage that they can be installed between supports. Groundwater variations are measured with piezometers (r).

An immovable bench mark is necessary for reliable measurements of deflections of adjacent ground, structures, and the excavation bottom.

3.5.4   Earth Dams

Objectives

An earth dam as actually constructed may differ from that designed because of the necessity of using earth materials for fabrication, which may vary in properties and distributions

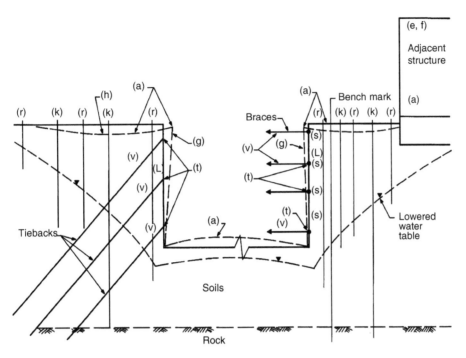

FIGURE 3.37
Instrumentation for excavation retaining structure constructed in soils. Legend: (a) precise leveling; (e) tiltmeter; (f) pendulum; (L) inclinometer; (g) convergence meter; (k) vertical extensometer; (h) strain meter; (r) piezometers; (v or w) strain gages; (t) load cells; (s) pressure cells.

from those assumed originally. Most important dam structures, therefore, are instrumented and their performance is monitored during construction, during impoundment, and while in service. Deformations in the foundation and embankment and seepage forces are monitored.

Instrumentation

Deformations

Most large dams undergo compression under their own weight and are often subjected to large foundation settlements. Cracks can develop across the core, near the crest soon after construction, and during reservoir filling. The cracks generally appear near the abutments as a result of differential settlement along the valley walls, or over irregularities in an underlying rock surface. As illustrated in Figure 3.38, external evidence of deformations is measured by optical survey (a). Internal evidence is monitored by settlement extensometers or cross-arm devices (k) and strain meters (p). Horizontal strain meters (p) permit the detection and location of internal cracking when it first occurs.

Pore Pressures and Seepage

Monitoring of pore-water pressures and seepage is necessary in the embankment, abutments, and foundation materials since these phenomena are normally the most critical factors of dam stability. Critical zones are at the toe, in front of, and below seepage cutoffs such as core trenches and grout curtains. Piezometers (r) measure pore pressures and the acoustic emissions device (g) may locate seepage paths and piping zones. Seepage may often be collected in drainage ditches and measured by weirs.

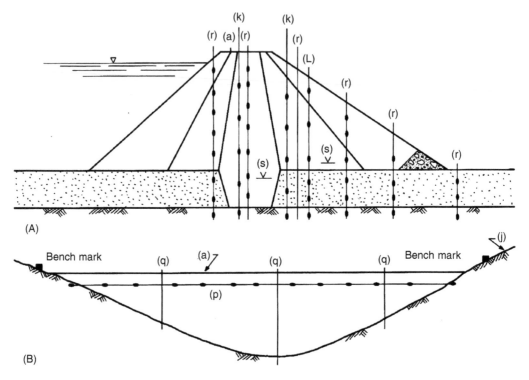

FIGURE 3.38
Instrumentation for an earth dam: (A) cross section. Legend: (a) precise leveling; (k) settlement extensometers or cross-arm devices; (L) inclinometer; (r) piezometer; (s) pressure cell. (B) Longitudinal section. Legend: (j) accelerometer; (q) acoustical emission devices; (p) horizontal strain meter.

Slope Stability

Inclinometers (L) monitor deflections in the slopes and pressure cells (s) monitor stresses.

Seismic Areas

Accelerographs (j) are installed on the dam and abutments to monitor earthquake loadings. At times seismographs have been used. Automatic-recording piezometers provide useful data because of the pore-pressure buildup occurring during seismic excitation.

3.5.5 Tunnels, Caverns, and Mines

Case 1: Closed Excavations in Rock

Objectives

Closed excavations in rock masses may or may not require support depending on rock quality, seepage forces, and the size of the opening. Support systems may be classified as temporary to permit construction to proceed safely and permanent to provide the final structure. In large openings there is the need to evaluate progressive deformation and the possibility of full-scale failure. In small openings, the most frequent problems involve partial or confined failures such as falling blocks and running ground. It is not enough only to provide adequate support to the opening itself; consideration must also be given to the possibility of surface subsidence and detrimental deflections of adjacent structures.

Design Criteria

Support requirements may be selected on the basis of experience and empirical relationships with rock quality or on the basis of finite element methods applied to assess stress conditions and deformations. Analytical methods require information on rock-mass stress conditions and deformation moduli, which are measured *in situ* (see Section 2.5.3). *In situ* test results may not be representative of real conditions for various reasons depending on the type of test, and are generally most applicable to rock of good to excellent quality in large openings.

A tunnel may become unnecessarily costly if all contingencies are provided for in design, especially with regard to the problem of identifying all significant aspects of underground excavation by exploration methods. The nature of tunnel construction allows design on a contingency basis in which the support systems are modified to suit the conditions encountered. This approach requires monitoring *in situ* conditions as excavation proceeds because of the relative nonpredictability of rock-mass response to an opening.

Construction Monitoring

Elements to be monitored during construction may include:

1. Closure of the opening and strain gradients in the rock mass
2. Support system strains and loads
3. Ground surface subsidence in urban areas over mines, caverns, and tunnels (surface subsidence may result from closure of the opening, from running ground into the opening, or from the lowering of groundwater causing compression in overlying soil formations)
4. Vibrations from blasting affecting surface structures or adjacent underground openings

Instrumentation (Figure 3.39)

Borehole extensometers (o) are installed to measure strain gradients and closure, and serve in a number of purposes. Rock-mass loads can be deduced for support design, and the modulus can be determined in good-quality rock when rock stresses are known from borehole stress relief or flat-jack tests.

Rock moduli computed from radial deflections may be lower than those computed from flat-jack tests. During tunnel driving, radial stresses diminish to zero at the edge of the opening while tangential stresses become concentrated (Kruse, 1970). Joints that have nearly tangential orientations have the greatest tendency to open. This condition influences borehole extensometer measurement as well as the compression generated by plate bearing loads. Borehole extensometer data are also used to estimate rock-bolt lengths.

Convergence meters (g) monitor closure. Deflectometers (m) installed in advance of the tunnel from a small pilot bore provide warning of mass deformations occurring in major shear zones; groundwater conditions are revealed during the drilling process and pilot holes serve to provide predrainage.

In situ stress meters (x,z) provide data for estimating roof and wall pressures that have to be retained by the support system. Pressure cells (s) installed between the tunnel lining

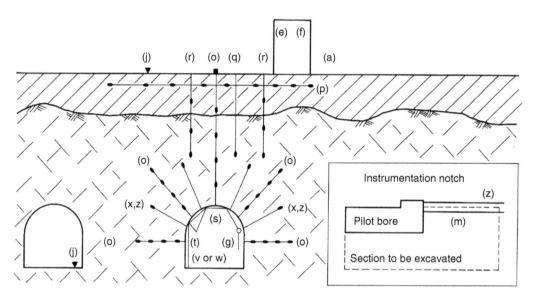

FIGURE 3.39
Instrumentation for tunnels, caverns and mines in rock. Legend: (a) precise leveling; (e) tiltmeter; (f) pendulum; (g) convergence meter; (j) vibration monitor; (m) deflectometer; (o) borehole extensometer; (p) multiple electric strain meter; (q) acoustical emission device; (r) piezometers; (s) pressure or stress cells; (t) load cells; (v) strain gages; (w) strain meter; (x) stress meter; (z) borehole residual stress measurements.

and the roof monitor stress changes occurring with time and forewarn of the necessity for additional supports. Strain gages (v or w) are installed on metal supports. Load cells (t) installed in the tunnel lining or on rock anchors monitor load changes. Load reductions in anchors indicate movements of the rock and a relaxation of the anchor load and the anchor's retention capacity. In this case, additional anchors, probably of greater length, are required.

Vertical extensometers (o), surface strain meters (p), and precise leveling (a) are used to monitor ground subsidence, and tiltmeters (e) and pendulums (f) monitor deflections in overlying structures.

Acoustical emission devices (q) provide warning of the impending collapse of mines or large caverns. Vibration monitoring (j) is required when blasting is conducted near an adjacent existing tunnel or overlying structures.

Case 2: Tunnels in Soil in Urban Areas

Objectives

The most important objective in instrumenting tunnels in urban areas is the monitoring of surface deflections and the prevention of damage to overlying structures.

Instrumentation

As illustrated in Figure 3.40, surface movements are monitored by precise leveling (a), but important data are obtained from tiltmeters (e) or pendulums (f) installed in structures. Borehole extensometers provide data on subsurface deformations in the vertical mode and inclinometers (L) in the lateral mode, serving as early warnings of excessive deformations. Pore pressures are monitored with piezometers (r), and load cells (t) installed in tunnel linings provide data on earth pressures.

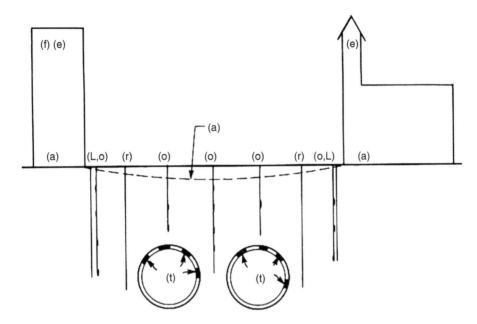

FIGURE 3.40
Instrumentation for tunnels in soil in urban areas. Legend: (a) precise leveling; (e) tiltmeter or (f) pendulum; (L) inclinometer; (o) borehole extensometer; (r) piezometer; (t) load cells.

3.5.6 Natural and Cut Slopes

Case 1: Rock Cuts

Objectives

The basic objective is to maintain slope stability, which is related to steepness, height, the orientation of weakness planes with respect to the slope angle, and cleft-water pressures. Failure in rock slopes usually occurs suddenly, with rapid movement.

Slope Stability Problems

Natural slopes may become unstable from weathering effects, erosion, frost wedging, or the development of high cleft-water pressures.

Open-pit mines are excavated with the steepest possible stable slope, and low safety factors are accepted for economic reasons. Instrumentation monitoring is generally accepted as standard procedure for deep pits.

Hillside cuts for roadways and other constructions are usually less steep than open-pit mine slopes in the same geologic conditions, and require a higher safety factor against failure. Evaluations are usually based on experience and empirical relationships, and instrumentation is normally used only in critical situations.

Instrumentation (Figure 3.41)

Internal lateral movements are monitored in terms of displacement vs. time with inclinometers (L) and deflectometers (m) which can be attached to alarm systems, or with borehole extensometers (o). Shear-strip indicators (n) and the acoustical-emissions device (q) provide indications of mass movements.

External lateral movements are monitored with the convergence meter (g) or tiltmeter (e) on benches, strain meters (h) on tension cracks, optical surveys (a) with the laser

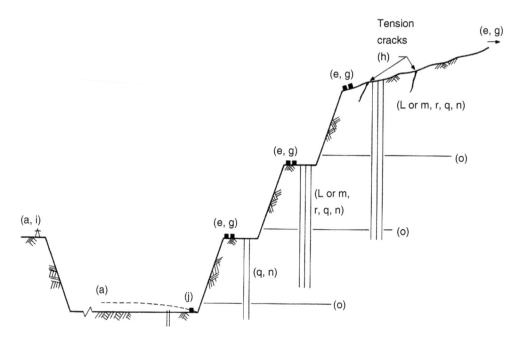

FIGURE 3.41
Instrumentation for rock cut. Legend: (a) precise surveying; (e) tiltmeter; (g) convergence meter; (h) strain meter; (i) terrestrial photography; (j) vibration monitoring; (L) inclinometer; (m) deflectometer; (n) shear-strip indicator; (o) borehole extensometer; (q) acoustical emissions device; (r) piezometers.

geodimeter that provide rapid measurements of the entire slope, and terrestrial stereopho-tography (i) that provides a periodic record of the entire slope face. GPS will find increas-ing use for long-term monitoring.

Bottom heave is monitored by the optical survey of monuments or "settlement points"; it often precedes a major slope failure.

Vibrations from blasting and traffic, which may affect stability, are monitored (j) and cleft-water pressures are monitored with piezometers (r).

Installations

Inclinometers, deflectometers, piezometers, and extensometers are often limited to early excavation stages to obtain data for the determinations of stable slope inclinations. They are expensive instruments to install and monitor, and not only will they be lost if failure occurs, but in mining operations many will be lost as excavation proceeds. In addition, they monitor only limited areas by section.

Tiltmeters, convergence meters, and optical surveys are both economical and "retriev-able," and provide for observations of the entire slope rather than a few sections; therefore, they provide the basic monitoring systems. In critical areas, where structures or workers are endangered if a collapse occurs during mining operations, MPBX extensometers and the acoustical-emissions device are used as early warning systems.

Case 2: Soil Slopes

Objectives

In soil slopes, the objectives are to detect movements when they first occur since in many instances slope failures develop gradually; and, when movement occurs, to monitor the rate of movement and accelerations, locate the failure surface, and monitor pore pressures. These data provide the basis for anticipation of total and perhaps sudden failure.

Slope Stability Problems

In potentially unstable natural or cut slopes, failure is usually preceded by the development of high pore-water pressures, an increase in the rate of slope movement, and the occurrence of tension cracks. Slope movement is not necessarily indicative of total failure, however, since movements are often progressive, continuing for many years. Stability evaluations require information on the failure surface location, pore pressures, and rates of movement as well as the geologic and climatic factors.

Instrumentation

Surface movements of the natural slope shown in Figure 3.42 are monitored by precise leveling with the laser geodimeter (a), convergence meters (g), and strain meters (h). The meters may be attached to alarm systems. GPS will find these meter to be increasingly useful for long-term monitoring.

Subsurface deformations are monitored with the inclinometer (L), shear-strip indicator (n), or steel-wire sensor to locate the failure surface. Acoustical emissions (q) may indicate approaching failure.

Pore-water pressures are monitored with piezometers (r). One can estimate the pore pressures required for total failure by knowing the failure surface location and soil shear strength.

Installations

Piezometers, inclinometers, extensometers, and other devices used in boreholes may not monitor the most critical areas, especially in slopes in a stable condition. Therefore, optical surveys, which provide information over the entire study area, are always important, although, because of terrain conditions, perhaps not always possible.

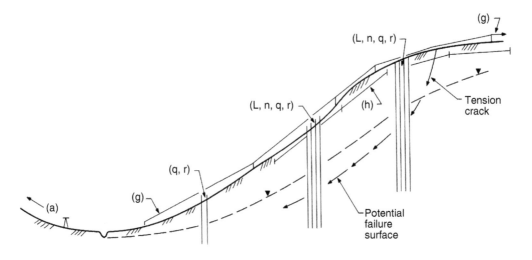

FIGURE 3.42
Instrumentation for a potentially unstable soil slope. Legend: (a) precise surveying; (g) convergence meter; (h) strain meter; (L) inclinometer; (n) shear-strip indicator; (q) acoustical emissions device; (r) piezometer.

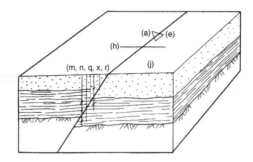

FIGURE 3.43
Instrumentation to monitor fault movements. Legend: (a) precise leveling and GPS; (e) tiltmeter; (h) strain
meter; (j) accelerometer; (m) deflectometer; (n) shear-strip indicator; (q) acoustical emission device; (r)
piezometer; (x) stress meter.

3.5.7 Fault Movements

Objectives

Earthquakes have been associated with fault rupture, which may be preceded by ground
warping, slippage along the fault, and increase in ground stresses and pore-water pressures. Fault monitoring is still in its experimental stages; eventually, some basis for predicting or anticipating rupture may be developed.

Instrumentation

Ground warping and fault movements on local and regional scales are monitored by several groups: NASA's Jet Propulsion Laboratory at Caltech, the U.S.G.S., and the California
Geological Survey. The source data include information from GPS (Section 3.2.2) and Insar
data (Section 1.2.3) as well as land-based and field geologic data (Chamot, 2003).

Surface indications of displacement along the San Andreas Fault are monitored by an
early warning system set up by NOAA's Earthquake Mechanism Laboratory at Stone
Canyon, California (Figure 3.43). The system consists of an interconnected 20 m triangular
array of mercury pools set in piers attached to Invar rods to monitor tilting (e) and strain
meters consisting of three 30-m-long extensometers (h) to measure creep (Bufe, 1972).
Accelerometers (j) set on the surface monitor ground motion accompanying fault activity.

Subsurface deformations may be monitored with deflectometers (m) and shear strips (n)
installed in boreholes across the fault, set to sound alarms if desired. Stress meters (x) monitor stress increase, as does the acoustic-emissions device (q) on a qualitative basis, when
set in or near the fault zone. Piezometers (r) set in the fault zone monitor water pressures
that may indicate stress changes in the rock mass.

References

Blackwell, G., Pow, D. and Klast, L., Slope Monitoring at Brenda Mine, *Proceedings of the 10th Canadian
Rock Mechanics Symposium*, Kingston, Ontario, Sept., 1975, pp. 45–79.

Brawner, C. O., Case examples of instability of rock slopes, *J. Assoc. Prof. Eng.*, British Columbia,
Feb. 26, 1975.

Broms, B. B., Landslides, *Foundation Engineering Handbook*, Winterkorn, H.P. and Fang, H.-Y., Eds.,
Van Nostrand Reinhold Publishers, New York, 1975, chap. 11.

Bufe, C. G., Strain and Tilt Measurements Near an Active Fault, Stability of Rock Slopes, *Proceedings ASCE, 13th Symposium on Rock Mechanics*, Urbana, IL, 1971, pp. 691–716, 1972.

Cording, E. J., Hendron, Jr., A. J., Hansmire, W. H., Mahar, T. W., MacPherson, H. H., Jones. R. A., and O'Rourke, T. D., Methods for Geotechnical Observations and Instrumentation in Tunneling, Vols. 1 and 2, Department of Civil Engineering, University of Illinois, Urbana, 1975.

Chamot, J., Earthquake Warning Tools, *Geotimes*, October 2003, p. 34

de la Cruz, R. V., Stress Measurements by the Retractable Cable Method, Proceedings of ASCE, Stability of Rock Slopes, *13th Symposium on Rock Mechanics*, University of Illinois, Urbana, 1971, pp. 856–882, 1972.

Dreyer, H., Long Term Measurements in Rock Mechanics by Means of Maihak Vibrating Wire Instrumentation, Field Measurements in Rock Mechanics, Vol. I, *Proceedings of the 4th International Symposium, Zurich*, April, A. A. Balkema, Rotterdam, 1977, pp. 109–135.

Dunnicliff, J., *Geotechnical Instrumentation for Monitoring Field Performance*, Wiley, New York, 1988, 577 pp.

FHWA, ROCK SLOPES: Design, Excavation, Stabilization, USDOT, FHA, Pub. No. FHWA-TS-89-045, September 1989.

Franklin, J. A., Monitoring rock structures, *Int. J. Rock Mech., Miner. Sci. Geomech.*, abstr., 14, 163–192, 1977.

Goble, G. G. and Rausche, F., Pile Load Test by Impact Driving, paper presented to Highway Research Board Annual Meeting, Washington, DC, January, 1970.

Haimson, B. C., Stress Measurements Using the Hydrofracturing Technique, Field Measurements in Rock Mechanics, *Proceeding of 4th International Symposium*, Zurich, April, Vol. I, A. A. Balkema, Rotterdam, 1977, pp. 223–242.

Hardy, Jr., H. R. and Mowrey, G. L., Study of Underground Structural Stability Using Near-surface and Downhole Microseismic Techniques, Field Instrumentation in Rock Mechanics, *Proceedings of 4th International Conference*, Zurich, April, A. A. Balkema, Rotterdam, 1977, pp. 75–92.

Hardy, Jr., H. R. and Leighton, F., Design, Calibration and Construction of Acoustical Emissions Equipment, *Proceedings of 1st Conference on Acoustical Emission/Microseismic Activity in Geologic Structures and Materials*, Pennsylvania State University, June, 1975, Transactions Technical Publishing Company, Clausthal, Germany, 1977.

Hartmann, B. E., *Rock Mechanics Instrumentation for Tunnel Construction*, Terrametrics, Wheatridge, CO, 1966.

Kennedy, B. A., Methods of Monitoring Open Pit Slopes, Stability of Rock Slopes, *Proceedings of ASCE, 13th Symposium on Rock Mechanics*, Urbana, IL, 1971, pp. 537–572, 1972.

Koerner, R. M., Lord, Jr., A. E. and McCabe, W. M., Acoustic emission behavior of cohesive soils, *Proc. ASCE, J. Geotech. Eng. Div.*, 103, 837–850, 1977.

Koerner, R. M., Lord, Jr., A. E. and McCabe, W. M., Acoustic monitoring of soil stability, *Proc. ASCE J. Geotech. Eng. Div.*, 104, 571–582, 1978.

Kruse, G. H., Deformability of Rock Structures. California State Water Project, Determination of the In-Situ Modulus of Rock, ASTM STP 477, American Society for Testing and Materials, Philadelphia, PA, 1970.

La Fond, K. J., Applications of Dynamic Pile Analysis, from Piletips, Assoc. Pile and Fitting Corp., Clifton, NJ, May–June, 1977.

Novosad, S., The use of modern methods in investigating slope deformations, *Bull. Int. Assoc. Eng. Geol.*, 17, 71–73, 1978.

Rausche, F., Goble, G. G. and Moses, F., A New Testing Procedure for Axial Pile Strength, OTC Paper 1481, *Offshore Technology Conference*, Houston, TX, Preprint, 1971.

Richart, F. E., Jr., foundation vibrations, *Proc. ASCE J. Soil Mech. Found. Eng. Div.*, 86, 1960.

Roberts, A., The measurement of stress and strain in rock masses, in *Rock Mechanics in Engineering Practice*, Stagg, K.G. and Zienkiewicz, O.C., Eds., Wiley, New York, 1969, chap. 6.

Sauer, G. and Sharma, B., A System for Stress Measurements in Constructions in Rock, Field Instrumentation in Rock Mechanics, Proceeding of 4th International Symposium, Zurich, Vol. I, A. A. Balkema, Rotterdam, 1977, pp. 317–329.

Scholz, C. H., Mechanism of creep in brittle rock, *J. Geophys. Res.*, 73, 3295–3302, 1968.

Sellers, J. B., The Measurement of Stress Changes in Rock Using the Vibrating Wire Stress Meter, Field Instrumentation in Rock, *Proceedings of the 4th International Symposium*, Zurich, Vol. I, 1977, A. A. Balkema, Rotterdam, pp. 317–329.

Sherard, J. L., Woodward, R. J., Gizienski, S. F. and Clevenger, W. A., *Earth and Earth-Rock Dams*, Wiley, New York, 1963.

Silveira, J. F. A. A Instrumentacao de Mecnica das Rochas em Tuneis. M, todos de Observacao e Historicos de Casos, *Proceedings of 1st Congress Brasileira de Geologia de Engenharia*, Rio de Janeiro, August, Vol. 1, 1976, pp. 131–154.

Tschebotarioff, G. P., *Foundations, Retaining and Earth Structures*, McGraw-Hill, New York, 1973.

USBR, *Earth Manual*, 2nd Ed., U.S. Bureau of Reclamation, Denver, CO, 1974.

Further Reading

ASTM Determination of Stress in Rock: A State of the Art Report, SPT 429, American Society for Testing and Materials, Philadelphia, PA, 1966.

Brawner, C. O., Stacey, P. F. and Stark, R., Monitoring of the Hogarth Pit Highwall, Steep Rock Mine, Atikokan, Ontario, *Proceedings of the 10th Canadian Rock Mechanics Symposium*, Kingston, Ontario, September, 1975.

Cotecchia, V., Systematic reconnaissance mapping and registration of slope movements, *Int. Assoc. Eng. Geol., Bull.*, 17, 5–37, 1978.

Gartung, I. and Bauernfiend, P., Subway Tunnel at Nurnberg—Predicted and Measured Deformations, Field Measurements in Rock Mechanics, *Proceedings of the 4th International Symposium*, Zurich, April, A. A. Balkema, Rotterdam, 1977, pp. 473–483.

Koerner, R. M., Lord Jr., A. E., McCabe, W. M. and Curran, J. W., Acoustic behavior of granular soils, *Proc. ASCE, T. Geotech. Eng. Div.*, 102, 1976.

Koerner, R. M., Reif, J. S. and Burlingame, M. J., Detection Methods for Locating Subsurface Water and Seepage, *Proc. ASCE, T. Geotech. Eng. Div.*, 105, 1301–1316, 1979.

Wilson, S. D., Investigation of embankment performance, *Proc. ASCE, J. Soil Mech. Found. Eng. Div.*, 93, 1967.

Catalogs

Boart Longyear Interfels GmbH, 48455 Bad Bentheim, Germany.
Geokon Inc., Lebanon, NH, USA.
Soil Instrumentation Ltd., East Sussex, England TN 221QL.
The Slope Indicator Co., Mukilteo, WA, USA.

Appendix A

The Earth and Geologic History

A.1 Significance to the Engineer

To the engineer, the significance of geologic history lies in the fact that although surficial conditions of the Earth appear to be constants, they are not truly so, but rather are transient. Continuous, albeit barely perceptible changes are occurring because of warping, uplift, faulting, decomposition, erosion and deposition, and the melting of glaciers and ice caps. The melting contributes to crustal uplift and sea level changes. Climatic conditions are also transient and the direction of change is reversible.

It is important to be aware of these transient factors, which can invoke significant changes within relatively short time spans, such as a few years or several decades. They can impact significantly on conclusions drawn from statistical analysis for flood-control or seismic-design studies based on data that extend back only 50, 100, or 200 years, as well as for other geotechnical studies. To provide a general perspective, the Earth, global tectonics, and a brief history of North America are presented.

A.2 The Earth

A.2.1 General

Age has been determined to be approximately 4 1/2 billion years.

Origin is thought to be a molten mass, which subsequently began a cooling process that created a crust over a central core. Whether the cooling process is continuing is not known.

A.2.2 Cross Section

From seismological data, the Earth is considered to consist of four major zones: crust, mantle, and outer and inner cores.

Crust is a thin shell of rock averaging 30 to 40 km in thickness beneath the continents, but only 5 km thickness beneath the seafloors. The lower portions are a heavy basalt ($\gamma = 3$ t/m^3, 187 pcf) surrounding the entire globe, overlain by lighter masses of granite ($\gamma = 2.7$ t/m^3, 169 pcf) on the continents.

Mantle underlies the crust and is separated from it by the Moho (Mohorovicic discontinuity). Roughly 3000 km thick, the nature of the material is not known, but it is much denser than the crust and is believed to consist of molten iron and other heavy elements.

Outer core lacks rigidity and is probably fluid.

Inner core begins at 5000 km and is possibly solid ($\gamma \approx 12$ t/m^3, 750 pcf), but conditions are not truly known. The center is at 6400 km.

A.3   Global Tectonics

A.3.1   General

Since geologic time the Earth's surface has been undergoing constant change. Fractures occur from faulting that is hundreds of kilometers in length in places. Mountains are pushed up, then eroded away, and their detritus deposited in vast seas. The detritus is compressed, formed into rock and pushed up again to form new mountains, and the cycle is repeated. From time to time masses of molten rock well up from the mantle to form huge flows that cover the crust.

Tectonics refers to the broad geologic features of the continents and ocean basins as well as the forces responsible for their occurrence. The origins of these forces are not well understood, although it is apparent that the Earth's crust is in a state of overstress as evidenced by folding, faulting, and other mountain-building processes. Four general hypotheses have been developed to describe the sources of global tectonics (Hodgson, 1964; Zumberge and Nelson, 1972).

A.3.2   The Hypotheses

Contraction hypothesis assumes that the Earth is cooling, and because earthquakes do not occur below 700 km, the Earth is considered static below this depth, and is still hot and not cooling. The upper layer of the active zone, to a depth of about 100 km, has stopped cooling and shrinking. As the lower layer cools and contracts, it causes the upper layer to conform by buckling, which is the source of the surface stresses. This hypothesis is counter to the spreading seafloor or continental drift theory.

Convection-current hypothesis assumes that heat is being generated within the Earth by radioactive disintegration and that this heat causes convection currents that rise to the surface under the mid-ocean rifts, causing tension to create the rifts, then moves toward the continents with the thrust necessary to push up mountains, and finally descend again beneath the continents.

Expanding Earth hypothesis, the latest theory, holds that the Earth is expanding because of a decrease in the force of gravity, which is causing the original shell of granite to break up and spread apart, giving the appearance of continental drift.

Continental drift theory is currently the most popular, but is not new, and is supported by substantial evidence. Seismology has demonstrated that the continents are blocks of light granitic rocks "floating" on heavier basaltic rocks. It has been proposed that all of the continents were originally connected as one or two great land masses and at the *end of the Paleozoic era* (Permian period) they broke up and began to drift apart as illustrated in sequence in Figure A.1. The proponents of the theory have divided the earth into "plates" (Figure A.2) with each plate bounded by an earthquake zone.

Wherever plates move against each other, or a plate plunges into a deep ocean trench, such as that exists off the west coast of South America or the east coast of Japan, so that it slides beneath an adjacent plate, there is high seismic activity. This concept is known as "plate tectonics" and appears to be compatible with the relatively new concept of *seafloor spreading*.

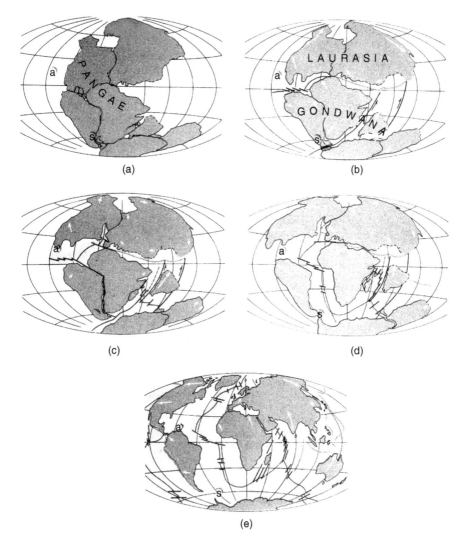

FIGURE A.1
The breakup and drifting apart of the original land mass, Pangaea: (a) Pangaea, the original continental land mass at the end of the permian, 225 million years ago; (b) Laurasia and Gondwana at the end of the Triassic, 180 million years ago; (c) positions at the end of the Jurassic, 135 million years ago (North and South America beginning to break away); (d) positions at the end of the Cretaceous, 65 million years ago; (e) positions of continents and the plate boundaries at the present. (From Dietz, R.S. and Holden, J.C., *J. Geophys. Res.*, 75, 4939–4956, 1970. With permission.)

A.4 Geologic History

A.4.1 North America: Provides a General Illustration

The geologic time scale for North America is given in Table A.1, relating periods to typical formations. A brief geologic history of North America is described in Table A.2. These relationships apply in a general manner to many other parts of the world. Most of the periods are separated by major crustal disturbances (orogenies). Age determination is based on fossil identification (paleontology) and radiometric dating.

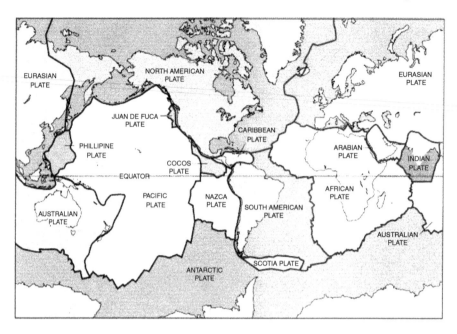

FIGURE A.2
Major tectonic plates of the world. (Courtesy of USGS, 2004.)

TABLE A.1

Geologic Time Scale and the Dominant Rock Types in North America

Era	Period	Epoch	Dominant Formations	Age (millions of years)
Cenozoic	Quaternary	Holocene	Modern soils	0.01
		Pleistocene	North American glaciation	2.5–3
	Neogene	Pliocene		7
		Miocene	"Unconsolidated" coastal-plain sediments	26
	Tertiary	Oligocene		37
	Paleogene	Eocene		54
		Paleocene		65
Mesozoic	Cretaceous		Overconsolidated clays and clay shales	135
	Jurassic		Various sedimentary rocks	180
	Triassic		Clastic sedimentary rocks with diabase intrusions	
Paleozoic	Permian		Fine-grained clastics, chemical precipitates, and evaporites. Continental glaciation in southern hemisphere	225
	Pennsylvanian		Shales and coal beds	280
	Mississippian	Carboniferous	Limestones in central United States. Sandstones and shales in east	310
	Devonian		Red sandstones and shales	400
	Silurian		Limestone, dolomite and evaporites, shales	435
	Ordovician		Limestone and dolomite, shales	500
	Cambrian		Limestone and dolomite in late Cambrian, sandstones and shales in early Cambrian	600
Precambrian	Precambrian		Igneous and metamorphic rocks	About 4.5 billion years

TABLE A.2

A Brief Geologic History of North America[a]

Period	Activity
Precambrian	Period of hundreds of millions of years during which the crust was formed and the continental land masses appeared
Cambrian	Two great troughs in the east and west filled with sediments ranging from detritus at the bottom, upward to limestones and dolomites, which later formed the Appalachians, the Rockies, and other mountain ranges
Ordovician	About 70% of North America was covered by shallow seas and great thicknesses of limestone and dolomite were deposited
	There was some volcanic activity and the eastern landmass, including the mountains of New England, started to rise (Taconic orogeny)
Silurian	Much of the east was inundated by a salty inland sea; the deposits ranged from detritus to limestone and dolomite, and in the northeast large deposits of evaporites accumulated in landlocked arms of the seas
	Volcanos were active in New Brunswick and Maine
Devonian	Eastern North America, from Canada to North Carolina, rose from the sea (Arcadian orogeny). The northern part of the Appalachian geosyncline received great thicknesses of detritus that eventually formed the Catskill Mountains
	In the west, the stable interior was inundated by marine waters and calcareous deposits accumulated
	In the east, limestone was metamorphosed to marble
Carboniferous	Large areas of the east became a great swamp which was repeatedly submerged by shallow seas. Forests grew, died, and were buried to become coal during the Pennsylvanian portion of the period
Permian	A period of violent geologic and climatic disturbances. Great wind-blown deserts covered much of the continent. Deposits in the west included evaporites and limestones
	The Appalachian Mountains were built in the east to reach as high as the modern Alps (Alleghanian orogeny)
	The continental drift theory (Section A.3.2) considers that it was toward the end of the Permian that the continents began to drift apart
Triassic	The Appalachians began to erode and their sediments were deposited in the adjacent nonmarine seas
	The land began to emerge toward the end of the period and volcanic activity resulted in sills and lava flows; faulting occurred during the Palisades orogeny
Jurassic	The Sierra Nevada Mountains, stretching from southern California to Alaska, were thrust up during the Nevadian disturbance
Cretaceous	The Rocky Mountains from Alaska to Central America rose out of a sediment-filled trough
	For the last time the sea inundated much of the continent and thick formations of clays were deposited along the east coast
Tertiary	The Columbia plateau and the Cascade Range rose, and the Rockies reached their present height
	Clays were deposited and shales formed along the continental coastal margins, reaching thicknesses of some 12 km in a modern syncline in the northern Gulf of Mexico that has been subsiding since the end of the Appalachian orogeny
	Extensive volcanic activity occurred in the northwest
Quaternary	During the Pleistocene epoch, four ice ages sent glaciers across the continent, which had a shape much like the present
	In the Holocene epoch (most recent), from 18,000 to 6,000 years ago, the last of the great ice sheets covering the continent melted and sea level rose almost 100 m
	Since then, sea level has remained almost constant, but the land continues to rebound from adjustment from the tremendous ice load. In the center of the uplifted region in northern Canada, the ground has risen 136 m in the last 6,000 years and is currently rising at the rate of about 2 cm/year (Walcott,1972)
	Evidence of ancient postglacial sea levels is given by raised beaches and marine deposits of late Quaternary found around the world. In Brazil, for example, Pleistocene sands and gravels are found along the coastline as high as 20 m above the present sea level

[a] The geologic history presented here contains the general concepts accepted for many decades, and still generally accepted. The major variances, as postulated by the continental drift concept, are that until the end of the Permian, Appalachia [a land mass along the U.S. east coast region] may have been part of the northwest coast of Africa [Figure A.1a], and that the west coast of the present United States may have been an archipelago of volcanic islands known as Cascadia. (From Zumberge, J.H. and Nelson, C.A., *Elements of Geology*, 3rd ed., Wiley, New York, 1972. Reprinted with permission of Wiley.)

The classical concepts of the history of North America have been modified in conformity with the modern concept of the continental drift hypothesis. The most significant modification is the consideration that until the end of the Permian period, the east coast of the United States was connected to the northwest coast of Africa as shown in Figure A.1.

A.4.2 Radiometric Dating

Radiometric dating determines the age of a formation by measuring the decay rate of a radioactive element.

In radioactive elements, such as uranium, the number of atoms that decay during a given unit of time to form new stable elements is directly proportional to the number of atoms of the radiometric element of the sample. This decay rate is constant for the various radioactive elements and is given by the half-life of the element, i.e., the time required for any initial number of atoms to be reduced by one half. For example, when once-living organic matter is carbon-dated, the amount of radioactive carbon (carbon 14) remaining and the amount of ordinary carbon present are measured, and the age of a specimen is computed from a simple mathematical relationship. A general discussion on dating techniques can be found in Murphy et al. (1979). The various isotopes, effective dating range, and minerals and other materials that can be dated are given in Table A.3.

In engineering problems the most significant use of radiometric dating is for the dating of materials from fault zones to determine the age of most recent activity. The technique is also useful in dating soil formations underlying colluvial deposits as an indication as to when the slope failure occurred.

TABLE A.3

Some of the Principal Isotopes Used in Radiometric Dating

Isotope				
Parent	**Offspring**	**Parent Half-Life (years)**	**Effective Dating Range (years)**	**Material That Can Be Dated**
Uranium 238	Lead 206	4.5 billion	10 million to 4.6 billion	Zircon, uraninite, pitchblende
Uranium 235	Lead 207	710 million		
Potassium 40[a]	Argon 40 Calcium 40	1.3 billion	100,000 to 4.6 billion	Muscovite, biotite hornblende, intact volcanic rock
Rubidium 87	Strontium 87	47 billion	10 million to 4.6 billion	Muscovite, biotite, microcline, intact metamorphic rock
Carbon 14[a]	Nitrogen 14	5,730±30	100 to 50,000	Plant material: wood, peat charcoal, grain. Animal material: bone, tissue. Cloth, shell, stalactites, groundwater and seawater.

[a] Most commonly applied to fault studies: Carbon 14 for carbonaceous matter, or K–Ar for noncarbonaceous matter such as fault gouge.

References

Hodgson, J.H., *Earthquakes and Earth Structure*, Prentice-Hall Inc., Englewood Cliffs, NJ, 1964.

Murphy. P.J., Briedis, J., and Peck, J. H., Dating techniques in fault investigations, geology in the siting of nuclear power plants, in *Reviews in Engineering Geology IV*, The Geological Society of America, Hatheway, A.W. and McClure, C.R., Jr., Eds, Boulder, CO, 1979, 153–168.

Zumberge, J.H. and Nelson, C. A., *Elements of Geology*, 3rd ed., Wiley, New York, 1972.

Further Reading

Dunbar, C.O. and Waage, K.M., *Historical Geology*, 3rd ed., Wiley, New York, 1969.

Guttenberg, B. and Richter, C.F., *Seismicity of the Earth and Related Phenomenon*, Princeton University Press, Princeton, NJ, 1954.

Walcott, R.L., Late quaternary vertical movements in eastern North America: quantitative evidence of glacio-isostatic rebound, *Rev. Geophys. Space Phys.*, 10, 849–884, 1972.

Index